Animal Cell Biotechnology
Volume 2

Contributors

G. D. J. Adams

G. D. Ball

R. Bomford

A. Brun

R. Camand

K. A. Cammack

Terence Cartwright

G. Chappuis

F. Colbère-Garapin

T. R. Doel

M. Duchesne

C. Duret

J. Fontaine

A. C. Garapin

J. B. Griffiths

F. Horaud

J. P. Jacobs

M. Lombard

Y. Moreau

Albert Osterhaus

T. M. Pollock

P. Precausta

R. A. J. Priston

M. Riviere

B. Roberts

C. Roulet

J.-P. Soulebot

R. E. Spier

C. Stellmann

John R. Stephenson

Fons UytdeHaag

Pieter van der Marel

C. N. Wiblin

Animal Cell Biotechnology

Volume 2

Edited by

R. E. SPIER

Department of Microbiology
University of Surrey
Guildford, Surrey
United Kingdom

J. B. GRIFFITHS

Vaccine Research and Production Laboratory
Public Health Laboratory Service
Centre for Applied Microbiology and Research
Salisbury, Wiltshire
United Kingdom

1985

ACADEMIC PRESS

(Harcourt Brace Jovanovich, Publishers)

London   Orlando   San Diego   New York
Toronto   Montreal   Sydney   Tokyo

COPYRIGHT © 1985 BY ACADEMIC PRESS INC. (LONDON) LTD.
ALL RIGHTS RESERVED.
NO PART OF THIS PUBLICATION MAY BE REPRODUCED OR
TRANSMITTED IN ANY FORM OR BY ANY MEANS, ELECTRONIC
OR MECHANICAL, INCLUDING PHOTOCOPY, RECORDING, OR
ANY INFORMATION STORAGE AND RETRIEVAL SYSTEM, WITHOUT
PERMISSION IN WRITING FROM THE PUBLISHER.

ACADEMIC PRESS INC. (LONDON) LTD.
24–28 Oval Road
LONDON NW1 7DX

United States Edition published by
ACADEMIC PRESS, INC.
Orlando, Florida 32887

British Library Cataloguing in Publication Data

Animal cell biotechnology.
    1. Cell culture
    I. Spier, R. E.   II. Griffiths, J. B.
    591.'0724   QH585

Library of Congress Cataloging in Publication Data
Main entry under title:

Animal cell biotechnology.

    Includes index.
    1. Cell culture.   2. Biotechnology.   I. Spier, R. E.
II. Griffiths, J. B.
QH585.A58   1984   615'.36   84-18567
ISBN 0–12–657552–5 (v. 2)

PRINTED IN THE UNITED STATES OF AMERICA

85 86 87 88   9 8 7 6 5 4 3 2 1

Contents

Part I Product Generation

1 Cell Products: An Overview

J. B. GRIFFITHS

2 Structure and Synthesis of Cellular Products

JOHN R. STEPHENSON

Part III   Product Testing

12   Product Testing: An Introduction

J. FONTAINE, A. BRUN, R. CAMAND,
G. CHAPPUIS, C. DURET, M. LOMBARD,
Y. MOREAU, P. PRECAUSTA, M. RIVIERE,
C. ROULET, J.-P. SOULEBOT, AND
C. STELLMANN

13   Pyrogenicity and Carcinogenicity Tests

C. N. WIBLIN

14   Field Tests

T. M. POLLOCK

Part IV Prospects

16 Genetic Engineering of Animal Cells

F. COLBÈRE-GARAPIN AND A. C. GARAPIN

17 The Biotechnological Future for Animal Cells in Culture

R. E. SPIER AND F. HORAUD

Contributors

Numbers in parentheses indicate the pages on which the authors' contributions begin.

G. D. J. Adams (251), Therapeutic Products Laboratory, Public Health Laboratory Service, Centre for Applied Microbiology and Research, Salisbury, Wiltshire SP4 0JG, United Kingdom

G. D. Ball (87), The Wellcome Research Laboratories, Beckenham, Kent BR3 3BS, United Kingdom

R. Bomford (235), Department of Experimental Immunobiology, The Wellcome Research Laboratories, Beckenham, Kent BR3 3BS, United Kingdom

A. Brun (291), Rhône Mérieux, Laboratoire IFFA, 69342 Lyon, France

R. Camand (291), Rhône Mérieux, Laboratoire IFFA, 69342 Lyon, France

K. A. Cammack (251), Therapeutic Products Laboratory, Public Health Laboratory Service, Centre for Applied Microbiology and Research, Salisbury, Wiltshire SP4 0JG, United Kingdom

Terence Cartwright (151), Laboratoire Roger Bellon, 37260 Monts, France

G. Chappuis (291), Rhône Mérieux, Laboratoire IFFA, 69342 Lyon, France

F. Colbère-Garapin (405), Unité de Virologie Medicale, Institut Pasteur, 75724 Paris, France

T. R. Doel (129), Vaccine Research Department, Animal Virus Research Institute, Pirbright, Woking, Surrey GU24 0NF, United Kingdom

M. Duchesne (151), Laboratoire Roger Bellon, 37260 Monts, France

C. Duret (291), Rhône Mérieux, Laboratoire IFFA, 69342 Lyon, France

J. Fontaine (291), Rhône Mérieux, Laboratoire IFFA, 69342 Lyon, France

A. C. Garapin (405), Unité de Virologie Medicale, Institut Pasteur, 75724 Paris, France

J. B. Griffiths (3), Vaccine Research and Production Laboratory, Public Health Laboratory Service, Centre for Applied Microbiology and Research, Salisbury, Wiltshire SP4 0JG, United Kingdom

F. Horaud (431), Unité de Virologie Medicale, Institut Pasteur, 75724 Paris, France

J. P. Jacobs (373), National Institute for Biological Standards and Control, London NW3 6RB, United Kingdom

M. Lombard (291), Rhône Mérieux, Laboratoire IFFA, 69342 Lyon, France

Y. Moreau (291), Rhône Mérieux, Laboratoire IFFA, 69342 Lyon, France

Albert Osterhaus (49), Rijksinstituut voor Volksgezondheid en Milieuhygiëne, NL 3720 BA Bilthoven, The Netherlands

T. M. Pollock (355), Epidemiological Research Laboratory, Central Public Health Laboratory, Public Health Laboratory Service, London NW9 5HT, United Kingdom

P. Precausta (291), Rhône Mérieux, Laboratoire IFFA, 69342 Lyon, France

R. A. J. Priston (71), Shell Research Laboratories, Sittingbourne, Kent ME9 8AG, United Kingdom

M. Riviere (291), Rhône Mérieux, Laboratoire IFFA, 69342 Lyon, France

B. Roberts (217), Glaxo Animal Health Ltd., Uxbridge, Middlesex UB9 6LS, United Kingdom

C. Roulet (291), Rhône Mérieux, Laboratoire IFFA, 69342 Lyon, France

J.-P. Soulebot (291), Rhône Mérieux, Laboratoire IFFA, 69342 Lyon, France

R. E. Spier (431), Department of Microbiology, University of Surrey, Guildford, Surrey GU2 5XH, United Kingdom

C. Stellmann (291), Rhône Mérieux, Laboratoire IFFA, 69342 Lyon, France

John R. Stephenson (13), Vaccine Research and Production Laboratory, Public Health Laboratory Service, Centre for Applied Microbiology and Research, Salisbury, Wiltshire SP4 0JG, United Kingdom

Fons UytdeHaag (49), Rijksinstituut voor Volksgezondheid en Milieuhygiëne, NL 3720 BA Bilthoven, The Netherlands

Pieter van der Marel[1] (185), Rijksinstituut voor Volksgezondheid en Milieuhygiëne, NL 3720 BA Bilthoven, The Netherlands

C. N. Wiblin (321), Vaccine Research and Production Laboratory, Public Health Laboratory Service, Centre for Applied Microbiology and Research, Salisbury, Wiltshire SP4 0JG, United Kingdom

[1]Present address: Intervet International, W. de Korverstraat 35, 5831 AN Boxmeer, The Netherlands.

Preface

"Animal Cell Biotechnology" is directed towards two groups of people. Those entering the field for the first time will find many details about how to set up and operate cell cultures in a variety of ways and at scales ranging between 0.001 and 10,000 litres. Others who are already engaged in the area will find that the comprehensive and detailed coverage of the selected topics written by experts in the various subjects will provide them with a well-referenced and well-indexed state-of-the-art report. This will enable them to expand their horizons further and to appreciate more fully those aspects of the subject with which they have not yet become deeply involved. It will also provide team leaders with the overview necessary to better coordinate and direct the various unit operations of manufacturing a product from animal cells in culture.

The contributing authors have been asked to assume that the reader will have been exposed at university or polytechnic level to two or more basic science subjects (mathematics, physics, chemistry and biology), although he/she may have qualified in the specialist disciplines of microbiology, chemical engineering, biochemistry, genetics, immunology or even biotechnology. In addition the writers have stressed, where appropriate, the underlying principles of the subject area and have provided many illustrations, diagrams, graphs and tables to present information which would otherwise take up much text and make for laborious reading. A balanced, thoughtful and fair assessment of alternative methods to achieve particular ends has been requested, yet in this burgeoning field the reader will have to contend with the clear enthusiasm of one or other of the various contributors. As editors, we are confident that there is a sufficient wealth of views for readers to have little difficulty in developing an appreciation of the strengths and weaknesses of alternative technologies and thereby be fortified in the choice of system for their specific application.

These volumes survey a new and as yet uncharted facet of biotechnology. They are designed both as an introduction and as an in-depth survey of the present situation. In view of the exciting scientific achievements but disappointing practical results derived from trying to express mammalian genes in prokaryotic organisms and the new capabilities in the area of genetic engineering of animal cells, we can anticipate that animal cell biotechnology is likely to retain its place at the cutting edge of biotechnology for some time to come. We trust that these volumes will help with the realisation of that future.

R. E. Spier
J. B. Griffiths

Contents of Volume 1

PART I

PRODUCT GENERATION

1

Cell Products: An Overview

J. B. GRIFFITHS

Vaccine Research and Production Laboratory
Public Health Laboratory Service
Centre for Applied Microbiology and Research
Salisbury, Wiltshire United Kingdom

1. INTRODUCTION

This part is the central core of animal cell biotechnology. Previous parts have described how the optimum environment and the most efficient bioreactor can be developed to produce the biomass. The biomass is the substrate from which a wide range of useful products can be generated. The remainder of this book describes how the products of the biomass are turned into purified products suitable for human and veterinary use. Despite product generation being the *raison d'être* of biotechnology, this section is perhaps the least exhaustive in the book. This is because the potential of generating cell products still exceeds the actual state of the art. The reason for this is discussed in this overview and in Chapter 17, this volume, which looks further into the future with regard to useful applications of animal cell cultures.

Animal Cell Biotechnology, Vol. 2

TABLE I   Products from Animal Cells

Viral Vaccines	Human and veterinary
Antibodies	Monoclonal antibodies
Interferons	Fibroblast and lymphoblastoid
Enzymes	e.g. Fibrinolytic
Whole cells	Cells and organelles
Insecticides	Insect viruses
Immunoregulators	e.g. Interleukins
Hormones	e.g. Insulin
Growth Factors	e.g. Platelet-derived growth factor (PDGF), epidermal growth factor (EGF), nerve growth factor (NGF)

The main categories of cell product are listed in Table I in the approximate order of magnitude in which they are currently produced. Some of these products, such as viruses for vaccines and insecticides, the immunoregulators and antibodies are fully described in the following chapters. The other products are briefly reviewed in this chapter.

2. VIRAL VACCINES

The human vaccines produced in cell culture are summarised in Table II. Cell cultures replaced whole animals (calf, sheep, mouse) for the manufacture of smallpox and rabies, and avian embryos for rabies, yellow fever and influenza. The first vaccine to be prepared in culture was the Salk polio vaccine in 1954, using primary monkey kidney cells. Since then the derivation of the human diploid cell line WI-38 made available a biologically safe substrate and provided the impetus, initially, for the widespread and large-scale production of measles, mumps and rubella vaccines. It has also allowed the development of better quality polio, rabies and yellow fever vaccines. Other vaccines produced in human diploid cells, but which are barely beyond clinical trial status, include those for adenovirus, varicella–zoster, cytomegalovirus, tickborne encephalitis, influenza, herpes simplex and respiratory syncytial virus. Even less advanced in their development are vaccines for parinfluenza, rota, Rift Valley fever, coxsackie, hepatitis A, rhino and rheo-viruses (common cold) and arenaviruses. The list of veterinary vaccines (Table III) is much larger and includes the largest single cell product manufactured today—foot-and-mouth disease virus (1.5×10^9 doses annually). Safety standards, particularly those relating to oncongenicity, are far

TABLE II   Human Vaccines Produced from Animal Cells

Principal vaccines	Limited application vaccines
Polio (Salk)	Herpes simplex
Polio (Sabin)	Cytomegalovirus
Measles	Varicella–zoster
Mumps	Respiratory syncytial virus
Rubella	Adenovirus
Rabies	Tickborne encephalitis
Yellow fever	
Influenza	

less exacting for veterinary products and thus allow a far more rapid, and cheaper, development of new vaccines.

Development of new viral vaccines has been slow in the last 10 years. Reasons for this are mainly economic in that: (1) quality control and testing procedures are now more stringent; (2) development costs cannot be met because their scale of use is too small, or because the main market is in countries that cannot afford them; (3) many of the new vaccines need to be either inactivated or prepared in a subunit form, which requires on the order of 10^5 times higher quantities of antigen than live vaccines; and (4) the productivity of many of these viruses in cell culture is extremely low. These factors mean that very efficient large-scale systems are necessary for an economical production of the virus. With the additional requirement to use either primary or diploid cells, which are anchorage dependent, cell culture systems cannot, at present, meet the demand. Most human vaccines are still produced in roller bottles, and microcarrier culture is used only for primary cells (e.g. for polio) (16). The vaccine manufacturing industry is almost marking time until either genetically engineered systems become commercially viable or heteroploid cells, which could be used in suspension culture, are accepted and licensed. These matters are dealt with in other sections. In this section the question of how the product is expressed by the cell and how this expression can be potentiated is addressed. One approach is to gain better knowledge at the molecular biology level of product generation. In Chapter 2, this volume, the host cell–virus relationships are reviewed. An example of how this approach can influence product development is the use of virus coded antigen on the surface of herpes simplex virus-infected cells. This antigen is present in greater quantities than can be obtained from the actual virus particles produced by the cell and investigators are currently testing the feasibility of using this material for a vaccine (15).

TABLE III   Some Principal Veterinary Vaccines[a]

Disease	Cell culture
Foot-and-mouth disease	BHK, calf and pig kidney
Newcastle disease	Pig kidney, chick embryo
Marek's disease	Chick embryo
Rinderpest (cattle plague)	Kidney
Rabies	CEF, BHK
Canine distemper	Dog kidney
Canine hepatitis	Pig/ferret kidney
Feline panleucopenia	Ferret kidney
Feline rhinopneumonitis	CEC
Parainfluenza 3	Bovine kidney
Bovine ephemeral fever	BHK
Infectious bovine rhinotracheitis	Kidney
Bovine viral diarrhoea	Embryonic kidney
Rift Valley fever	CEF
Theileriasis	Bovine lymphoblasts
African horse sickness	Monkey kidney
Equine encephalitis	CEC
Equine rhinopneumonitis	Pig kidney
Swine fever (hog cholera)	Kidney
Pseudorabiles (Aujeszky)	Pig kidney
Teschen disease	CEF
Bluetongue	Sheep kidney
Sheep pox	BHK
Louping ill	Kidney
Contagious pustular dermatitis	Cat kidney
Wesselsbron disease	Lamb kidney
Fowl pox	CEF
Infectious bursal disease	CEC
Duck plague	CEF
Mink enteritis	Mink kidney
Laryngotracheitis	CEC
Transmissible gastroenteritis	Dog kidney

[a] Based on data of Ozawa (9) and Spier (12).

3. ENZYMES

A cell contains many thousands of enzymes but at the present time it is only the thrombolytic enzymes which are receiving any serious attention as a product of animal cells. Thromboembolism is now one of the major killing diseases, particularly in the Western world, and the market for an effective diagnostic aid and therapeutic treatment is enormous. At present treatment is confined to two fibrinolytic enzymes: streptokinase, a bacterial protein, and urokinase, a protein isolated from human urine, but also secreted by some kidney cell lines (2). However, these enzymes, as well as destroying the blood clot, also degrade the blood coagulating factors and thus bring about a serious risk of haemorrhage during treatment. Plasminogen is an inactive proenzyme, present in blood, which when activated by plasminogen activators (either tissue-type or urokinase-type) becomes the enzyme plasmin. Plasmin acts only on the fibrin within the clot and not on circulating blood factors and is thus the enzyme of choice for clinical use.

Tissue plasminogen activators (t-PA) have been isolated from normal endothelial and even epithelial (1,3) cell lines and from tumour tissues (5). A melanoma cell line (Bowes) has been used to produce a t-PA which was subsequently tested in patients (17). Melanoma cells produce at least a 10-fold higher concentration than normal cells but, being derived from a tumour, have no potential as a substrate for a therapeutic agent. Normal human breast epithelial cells produce about 1 mg of highly purified t-PA per 10^{10} cells (3). The clinical trial described above used 7.5 mg per patient per day, although probably 1–2 mg would be sufficient. Obviously a future therapeutic agent relies on a genetically engineered product and such studies are well advanced (10).

Substantial quantities of this enzyme are needed to develop the second-generation product. This is for development of a purification process, for efficacy tests in animals, for immunological and biochemical characterisation studies, for genetic sequencing and for the cloning experiments. To produce the required quantities large-scale culture systems are being used, e.g. continuous culture systems for melanoma cells and, in the author's laboratory, high-density perfusion microcarrier cultures (15 g microcarrier per litre supporting $5–10 \times 10^6$ cells/ml) for epithelial cells. In addition, ways of potentiating the yield are being investigated, particularly with the use of steroid hormones and other regulatory compounds (e.g. serotonin, bradykinin, histamine, Stromba).

Many other cellular enzymes are being investigated and used, e.g. as diagnostic aids for blood cells. These studies are not relevant to biotechnology at the moment but some of them may lead to future production processes. This topic has been reviewed by Katinger and Bleim (6).

4. HORMONES AND GROWTH FACTORS

There are available a few cell lines that secrete hormones, e.g. luteinizing, follicle-stimulating, growth, ACTH (14), but these are mainly derived from endocrine tumours (Table IV). For the reasons discussed in Volume 1, Chapter 2, Section 1, normal cells growing in culture do not normally show such a

TABLE IV   Hormones from Cultured Cells

Hormone	Cell source
Growth hormone	Rat pituitary tumour
Prolactin	
Insulin	Foetal rat pancreas
	Hybrid cells
ACTH	Pituitary cells, HeLa V3
Thyroid	
Parathyroid	Parathyroid adenoma
Adrenal	Adrenal cortex tumours
Luteinizing	Anterior pituitary (human, rat, sheep)
Follicle-stimulating	
Thyrocalcitonin	Medullary carcinoma
Erythropoetin	
(TSH)	Haemopoietic cells
Thyrotropin (TSH)	Sheep pituitary
Gonadal	Placental and follicle cells

specialised function. Hormones can be expressed by epithelial/endothelial cells cultured in tissue-like systems (e.g. capillary fibres), by some cancer cells (13) or by fusion of a hormone-secreting cell with a rapidly growing cell (analogous to hybridoma production). An example of this latter approach is the fusion of mouse A9 cells with pig pancreatic cells to produce an insulin-secreting hybrid cell line. An interesting source is the ectopic hormone production from tumours of tissues other than the appropriate endocrine organ (4). Examples of this phenomenon are given in Table V. The ability to express this genetic capability will have been influenced during differentiation of the cells from which these tumours arise and by environmental controls. The tumour cells will have escaped from these influences due

TABLE V   Ectopic Hormones

Hormone	Source
ACTH	Bronchus, thymus
ADH	Bronchus
Gonadotrophin	Liver, bronchus
Erythrocytosis	Kidney, brain
Gastrin	Pancreas
Parathormone	Kidney, bronchus
Insulin	Bronchus, kidney
Corticotrophin	Bronchus, pancreas
TSH	Hydatiform mole

TABLE VI   Growth Factors

Growth Factor	Action
Fibroblast	Growth of mesodermal cells
Epidermal	Growth of cultured cells
Nerve	Nerve cell development
Platelet-derived[a]	Growth of endothelial cells
Endothelial CGS	Maintain endothelial cells
MSA (insulin-like)[a]	Growth of mesodermal cells
T cell (TCGF)[a]	T lymphocyte culture
BCGH[a]	B lymphocyte culture
Calmodulin	Calcium activation
Fibronectin	Attachment and growth
Laminin[a]	Attachment of epithelial cells
Transferrin	Transport factor

[a] Factors at present derived from cultured cells.

either to loss of sensitivity to regulatory controls or to genetic changes allowing these genes to become switched on and be free to express the hormone. This example of unlocking the cells' suppressed ability to produce a useful product gives encouragement that further studies of growth regulation and differentiation will bring about new production capabilities from animal cells.

At present, commercial production of a hormone from animal cells cannot be foreseen. The only cells capable of significant hormone expression are tumour cells or transformed cells such as the GH3 (rat pituitary). However, these might, like plasminogen activator, provide material for genetic engineering developments. Human growth hormone has been cloned in *Escherichia coli* but still the main commercial source for therapy is extraction from human pituitary glands.

Of great significance for developing cell systems capable of more differentiated and specialised function are the growth factors. These compounds are being isolated from many types of tissue and body fluids and are specific for either their growth-promoting activity or other specialised cellular function. These factors are listed in Table VI and have been discussed in Volume 1 (Chapters 2 and 4).

5. CELLS AND CELL COMPONENTS

Cell production (Table VII) is sometimes needed to provide a large source of cellular organelles such as chromosomes (or DNA), messenger RNA, ribosomes, mitochondria and cell membrane components (antigens, attach-

TABLE VII   Whole Cell and Other Products

Whole cells	Assays
	Bone marrow, lymphocyte and skin grafts
Cell constituents	Chromosomes, DNA, mRNA, organelles
	Tumour antigens
	Attachment factors
	Fibronectin
	Laminin
	Transferrin
Other products	Heparin
	Tumour angiogenesis factor
	Albumin
	Blood clotting factors
	Chondroitin sulphate

ment proteins). In addition, whole cells can be used for chemical transformation, enzyme degradation, or synthesis of metabolites (e.g. steroids) and for use in assay systems for screening and toxicology. A potential use is to provide material for bone marrow and lymphocyte replacement and skin grafts, particularly for burn patients. Probably the most widespread use of cells is in research and development. Many of the products listed under other categories, such as the immunoregulators (Chapter 2, this volume), are used as research probes for understanding cell function in both the normal and diseased state. The other role of animal cells is to provide for the development of genetically engineered products. Whether these products will be derived from large-scale cell cultures in the future and the potential for genetically engineered cell lines are discussed in Chapter 17 of this volume.

6. CONCLUSION

The production of t-PA provides a good example of the way in which the low productivity of cultured cells is limiting their use in biotechnology. Typical production values are in the order of 0.01 to 0.1 μg/million cells/day for enzymes and hormones to 5–50 μg of antibody from hybridomas. Despite the high specific activity of these products and the relatively low dose rate needed, a single therapeutic dose will require between 10^9 and 10^{12} cells for its production. It is also typical in that although the future depends upon a genetically engineered product, there is a large requirement for the product now. The product from cells has to be proven before it is feasible to consume the resources necessary to clone it. The approach to this low productivity is (1) to increase the scale and efficiency of the culture system, (2) to potentiate the yield by gaining an understanding of the product's synthesis in the cell and (3) to derive new cell lines and screen for product generation.

A factor to consider is that cells do not necessarily manufacture the product throughout their whole life cycle. Very often only a certain proportion of a population may be active or capable of activity and therefore cloning and selection of clones may be a productive procedure.

It must be stressed that conditions for the manufacture of any product have to be empirically defined as they will be unique to that cell, system and product. The antigenic variation between virus produced in monolayer and suspension strains of the same cell line is one such example. Optimal environmental conditions for growth might be different from those for product generation (e.g. oxygen) (7). Some products are expressed from growing cells, others from stationary cells. The problems of keeping non-growing cells physiologically active are legendary and recourse is usually made to perfusion systems. Selective gene expression by cells has been discussed. This also applies to viruses. Herpes simplex, for example, contains far more genetic information than is ever expressed in culture. The interpretation of this is that the information was gained in evolution for survival in many complex multicellular hosts. A unique relationship with a particular host will have evolved, and only in this host will their full genetic potential be fulfilled. Cell cultures are not a native habitat, and it is unlikely that the entire genetic contents of herpes simplex virus will be expressed in single cell suspensions of undifferentiated cells (11).

Another factor to be considered is whether downstream processing can cope with a large volume of low-titre product, or if smaller volumes of high concentration are necessary. A product such as plasminogen activator tends to stick to glassware, filters, etc., and high titres have to be produced otherwise purification losses are intolerable. The complexity of the purification process is obviously a key factor in deciding the culture volume that should be generated. The point is that cell growth and product generation parameters cannot be designed in isolation—they are steps in a long production chain which starts at Chapter 1 (Volume 1) and finishes at Chapter 14 (this volume).

REFERENCES

1. Atkinson, A., Electricwala, A., Latter, A., Riley, P. A., and Sutton, P. M. (1982). New tissue sources and types of fibrinolytic enzymes. *Lancet* (July) 132–133.
2. Balow, G. H., Lazar, L., and Tribbey, I. (1977). Production of plasminogen activity by tissue culture techniques. *In* "Thrombosis and Urokinas" (R. Paoletti and S. Sherry, eds.), p. 75. Academic Press, New York.
3. Electricwala, A., Sutton, P. M., Griffiths, J. B., Riley, P. M., Latter, A., and Atkinson, A. (1983). Purification and properties of the plasminogen activator secreted by guinea pig keratocytes in culture. *In* "Clinical Aspects of Fibrinolysis and Thrombolysis," pp. 85–91. South Jutland Univ. Press, Denmark.
4. Frohman, M. D. (1981). Ectopic hormone production. *Am. J. Med.* **70**, 70–74.

5. Gronow, M., and Bleim, R. (1983). Production of human plasminogen activators by cell culture. *Trends Biotechnol.* **1**, 26–29.

6. Katinger, H. M. D., and Bleim, R. (1982). Production of enzymes and hormones by mammalian cell culture. *Adv. Biotechnol. Processes* **2**, 61–95.

7. Mizrahi, A., Vosseller, G. V., Yagi, Y., and Moore, G. E. (1972). The effect of dissolved oxygen partial pressure on growth, metabolism and immunoglobulin production in a permanent human lymphocyte cell line culture. *Proc. Soc. Exp. Biol. Med.* **139**, 118–122.

8. Munro Neville, A. (1972). Ectopic production of hormones by tumours. *Proc. Soc. Med.* **65**, 55–60.

9. Ozawa, Y. (1979). Common problems in tissue culture. *In* "Practical Tissue Culture Applications" (K. Maramorosch and H. Hirumi, eds.), pp. 67–75. Academic Press, New York.

10. Pennica, D., Holmes, W., Kohr, W., Harkins, R., Vehar, G., Ward, C., Bennett, W., Yelverton, E., Seeburg, P., Heyneker, H., and Goeddel, D. (1983). Cloning and expression of human tissue-type plasminogen activator cDNA in *E. coli. Nature (London)* **301**, 214–221.

11. Roizman, B. (1969). The herpes viruses—a biochemical definition of the group. *Curr. Top. Microbiol. Immunol.* **49**, 1–79.

12. Spier, R. E. (1983). Opportunities for animal cell biotechnology. *In* "Proceedings of Biotechnology," pp. 317–335. Online Publications Ltd., Northwood Hills, U.K.

13. Tashjian, A. H. (1969). Animal cell cultures as a source of hormones. *Biotechnol. Bioeng.* **11**, 1109.

14. Tashjian, A. H., Yasumura, Y., Levine, L., Sato, G. H. and Parker, M. L. (1968). Establishment of clonal strains of rat pituitary tumour cells that secrete growth hormone. *Endocrinology* **82**, 342–352.

15. Thornton, B., Griffiths, J. B., and Walkland, A. (1982). Herpes simplex vaccine using cell membrane associated antigen in a animal model. *Dev. Biol. Stand.* **50**, 201–206.

16. van Wezel, A. L., Van Steemis, G., Hannik, C. A., and Cohen, H. (1978). New approach to the production of concentrated and purified inactivated polio and rabies tissue culture vaccines. *Dev. Biol. Stand.* **41**, 159–168.

17. Weimar, W., Stibbe, J., Billiau, A., DeSomer, P., and Collen, D. (1981). Specific lysis of an iliofemoral thrombus by administration of extrinsic (tissue-type) plasminogen activator. *Lancet*, 1018–1020.

2

Structure and Synthesis of Cellular Products

JOHN R. STEPHENSON

Vaccine Research and Production Laboratory
Public Health Laboratory Service
Centre for Applied Microbiology and Research
Salisbury, Wiltshire, United Kingdom

1. INTRODUCTION

The purpose of this chapter is to outline the mechanisms whereby viruses infecting animal cells are manufactured, and also to outline the production of cellular products involved in the control of viral infections. By concentrating

Animal Cell Biotechnology, Vol. 2
Copyright © 1985 by Academic Press Inc. (London) Ltd.
All rights of reproduction in any form reserved.

on the molecular aspects of these processes, it is hoped that readers without a formal education in the biological sciences will appreciate the inherent problems associated with biological systems as well as the considerable advantages they offer. It will become apparent from the following discussion that there are three parameters which differentiate biological systems from their chemical counterparts, i.e. their large size, their considerable complexity and the fact that growing organisms mutate. Although these properties can pose significant problems for the biotechnologist, especially in ensuring the integrity and uniformity of the product, the ability of biological systems to efficiently produce complex products with the correct stereoisomerism frequently outweighs the problems encountered.

This review is intended to give a brief overview of the biological processes involved in the production of these animal cell products, but will not include a discussion of antibody production or the expression of eukaryotic genes in bacterial cells, as these will be dealt with in other chapters of this volume. A considerable proportion of this chapter will be devoted to discussing the replication of animal viruses, for not only have they been used extensively to elucidate the mechanisms of eukaryotic protein and nucleic acid synthesis, but the production of vaccines against virus diseases occupies a sizeable section in the field of biotechnology. Therefore, as far more is known about the replication of animal viruses than about similar processes in the whole cell, the synthetic processes in the cell will be discussed by reference to the replication strategy of the various virus families and, where possible, comparing them with what is known about the processes in the whole cell. As most of this chapter has been compiled from several review articles, these will be indicated at the end of each section. Although there is no intention of supplying a complete bibliography, reference is made at the appropriate point in the text to experiments of particular relevance if they are not covered by the review articles.

2. REPLICATION OF VIRUSES AND VIRAL COMPONENTS

2.1. Taxonomy

The replication strategies employed by animal viruses are extremely diverse and it is essential, therefore, to understand their principles when attempting to grow viruses or viral components *in vitro*, or to extract them from infected cells. For the purpose of this chapter, the virus types have been split into taxonomic groups according to their genomic structure. However, as the synthesis of enveloped viruses contains a number of processes which distinguish them from the non-enveloped viruses, it is useful to combine several taxonomic groupings when discussing many aspects of protein synthesis. Taxonomic relationships between the DNA viruses and RNA viruses are illustrated in Tables I and II and the relevant characteristics of the various taxonomic groups are listed.

2.2. Virus Attachment and Uncoating

When compared to the large amount of research performed on virus replication, this area of virology has received little attention. This may seem surprising, as the mode of attachment of viruses determines their cell tropism and in all probability their ultimate pathogenicity. As the mechanisms of attachment for all viruses are broadly similar, they will all be discussed together. However, significant differences exist between the mechanisms whereby enveloped and non-enveloped viruses enter their target cells, so these will be discussed separately. Originally, it was thought that the mechanisms of virus penetration and uncoating could be separated, but recent research seems to indicate that at least for the smaller enveloped viruses, these process may be inextricably linked. Therefore, penetration and uncoating will be discussed together in the following section.

2.2.1. Attachment of Virus Particles to Receptors on the Cell Surface

Under physiological conditions, the external surface of both the host cell and a virus particle are negatively charged; thus, the reaction between a virus and its host cell receptor must be capable of overcoming this natural repulsion. Experiments designed to identify such components are fraught with difficulties. Firstly, viruses bind to several non-physiological surfaces, such as glass, nitrocellulose and carbon, and these may be inhibited by neutralizing antiserum in the same way as binding to a cell surface. Secondly, it is difficult to distinguish between a receptor which binds a virus non-specifically and one that will bind a virus particle and lead to a productive infection. This latter point is further exacerbated by the fact that the particle to infectivity ratio of all animal viruses is greater than unity and it is impossible to distinguish infectious and non-infectious particles by visual techniques such as electron microscopy (EM). However, this may be a reflection of a situation in which more than one particle is required to infect a cell. In addition, relatively few viruses grow well enough in tissue culture for detailed studies on attachment to be performed, so this discussion is limited to those virus–cell systems which have been best characterised.

In theory, any normal cell-surface component may serve as a virus receptor. However, lipids have not been shown to act as productive receptors, except in artificially produced liposomes. In general, animal viruses appear

TABLE I Animal DNA Viruses

Nucleic acid type	Envelope	Taxonomic group	Core morphology	Core polypeptides	Virion morphology	Envelope proteins	Examples
Double-stranded circular 5.5 kbp[a]	None	Papovaviridae	—	—	Icosohedral 45–55 nm[b]	45K[c](major) 35K, 23K (minor) + histones H2A, H2B, H3, H4	SV40, polyoma, BK virus; Human papilloma
Double-stranded linear 35–45 kbp	None	Adenoviridae	8–12 spherical subunits 66 nm[b]	pVII (major) p48.5K (minor)	Icosohedral 80 nm	pII (hexon) pIII (penton) pIV (fibre)	Human adenoviruses and adenoviruses from several other vertebrates
Double-stranded linear; bisegmented; $1.3–2.6 \times 10^2$ kbp	Multicomponent complex lipoprotein shell	Herpesviridae	Icosohedral 100 nm	157K, 55K, 47K, 39K, 36K, 30K	Spherical 120–280 nm	g A/B, g C, g D, g E	Herpes simplex, Cytomegalovirus, Epstein–Barr virus, Varicella zoster
Double-Stranded linear; $2–4 \times 10^2$ kbp	Lipoprotein bilayer	Poxviridae	Biconcave	At least 12 major species	"Brick-shaped" 200–300 nm	p 4 C, p 6 a (g)[d], p 6 b (g), p 11 b, p 10, p 12	Vaccinia, Smallpox
Double-stranded linear 1.6×10^2 kbp	Linoprotein bilayer	Iridoviridae	Icosohedral 136 nm	VP48	Spherical 160 nm	VP 58, VP 44, VP 63	Frog virus 3, African swine fever
Partially double-stranded circular 3.2 kb	Simple lipoprotein bilayer	Hepatitis B	Spherical 28 nm	5K (major), 16K (major), 68K (minor)	Spherical 42 nm	23K, 26K–28K (g), 29K–30K (g)	Human hepatitis B virus
Single-stranded linear 4.5–5.5 kb	None	Parvoviridae			Icosohedral 20–25 nm	72–92K, 64–80K, 56–69K	Kilham rat virus, Minute virus of mice

[a] kbp, size of genome in kilobase pairs.
[b] Diameter in nanometers as seen by electron microscopy.
[c] Apparent molecular weight on SDS–PAGE, in kilodaltons.
[d] (g) Denotes that this polypeptide is glycosylated.

to bind to naturally occurring glycoproteins on the cell surface, although at present it is difficult to determine the relative roles of the protein and carbohydrate moieties on these molecules. However, it has been shown that the orthomyxoviruses, the paramyxoviruses and polyoma virus bind to N-acetylneuraminic acid (NeuAc) *in vitro,* but experiments with influenza on erythrocytes at various pH levels suggest that this virus is only recognizing the spatial arrangement of electrostatic charge, rather than the sugar residues per se. Conversely, Sendai and polyoma viruses appear to show a conformational requirement for a $2 \rightarrow 3$ sugar linkage.

Although identification of cellular virus-receptors is fraught with difficulties, membrane proteins with such properties have been identified for both enveloped viruses (e.g. VSV, murine leukaemia virus and Sindbis virus) and non-enveloped viruses (e.g. adenovirus). Of special interest is the observation that membrane products of immunological importance can act as virus receptors. It has been shown for SFV, for example, that virus coat protein can bind to the almost ubiquitous and highly conserved transplantation antigens. However, other studies have indicated that these proteins are not the exclusive receptors for this and other families of viruses. Of greater importance is the observation that virus-specific, non-neutralizing immunoglobulin will bind to the F_c receptor of monocytes or macrophages, and thus cause an infection in a situation where virus by itself would not (39). This phenomenon of immunological enhancement has been described for several viruses including rabbitpox and dengue and is thought to be responsible for dengue shock syndrome (22) and early fatalities in rabies infection. Thus, in view of the apparent similarity of host cell receptors, it is tempting to assume that all viruses use similar proteins, and indeed this is partly true. However, similar viruses such as the coxsackie B group and poliovirus types 1, 2 and 3 do not apparently share similar receptors, whereas viruses of dissimilar structure such as adenovirus share common receptors with the coxsackie B group and viruses such as influenza and polyoma also apparently share similar receptors. However, this may reflect a similarity in the local conformation for the receptor molecule on the virus, which, contrary to popular belief, may have little to do with taxonomic status or serotype specificity.

The physical parameters for virus attachment to cell receptors have been discussed at length by several workers. The cell receptors appear to be distributed evenly over the cell surface and occur at a frequency, at least for tissue culture cells, of about 10^4–10^5 molecules per cell. The effect of temperature on virus attachment is the most studied parameter and is the most variant. Viruses such as influenza and Mengovirus are little affected by reductions in temperature, but other similar viruses such as EMC do appear to be temperature-dependent.

As the protein composition of most virus surfaces is much simpler than

TABLE II Animal RNA Viruses

Nucleic acid type	Molecular weight (daltons)	Envelope	Taxonomic group	Core morphology	Virion morphology	Internal polypeptides	External polypeptides	Examples
	$2-3 \times 10^6$	None	Picornaviridae	—	Icosohedral (27–28 nm)[a]	VPg	VP1, VP2, VP3, VP4	Poliovirus, rhinovirus, foot and mouth disease virus, hepatitis A
	2.5×10^6	None	Caliciviridae	—	Icosohedral (35–40 nm)	VPg	VP1	Vesicular exanthema of swine viruses. Feline calicivirus
Positive sense[b] linear contiguous	$3-4 \times 10^6$ (two molecules per virion + tRNA)	Proteolipid bilayer	Retroviridae	Icosohedral (50 nm)[a]	Spherical (100 nm)	p10, p21 p27, p14 p100	gp36[d] gp52	Avian myoblastosis Visna virus Human T-cell leukaemia virus
	4.4×10^6	Proteolipid bilayer	Alphaviruses Rubiviruses Pestiviruses	Icosohedral (30–40 nm)	Spherical (70 nm)	C	E_1, E_2 (E_3)[c]	Sindbis virus, Venezuelan equine encephalitis virus Rubella, bovine diarrhoea virus
	4.0×10^6	Proteolipid bilayer	Flaviviruses	Icosohedral (25–35 nm)	Spherical (50–60 nm)	C	E (g)[d] (M) (46–51K)	Yellow fever TBE Dengue
	$5.5-7 \times 10^6$	Proteolipid bilayer	Coronaviridae	Coiled helix (20–30 nm)	Roughly spherical with "corona" (60–220 nm)	50–60K	20–30K (g) 80–200K (g)	Infectious bronchitis virus Human coronavirus Mouse hepatitis virus

Negative sense linear contiguous	$4–5 \times 10^6$	Proteolipid bilayer	Rhabdoviridae	Helical 40–60 nm × 50–100 nm	"Bullet" shaped 70 × 100 nm	L, N, NS, M	G	Vesicular stomatitis virus, Rabies virus
	$5–6 \times 10^6$	Proteolipid bilayer	Paramyxoviridae	Linear "herring-bone" structure	Pleomorphic sphere (150–200 nm)	N, P, M, L	H or HN, F	Sendai virus, Newcastle disease virus, Measles virus, Mumps virus, Canine distemper virus
Negative sense segmented linear	4×10^6	Proteolipid bilayer	Orthomyxoviridae	Helical	Spherical (100 nm)	N, M, P1, P2, P3	H, NA	Human influenza, Swine influenza, Fowl plague virus
	$3–5 \times 10^{6e}$	Proteolipid bilayer	Arenaviridae	Filamentous	Pleomorphic (80–150nm)	54–68K 200 K	34–44K (g) (G1) 52–65K (g) (G2)	Lymphocytic choriomeningitis virus, Lassa fever virus, Pichinde virus
	$4.8–6 \times 10^{6e}$	Proteolipid bilayer	Bunyaviridae	Helical	Spherical (80–110 nm)	N, L	G_1, G_2	Bunyumwera virus, Crimean haemorrhagic fever virus, Nairobi sheep disease
Linear double-stranded segmented	1.5×10^{7e}	None	Reoviridae	Icosohedral (65 nm)	Icosohedral (76 nm)	λ1, λ2, λ3, μ2 σ2	μ1C, σ1 σ3	Reovirus, Rotavirus
	2×10^{6e}	None	IBDV group	—	Icosohedral (60 nm)		VP1, VP2, VP3, VP4	Infectious bursal disease virus, Infectious pancreatic necrosis virus

[a] Diameter in nanometers as determined by electron microscopy.

[b] That is, genomic RNA can function as a messenger RNA.

[c] Polypeptides in parentheses may only occur in some members of the group.

[d] A "g" prefix or a "g" suffix in parentheses denotes that this protein is glycosylated.

[e] These values represent the combined molecular weights of all the segments.

that of the cell, it would seem that identification of the virus protein responsible for attachment (virus receptor) would be relatively straightforward. However, all three of the most common methods used—antibody binding, binding of isolated proteins and digestion with enzymes—abound with problems. Firstly, it has been shown by several authors that neutralizing antibody against viruses does not always prevent binding, penetration or even uncoating. In addition, neutralizing monoclonal antibodies against influenza virus have recently been shown to act on the virion transcriptase (*40*). Secondly, isolated virus proteins seldom retain their native conformation when removed from virus particles; and finally, so-called receptor-destroying enzymes are seldom pure and in many studies have not led to a clear understanding of receptor structure. In spite of these strong reservations, it has been possible to identify receptors for a few viruses. The VP4 polypeptide from poliovirus has been shown to act as a receptor, but other picornaviruses use other polypeptides as receptors. The architecture of the adenoviruses indicates that the site of attachment is the penton fibre and this observation is supported by some biochemical data. In another non-enveloped virus, reovirus, the σ1 protein was shown by competitive antibody binding to be important in attachment. Electron microscopy of enveloped viruses such as influenza and Sendai virus show the virus attaching by the envelope spikes of HA and neuraminidase. Several lines of study indicate that the HA is the binding protein for influenza virus, but it is difficult to completely eliminate a similar function for the neuraminidase. With parainfluenza viruses such as Sendai and SV5, both genetic studies and antibody binding experiments strongly indicate the role of the HN protein in attachment, although the F protein is essential for penetration (see paragraph below). Similar studies on VSV, SFV and several retroviruses also implicate the major envelope glycoprotein of these viruses as the virus receptor. But, to date, there is no conclusive evidence to indicate which amino acid sequence in any virus envelope or capsid protein is directly involved in the binding of the virus to its host cell. Work on virus receptors has been of interest over several years as, at least in theory, isolated virus proteins could be used prophylactically to block cell receptors. As there are as many as 10^5 receptors per host cell, such a large amount of foreign protein on the surface may well interfere with normal cell functions. Such an effect, however, while probably of no use for chemotherapy, might be a useful spin-off for a subunit vaccine.

2.2.2. *Penetration and Uncoating of Non-enveloped Viruses*

All studies on virus penetration suffer from all the limitations listed above in discussing mechanisms of attachment. As electron microscopy is the main tool used in this area, these limitations must be borne in mind when assessing relevant experiments. A further consideration must also be remembered, and that is that most cells possess several mechanisms whereby they ingest extracellular compounds and any, or all, of these may be employed by any virus. "Translocation" or direct penetration is a phenomenon described mainly for non-enveloped viruses, such as poliovirus and adenovirus, whereby the virus passes directly through the membrane, without evidence of fusion or envelopment in a vesicle. No satisfactory mechanism has been described at present to explain this phenomenon, although one could imagine a situation whereby on binding to cell surface receptors, the virus would cause an aggregation of these receptors which would then form a hydrophilic tunnel through the cell membrane, through which the virus could pass directly into the cytoplasm. A second mechanism for virus entry has been postulated, that of endocytosis (sometimes referred to as viropexis when used to describe virus entry). This is the mechanism whereby the cell naturally takes up a variety of materials from its environment, and therefore it is difficult to determine whether this is a process relevant to infection or just reflects a natural scavenging by the cell. As this mechanism results in a fusion of the endocytotic vesicle with lysosomes and would thus expose the virus to a battery of hydrolytic enzymes, many objections have been raised to the relevance of this mechanism. However, recent studies on the uptake of various hormones, yolk proteins etc. have demonstrated the ability of several endocytosed moieties to avoid exposure to lysosomal enzymes. It should be remembered that endocytosis describes a wide variety of mechanisms, at least some of which are present in all cells. Indeed, Allison and Davies (3) divided the process into three classes, naming them phagocytosis, macropinocytosis and micropinocytosis, each with different energy requirements and drug sensitivities. It should also be realized that these processes have a time scale of seconds rather than minutes and even a reduction in temperature of 20–30°C will only slow the mechanism by an order of magnitude—still too fast for normal experimental procedures to detect. For these reasons, both enveloped viruses (such as SFV and influenza) and non-enveloped viruses (such as SV40 and poliovirus) have been reported to penetrate cells by this mechanism. However, as may be seen in the discussion below, workers with enveloped viruses may be observing only half the process.

Attempts to study the mechanism of uncoating have been hampered by the fact that once inside the cell, most viruses lose their distinct morphology and thus become difficult to monitor. Several viruses such as VSV and SV40 have been studied by degrading purified virions and observing the effect on the virion-associated transcriptase. However, the relevance of these studies to uncoating inside the host cell is not clear. Detailed studies with poliovirus have shown that penetration results in a loss of VP4, which results in an increase in RNase sensitivity and therefore presumably an increase in the permeability of the core. Such a permeable core would then be able to open out and function as a messenger RNA. In addition, other studies have demonstrated two conformational forms of the poliovirus nucleocapsid which are interchangeable, but only one of which results in an infection.

In spite of these difficulties, adenoviruses have a distinct morphology and their penetration and uncoating have been followed in some detail. When these viruses cross the plasma membrane, they change from a hexagonal to a spherical conformation, which results in a concomitant increase in DNase sensitivity. These subviral particles then decrease in size as they pass through the cytoplasm in association with microtubules and finally enter the nucleus via the nuclear pores, by an unknown energy-dependent process. Once in the nucleus, transcription can begin. Some biochemical studies have indicated that the subviral particles lack polypeptides IVa2, VII and V and that the integrity of polypeptide VII is crucial for the uncoating process. By contrast, other non-enveloped DNA viruses like the papovaviruses remain DNase resistant until they are inside the nucleus.

2.2.3. *Penetration and Uncoating of Enveloped Viruses*

The mechanisms whereby all small enveloped viruses enter the cell are probably very similar. As the best characterized system has been that of the entry of the alphavirus SFV into BHK cells, this system will be described in some detail and the differences exhibited by other groups of viruses discussed later. At present, very little is known of the mode of entry of the larger enveloped viruses such as the poxviruses and herpesviruses, so these will not be included in this discussion.

After SFV particles attach to the tips of the microvilli, they travel down them until they reach the main body of the cell. After a few minutes, the virus reaches a "coated pit". This is a specialized region of the cell, coated on the inside with the protein clathrin. The cell membrane then folds inwards and pinches off to form a "coated vesicle". The clathrin on the inner or cytoplasmic side of the membrane now forms a polygonal structure around the vesicle and it moves into the cytoplasm. As the vesicle moves through the cytoplasm, it loses its clathrin coat and fuses with a large smooth-surfaced vesicle, an endosome. The virus-containing endosome now fuses with a lysosome which contains many of the degradative enzymes of the host cell. However, instead of being destroyed by these enzymes, the virus utilises the low pH in the interior of these lysosomes to trigger the next step in penetration. The reduction in pH apparently induces a conformational change in the E1 protein which then allows the virus membrane to fuse with the membrane of the lysosome, thus releasing the nucleocapsid into the cytoplasm and allowing the first stage of replication to begin. Work with SFV on free membranes *in vitro* demonstrated that this fusion was instantaneous, indicating that his process would allow the rapid transfer of the nucleocapsid to the cytoplasm and allow the first stage of replication to begin. Thus, these viruses use both fusion mechanisms and endocytosis to enter their host cell. Recent work in several laboratories suggests that VSV and influenza penetrate cells by the same route. Moreover, a fusion activity has been demonstrated for influenza virus when the pH is lowered (25). However, the situation with influenza may not be identical to that with SFV as several workers have reported entry of influenza by uncoated vesicles. Although the uptake of virus has been observed at temperatures too low for simple membrane fusion to occur (55), it must be borne in mind that measurements on membrane transition temperatures have been made only with pure lipid membranes and the high proportion of proteins in viral membranes probably fundamentally alters their physical characteristics. With the paramyxoviruses, the mechanism of entry may be much simpler. One of their envelope glycoproteins, the F protein, exhibits a fusion activity at physiological pH and its integrity and correct cleavage have been shown to be essential for pathogenesis. Therefore, on binding to the cell surface, these viruses simply fuse with the plasma membrane and release the nucleocapsid directly into the cytoplasm.

For additional information, see Dimmock, 1982 (15); Dales, 1973 (13); Lonberg-Holm and Philipson, 1980 (31); Meager and Hughes, 1977 (33); Simons et al., 1982 (50); Acheson, 1980 (1) and Flint and Broker, 1980 (18).

2.3. Replication of the Viral Genome

If we assume that all vertebrate viruses originally derive from vertebrate cells, it may seem surprising that such a great variety of replication strategies is observed in these viruses. However, it is tempting to speculate that this variety in the structure and replication of viral genomes may be a reflection of the structural variety resident in the genome of the animal cell itself. Because of this great diversity, it would be impossible here to give even the briefest details of the replication strategies of all known virus families. Therefore, only those viruses whose replication mechanisms are best known have been described and, where appropriate, reference made to families where similar mechanisms probably operate.

2.3.1. The DNA Viruses

For the sake of brevity, only those viruses with a proven role in infectious disease will be discussed here. Although the papovaviruses cause only mild or rare diseases in humans, they are discussed in detail, since they have been studied at length and much is known of their molecular architecture.

Unfortunately, the viruses with the greatest pathological importance such as herpes and smallpox are too complicated at present for a detailed knowledge of their molecular structure to be available.

Papovaviruses DNA Replication.   Although this group of viruses have a circular genome, the organisation of their genetic material into mini-chromosomes along with cellular histones has long made them a popular model for the study of eukaryotic gene expression and replication. As soon as the infecting viral genomic material enters the nucleus, initiation of replication commences. This always starts at a specific site on the genome charac-terised by a 27-base-pair palindrome flanked on one side by 17 A/T base pairs. Initiation requires the binding of T antigen to the sequence GCCTC and then proceeds in both directions simultaneously. Although this system is a useful model for eukaryotic DNA replication, it is thought that the initia-tion process is not similar to that in the host cell, but rather represents a mechanism for uncoupling viral DNA synthesis from the cell cycle. Chain elongation occurs in association with two host DNA polymerases, α and γ and proceeds in a similar fashion to that of chromosomal DNA. The leading strand grows in a $5' \rightarrow 3'$ direction, that is in the same direction as the replication fork is moving, but as all polymerases synthesize nucleic acids in a $5' \rightarrow 3'$ direction, the other strand (the lagging strand) must grow in an opposite direction. This is accomplished by the synthesis of short fragments of DNA (Okazaki fragments) about 40–300 nucleotides in length, whose synthesis is initiated by small oligoribonucleotide primers. The primers are then removed by an as yet undescribed enzyme and then the gaps are filled by a DNA polymerase. Thus the lagging strand can grow in a $3' \rightarrow 5'$ direc-tion in a semi-discontinuous fashion. The nascent DNA strands are rapidly associated into nucleosomes, with the histones from the parental strands reforming on the leading daughter strand and new histones attaching to the lagging daughter strand. The mechanism whereby replication terminates and the strands separate is poorly understood. However, it appears that termination occurs at a point on the genome diametrically opposite to the origin, but does not appear to have an exact sequence specificity. Although little experimental detail is known, a plausible hypothesis has been pro-posed. This involves a continuation of replication until the two forks meet, at which point the parental strands are no longer held together by hydrogen bonds. This results in a dimer with a small gap in each nascent strand which is filled and ligated. The two resulting covalently closed circles can then be resolved into two monomeric circles by host cell topoisomerases.

Replication of Adenovirus DNA.   Upon entry into the nucleus, the final stages of uncoating are completed, leaving the genomic DNA almost free of protein. As both ends of the linear double-stranded molecule appear identical, replication can proceed from either end. However, as with all linear nucleic acid molecules, a major problem in replication is the accurate transcription of the terminal nucleotides. This virus appears to have overcome this problem by using a protein covalently linked to the 3' end of the DNA as the primer for DNA synthesis. As with all DNA polymerases, the daughter strands are synthesized in a $5' \rightarrow 3'$ direction, probably by the DNA polymerases from the host cell. However, at least two more virus-specific proteins appear to be associated with DNA replication. *In vitro*, replication has been shown to be dependent on the presence of a protein of M_r 55,000, and a DNA binding protein of M_r 72,000 has also been shown to be necessary for DNA replication. The exact role of this protein is not clear at present, but the single-stranded daughter molecules appear to be completely coated with it, and only the very ends of linear double-stranded molecules appear to bind it. These observations are consistent with a role as an unwinding protein. Replication continues along the complete length of the parental molecule to form a single-stranded daughter molecule, which is then converted to a double-stranded molecule by a similar mechanism. There is no evidence at present for the existence in the adenovirus genome of a chromatin-like structure with host cell histones or of Okazaki fragments occurring during DNA replication, as seen in the replication of the papovaviruses.

The Replication of Herpesvirus DNA. Predictably, as these are large complex viruses, it is impossible to discuss in detail what is known about their replication or give credence to all the postulated models. Only the main points are outlined below and reference must be made to the relevant literature for further details.

Before understanding the replication of HSV DNA, it is essential to appreciate that the L and S segments can invert relative to each other and that all four isomers will appear in a cell infected with only one isomer. However, it may not follow that all four isomers are functionally equivalent either in their recombinational frequency or in their transcriptional products, even though their genetic material is essentially identical.

On entry into the nucleus, the DNA circularizes by uncovering complementary terminal sequences after digestion with an exonuclease. Presumably this must occur after several rounds of transcription as at least two virion proteins are implicated in DNA replication. Thus, the circle would have one normal junction between the L and S segments and one modified junction. Initiation occurs at unique internal sites on either the L or S segment, or both. It is possible that, as with the papovaviruses, replication proceeds in both directions. This will proceed until nicks in the parental strand convert the replication complex into a "rolling circle", yielding head to tail multimers of the viral DNA. Excision of monomeric viral DNA can then occur at the modified junctions. The missing sequence at the modified junction can be generated as follows. The opposite strand to the template is first nicked. Then after branch migration it base pairs to part of the other terminal reiterated sequence and then is repaired. The linear form can be regenerated, either by nicking of the complementary strand, followed by branch migration to yield the same isomer as the parent, or by nicking of the template strand and subsequent branch migration which would lead to the inversion of the component with the repaired terminals. Therefore, herpesviruses have evolved yet another mechanism to ensure the complete replication of a linear genome.

For additional information, see Acheson, 1980 (1); Griffin, 1980 (21); Flint, 1980 (17); Flint and Broker, 1980 (18) and Spear and Roizman, 1980 (52).

2.3.2. The RNA Viruses

As Table II indicates, the RNA viruses demonstrate a wealth of variety in their genomic structure. For the sake of brevity and clarity, however, only four groups will be discussed which illustrate the main mechanisms of RNA virus replication. One factor should be noted and that is the comparatively narrow range of genome sizes. It is possible that because of the high mutation rate observed in RNA genomes (43) (because they have no equivalent of the DNA repair enzymes), molecules larger than 15 kilobases could not be replicated accurately enough to ensure successful transmission of all the necessary genetic information to the next generation.

Replication of Picornaviruses.   In spite of the wealth of detail on the structure and synthesis of the capsid proteins of this group of viruses, comparatively little is known about the mechanism of picornavirus RNA replication. However, an RNA polymerase complex has been isolated from infected cells. A surprising characteristic of this replicase complex is that it is associated with smooth host cell membranes, even though the virus contains no lipid and apparently no glycoproteins. This, however, would correlate with two unpredicted observations. Firstly, there appears to be a significant increase in membrane synthesis during poliovirus replication, and secondly, the majority of virus-specific synthesis occurs on membrane-bound ribosomes, even though there is no lipid envelope or membrane proteins in the virus. This latter observation can be explained by assuming that nascent progeny RNA molecules of positive polarity would preferentially associate with the ribosomes nearest at hand, i.e. those on patches of rough endoplasmic reticulum. However, one central problem of RNA replication remains unresolved, that is what destines a nascent positive sense RNA molecule for protein synthesis, further minus sense template synthesis or assembly into progeny virions. One recent observation that polysome bound virion RNA molecules do not have VPg on their 5′ end would suggest that proteolytic removal of this protein would partially determine the fate of these molecules in that only molecules without a proteinaceous 5′ terminus can act as messenger.

For additional information, see Agol, 1980 (2) and Putnak and Phillips, 1981 (42).

Replication of Retrovirus Genome.   The replication of these viruses has been studied in some detail. The initial stage in the replicative cycle is the conversion of the virion RNA into a double-stranded linear DNA molecule— the "provirus". The reverse transcriptase contained in the virus particle commences transcription at the 3′ end of a tRNA bound near the 5′ end of one of the two virion RNA strands. However, the template is exhausted within 100–200 bases of the point of initiation and transcription can only continue by the polymerase switching strands. The 5′ and 3′ ends of the viral RNA have a short nucleotide sequence (R-region) in common and, after the 5′ end of the template RNA has been digested by the RNase H activity associated with the reverse transcriptase, the nascent cDNA binds with its complementary sequence on the 3′ end of the other RNA molecule. Transcription will then occur throughout the length of the new template. However, sequence analysis of the proviral DNA indicates that a second strand switch must occur. Thus, when the reverse transcriptase reaches the 5′ end of its new template, the R-region of the template is again digested and the nascent cDNA binds to the 3′ R-region, either of itself or its sister template, and then transcription proceeds. The second strand of the proviral DNA appears to be synthesized by the same enzyme once the template RNA has been removed by the RNase H activity. The next stage in replication is the conversion of the linear duplex DNA to a covalently closed circle. This occurs in the nucleus and probably involves several, at present unknown, cellular enzymes. This is probably expedited by the inverted repeat sequences at the end of the linear DNA, which can form single circles, dimeric or trimeric structures. This circularization is not a precise mechanism, since variable amounts of the repeated sequences from either end can be lost. The penultimate step in replication is the integration of the circular DNA into the host chromosome. This mechanism is not at present well understood, but it appears that almost the entire genome is integrated, at a large but finite number of sites in the host chromosome. Although some small sequences may be missing from the ends of the proviral DNA, this DNA is generally inserted with its inverted repeat sequence at the ends. Once integrated into the host chromosomal DNA, the proviral DNA is transcribed by cell DNA polymerase II to form progeny virion RNA, which is capped, a poly A tail added and then transported to the cytoplasm for incorporation into progeny virus. Finally, the integrated proviral DNA will be replicated along with the host chromosomal DNA and thus be transferred to subsequent generations of host cells. The integrated DNA can be excised from the host chromosome, but as this mechanism is not precise (as is the integration step), host sequences can be incorporated into proviral DNA. Thus, these viruses have a mechanism for transferring host sequences from cell to cell. Such a process may have resulted in the integration of oncogenic host DNA sequences into some of these viruses.

For additional information, see Coffin, 1982 (12); Varmus and Swanstrom, 1982 (58) and Bishop and Varmus, 1982 (7).

Replication of Alphavirus RNA.   Although the alphaviruses are similar to the picornaviruses in that they contain a single-stranded, positive sense RNA molecule in their virions, they are fundamentally different in their mode of replication. The virion RNA of the infecting virus has two functions, one as a template for negative strand synthesis and the other as a messenger for the non-structural polypeptides, including the replicase (or replicases). As there is no polymerase function associated with purified virions, the first step must be the synthesis of the RNA replicase. This is situated at the 5' end of the genome and therefore the enzyme can be synthesized directly from the input RNA. The 5' and 3' termini of the input RNA have inverted complementary sequences and can anneal together to form "pan-handle" structures. This terminal double-stranded region possibly acts as an initiation signal for the binding of the replicase at the 3' end. Negative strand synthesis then proceeds in a 3' → 5' direction along the parental strand and, like picornaviruses, appears as a replication complex on the smooth membranes of the host cell. The 3' poly A tail is also copied into a poly U tract on the negative strand RNA and thus these sequences are not added post-transcriptionally as they are for cell messengers. The 42S negative sense molecule can then act as template for both progeny, positive sense 42S virion RNA and the 26S messenger RNA coding for the coat proteins. The virus-specific polymerase consists of three separate polypeptides, NSp60, NSp82 and NSp89, which are all cleaved from a large polyprotein. Several independent lines of evidence have indicated how both the 26S and 42S positive sense RNAs are synthesized from the 42S template. Briefly, both NSp60 and NSp82 form the "core" polymerase which functions in the synthesis of both 42S and 26S RNA. If NSp89 also binds to this core, but is inactive, only 42S RNA is made. However, if NSp89 is activated, this allows the polymerase to bind to an internal initiation site and only 26S RNA is made. This modulation of NSp89 is thought to occur by a rapid but transient interaction of the nucleocapsid polypeptide with NSp89. Thus with excess nucleocapsid polypeptide, NSp89 is inactivated and only 42S RNA is made, but when levels fall, NSp89 is activated and thus the 26S RNA, the messenger for structural proteins, is synthesized. Therefore, by this elegant mechanism, the virus ensures that the synthesis of progeny virion RNA and the proteins to encapsulate it are synchronized. No mention has been made here of the synthesis of defective genomes and their putative role in persistent infections, even though they have thrown much light on the mechanisms discussed above. This aspect of alphavirus RNA synthesis has been reviewed elsewhere (26).

For additional information, see Kennedy, 1980 (26), (27).

Replication of the RNA from Negative Strand Viruses.   As the majority of data from this type of virus have been collected from experiments on paramyxoviruses and rhabdoviruses, this discussion will be centred on them. However, a paragraph has been included to indicate the areas in which the replication of the orthomyxoviruses differs.

The genomic RNA of negative strand viruses serves two mutually exclusive functions, that of acting as template for messenger RNA synthesis and that of acting as template for the synthesis of full-length complementary RNA from which progeny virion RNA molecules can be synthesized. The alternation between these two roles is the key function in the control of virus replication. The virion RNA polymerase first of all binds to the 3' end of the virus negative strand RNA and transcribes a short sequence of bases to generate a "leader" RNA molecule, 48 nucleotides in length. At the junction of the leader sequence and the coding region for the first gene (that for N protein) lies the leader RNA termination site or alternator, which is thought to determine whether transcription or replication occurs. If low levels of N protein are present, the leader termination signal is recognized. This, in turn, allows RNA polymerase molecules to bind to the promoters at the start of each virus gene. The genes are then transcribed to produce monocistronic messenger RNAs. When the level of N protein rises and N protein molecules bind to the nascent leader RNA sequence, the alternator is suppressed and RNA synthesis occurs along the entire genome, with the concomitant binding of N protein molecules to form a nucleocapsid structure. In a similar way, progeny negative sense RNA would be transcribed from the newly-formed positive sense antigenomic nucleocapsid. In this latter case, negative sense leader RNAs are also found and are thought to control progeny RNA synthesis in a similar manner. However, in this case, they do not allow the synthesis of "negative sense messengers", but prevent synthesis of progeny genomes if there are insufficient levels of N protein to form nucleocapsids. This could be an important control mechanism, as an excess of negative sense "naked" RNA would hybridize to viral messengers and inhibit protein synthesis as well as induce an antiviral state by using these double-stranded RNA molecules to induce interferon, or other related antiviral products.

In general, the synthesis of negative strand segmented genomes (such as those in influenza virus) follows a similar pattern. However, even though each segment appears to code for only one gene, each physiological messenger RNA is not a complete transcript of the virion RNA. Antigenomic template RNA molecules are found in infected cells and these are complete transcripts of the virion nucleic acid. It is assumed that the switch from messenger synthesis to template synthesis is similar to that described above for rhabdoviruses, but the temporal control of protein synthesis discussed below should also be borne in mind. The other major difference seen in the orthomyxoviruses is the role of the host nucleus. It has been known for some time that the replication of influenza virus is dependent on an intact cell nucleus and that this is the initial target for virus RNA (53). Recently, these earlier observations have been explained by the demonstration that influenza messengers require the transfer of sequences from the 5' termini (including cap structures) of cell messengers in order to function. As these enzymes are only found in the nucleus, these events would explain the nuclear requirement for influenza replication.

For additional information, see MacCauley and Mahy, 1983 (32); Kolakofsky and Blumberg, 1982 (28); Banenjee et al., 1977 (5); Choppin and Compans, 1975 (11); Wagner, 1975 (59) and Skehel and Hay, 1978 (51).

2.4. Synthesis of Virus-specific Proteins

As there appears to be little evidence of translational control in animal viruses and therefore all control mechanisms appear to operate at the level of transcription, RNA transcription and translation will be discussed together. In order to illustrate general principles, only those virus groups whose protein synthesis has been studied in some depth will be discussed. Comparatively little work has been done on many clinically important virus groups such as the poxviruses, arenaviruses, bunyaviruses, herpesviruses and hepatitis B virus and what information is available suggests that their protein synthesis mechanisms are very like those of the virus groups discussed below.

2.4.1. Synthesis of Papovavirus Proteins

Synthesis of papovavirus proteins in a lytic infection is temporally regulated into "early" and "late" phases, separated by the onset of virus DNA synthesis. The messenger RNAs for the early protein are transcribed from the "E" strand of the circle viral DNA molecules. The initial transcripts cover all the nucleotides from the "A" region of the genome starting near the origin of the replication. This precursor molecule is then capped at the 5' end and polyadenylated at the 3' end, probably by the same enzymatic mechanisms used for the host cell. There is some evidence that the poly A tail may be transcribed from a poly T sequence in the viral DNA, but in general, it is thought that the 3' poly A tail is added by cellular enzymes. The other early proteins, the U and S antigens and transplantation antigens are probably not virus coded and their synthesis will not be discussed here.

In order to form the messenger for the small T polypeptide, only a small deletion is made in the primary transcript after the termination codon for the small T polypeptide and then the 5' and 3' fragments are joined together. The messenger for the large T polypeptide has a much larger deletion, which includes the translation termination codon found in the small T messenger. The 5' fragment and 3' fragment are then spliced to form the messenger for the large T polypeptide. The spliced messengers are then transported from the nucleus for translation in the cytoplasm. The synthesis of messenger for large T antigen is regulated by a negative feedback mechanism triggered by the levels of large T antigen in the cell.

Once DNA replication commences, both late and early messenger RNAs are transcribed. The messengers for the late proteins are transcribed from the L strand, again commencing at the origin of replication, but covering the other half of the DNA to that transcribed during the early phase. Transcription of late messengers is about 10–20 times more efficient than that of the early messengers. Very little is known about the termination of the L strand transcripts, but it appears to be very inefficient as transcripts up to four times the length of the genome can be found. These transcripts are then capped and spliced in a way similar to that described for the early messenger precursors. The three structural polypeptides VP1, VP2 and VP3 are all translated from separate messengers, albeit with virtually identical leader sequences on the 5' end. VP2 is made from the largest messenger RNA, which is transcribed from the entire late region and is encoded at its 5' end. VP3 is made from a similiar messenger, but with a small deletion near its 5' end. VP1 is made from the 16S messenger with the same 5' leader sequence as VP2 and the same 3' sequences, but contains a deletion covering most of the sequences for VP2 and VP3. Thus, both the 19S messengers contain the sequences for VP1, but neither message is capable of translating them into protein. Transcription of the L strand is also regulated by large T antigen, but this time by a positive feedback mechanism.

For additional information, see Acheson, 1980 (1) and Tooze, 1973 (57).

2.4.2. Synthesis of Picornavirus Polypeptides

All virus-specific proteins in picornavirus-infected cells are synthesized from an RNA molecule which is indistinguishable in sequence from virion RNA, but this active messenger RNA is unlike virion RNA in the sense that it lacks the 5' terminal protein VPg. Even though the virus genomic RNA codes for several proteins, it behaves as a monocistronic messenger, since its primary translation product is a single large precursor polypeptide. The primary processing of this molecule is probably performed by cellular proteases to form the proteins NCVP1a, NCVP3a and NCVP1b, which represent the products of viral genes CR1, CR2 and CR3 respectively. All further processing is probably mediated via virion proteases and occurs in the cytoplasm, with the exception of the cleavage of NCVP0 to VP2 and VP4, which takes place in the completed virus particle. NCVP1a is cleaved via several intermediates to form the four capsid proteins. NCVP3b is cleaved via a single intermediate to form NCVPX, the putative viral protease. NCVP1b is first cleaved to form the viron-associated protein VPg and the precursor to the other four stable non-structural proteins. These latter proteins are assumed to form the replicase complex. One of the notable features of picornavirus replication is the rapid shut-off of protein synthesis. The mechanism of this process is unknown, but one theory which has some experimental support exploits the fact that some viral messengers are capable of more efficient translation at elevated salt levels than are cellular messengers. This theory proposes that early in infection, the accumulation of viral proteins on the cell surface stimulates an influx of cations which is preferentially favourable for the translation of viral proteins (10). Another theory, which has much experimental data to support it, involves the inhibition of a cellular "cap binding protein" (a protein responsible for attaching "capped" cellular messengers to ribosomes) by a viral component (16).

For additional information, see Putnak and Phillips, 1981 (42) and Agol, 1980 (2).

2.4.3. Synthesis of Retrovirus Polypeptides

Even though the virion RNA of these viruses is "messenger sense", it does not appear to code directly for virion proteins in the infected cell; instead it must first be transcribed into DNA which is then integrated into the host chromosome DNA as proviral DNA from which the transcription of all viral messenger RNA occurs (see section above on replication). The major internal structural polypeptides are formed from a precursor protein (Pr76^{gag}) synthesized from a messenger RNA which is apparently indistinguishable from full-length genomic RNA and represents a complete transcript of the proviral DNA. The coding sequences for Pr76^{gag} lie near the 5' end of the viral RNA and are preceded by at least three possible initiation codons and a ribosome binding site. The sequence of viral core proteins on the precursor is NH_2-p19-p10-p27-p12-p15-COOH, and thus translation is normally terminated on this messenger after p15. A putative host cell protease first cleaves off p15, which then acts as a *gag*-specific protease to cleave the remainder of the molecule to form p12, p23, and p27. Polypeptide p23 is then further processed to form p19 and p10. In addition, some RNA tumour viruses produce a glycosylated *gag* gene product which appears on the surface of the cell membrane. Its function is not known, but some evidence indicates it may be synthesized from a different population of messengers than its non-glycosylated counterpart. Synthesis of the reverse transcriptase occurs from the same RNA molecule as Pr76^{gag} by an apparently haphazard mechanism involving an occasional readthrough of the *gag* gene. Thus, the initial product of translation is a polyprotein of M_r180,000 termed Pr180$^{gag\text{-}pol}$. The situation is not that simple, however, as the initiation condon of *pol* and the termination codon of *gag* are in different translational reading frames. Recent evidence seems to indicate that the message for Pr180$^{gag\text{-}pol}$ is not the same as for Pr76^{gag}, but contains a deletion in the intercistronic junction; subsequent splicing then brings the two genes into phase. The precursor is first cleaved in the same place as Pr76^{gag}, i.e. between p15 and p12, which is then cleaved to release p15 and the β subunit of the reverse transcriptase, which itself is further cleaved to produce the α subunit and p32 (i.e. ribonuclease H). The cleavage of the β subunit does not occur in the cell, but in the mature virion. Unlike the products of the *gag* and *pol* genes, the *env* gene products are synthesized from a subgenomic RNA derived from a segment of the 3′ end of the genome spliced onto a sequence from the 5′ end of complete viral genomic RNA. After the translating ribosome has traversed the initial 192 nucleotides, the N-terminal amino acid chain binds the translating ribosome to the host cell membranes. This hydrophobic N-terminal sequence is termed a "signal" sequence (9) and enables the subsequent nascent amino acids to be passed through the membrane and thus reach the outside of the cell. Recent studies have described a cellular ribonucleoprotein particle containing a highly conserved 7S RNA molecule which is responsible for attaching the signal sequence to the ribosome (60). The precursor molecule (P 63^{env}) rapidly loses its signal sequence (probably before completion) to form P 57^{env}, but is probably held in the membrane by other hydrophobic sequences further along the amino acid chain. This latter molecule is then extensively glycosylated (probably as it passes through the cell membrane) and is cleaved to form the two subunits of the mature envelope protein gp85 and gp37. This mechanism whereby membrane proteins can be synthesized directly into the cell membrane is used for the synthesis of most cellular membrane proteins. Furthermore, secretory proteins are also made in this way, but have no second hydrophobic tail and are thus released into the extracellular environment. In viruses with a *sarc* gene, this is situated at the 3' end of the *env* gene and is translated from its own monocistronic messenger. In these viruses the messenger for the *env* gene products contains sequences for both *env* and *sarc*, but again only the 5' gene (i.e. *env*) is translated.

For additional information, see Varmus and Swanstrom, 1982 (*58*); Dickson *et al.*, 1982 (*14*); Bishop, 1978 (*6*) and Temin, 1971 (*56*).

2.4.4. Synthesis of Alphavirus Proteins

Alphavirus proteins are synthesized from two discrete messenger RNAs in the infected cell. Other intracellular species of RNA are structural isomers of these two major virus-specific RNA species. The viral genome acts as messenger for the non-structural proteins, but even though this RNA molecule contains the sequences for the structural genes, they are never translated, as this molecule has a termination codon immediately preceding the gene for the core polypeptide. Thus, it conforms to the general pattern of eukaryotic messengers, in that they can terminate protein synthesis internally, but will only initiate at the 5' terminus of a messenger RNA molecule. Translation of this molecule results in a large precursor polypeptide with an approximate M_r of 200,000, which is then cleaved via a variety of intermediates to form the major non-structural polypeptides, at least some of which form the viral replicase. These cleavages are possibly the result of the activity of cellular proteases. The second intracellular messenger is a 26S RNA molecule and codes for both the non-glycosylated core polypeptide and the glycosylated membrane polypeptides. The synthesis of proteins with such apparently disparate functions is achieved as follows. The capsid protein is encoded at the 5' end of the 26S messenger RNA and is translated first. The ribosome then continues along the RNA molecule and instead of producing a polyprotein, the capsid is cleaved off immediately, probably by self-proteolysis. The next sequence to be translated produces a hydrophobic signal peptide similar to that described for the retrovirus envelope proteins above, and this enables the ribosome to bind to cellular membranes. The precursor to the envelope polypeptides E3 and E2 (p62) is then secreted through the membrane and anchored in the cell membrane by a hydrophobic sequence near its C terminus; the N-terminal signal sequence is then cleaved off. The ribosome subsequently proceeds into the coding region for E1 (which is also embedded in the membrane at its C terminus) and upon reaching the termination codon, p62 and E1 are cleaved. Both polypeptides are then glycosylated and the final cleavage of p62 to E3 and E2 completed. A crucial problem in the replication of an alphavirus is what determines whether a 26S or 42S positive sense RNA is transcribed from the negative sense 42S template RNA. The mechanism whereby the synthesis of 42S RNA and 26S RNA is regulated by the intracellular levels of core polypeptide is discussed above in the section on RNA replication. Details of the glycosylation mechanism are included in the section on modification below.

For additional information, see Lodish *et al.*, 1981 *(30)*; Simons *et al.*, 1980 *(49)*; Simons *et al.*, 1982 *(50)* and Schlesinger and Kääräinen, 1980 *(48)*.

2.4.5. *Synthesis of Negative Strand Virus Proteins*

All negative strand viruses appear to synthesize their proteins by similar mechanisms in that they all appear to be translated from monocistronic messengers. Therefore this section is based on studies with rhabdoviruses, but indicates the areas where the orthomyxoviruses and paramyxoviruses differ.

Synthesis of rhabdovirus proteins occurs from monocistronic messengers with 5' caps and 3' poly A tails transcribed from virion RNA as described in the section on replication and is concomitant with a decrease in cellular protein synthesis. There is no evidence of an early and late phase as found with the DNA viruses, although more NS protein is synthesized early in infection than later on. Although all virus proteins are synthesized at all times, they are not synthesized in equal amounts, the N protein (major nucleocapsid protein) being predominant and the L protein (putative replicase) being the least abundant. As these proportions are not reflected exactly in mRNA levels, there is some evidence for translational control. The synthesis of the orthomyxoviruses and the paramyxoviruses is broadly similar; however, the levels of active messengers in infected cells appear to reflect the levels of *in vivo* protein synthesis, indicating that only transcriptional control is operating. In addition, orthomyxoviruses not only differ in their nuclear requirement (see section on replication) but also demonstrate temporal control of transcription. Furthermore, this control mechanism appears to be defective during abortive infections *(54)*. The mechanisms of membrane protein synthesis described for the retroviruses and alphaviruses also appear to apply for both the orthomyxoviruses and paramyxoviruses.

For additional information, see Wagner, 1975 *(59)*; Choppin and Compans, 1975 *(11)* and Lodish *et al.*, 1981 *(30)*.

2.5. Processing and Modification of Viral Products

The processing of viral nucleic acids has been dealt with in the sections above on replication and protein synthesis. In addition, processing of viral proteins by proteolytic cleavage has been covered in the section on protein synthesis. The correlation between such proteolytic cleavage and viral pathogenicity has been recently reviewed by Rott *(46)*. Apart from proteolytic cleavage of precursor molecules, the other major post-translational modification showed by viral and cellular proteins is that of glycosylation. Glycosylation of most viral polypeptides and cellular proteins occurs on the rough endoplasmic reticulum and commences before translation is complete. Sugar residues are either added to asparagine residues via an N-glycosidic linkage or to threonine, hydroxyproline or hydroxylysine via an O-glycosidic linkage. The former linkages are normally associated with the amino acid sequence asn-X-thr(ser) and the latter with the sequence gly-X-hyl-gly. Most virus glycoproteins are of the N type, but recently, one of the coronavirus glycoproteins has been shown to contain the O type of linkage (38). The O-type linkages appear to be made by glycosyltransferases, which transfer sugar residues from nucleoside monophosphates to the required amino acid in the protein or to an already growing oligosaccharide chain. The synthesis of the N-type linkages is more complex. A "core" branched oligosaccharide is first generated from mannose, glucose and N-acetyl-glucosamine residues linked to a dolicol phosphate lipid carrier. The structure of this oligosaccharide is heavily conserved in both viral and cellular glycoproteins. The dolicol phosphate carrier then transfers the oligosaccharide en bloc to the nascent polypeptide chain, and immediately following this transfer, one or two glucose residues are removed. Further processing occurs after the completed protein has been transferred to the Golgi apparatus. Firstly, the remaining glucose residues and six of the nine mannose residues are removed; then the "peripheral" sugars—three N-acetyl-glucosamines, three galactoses, one sialic acid and one fucose—are added in a stepwise fashion. Finally, the finished glycoprotein is transported to the cell surface.

Phosphorylation is the next most widespread form of post-translational modification amongst both viral and cellular proteins and is especially common amongst viral proteins associated with nucleocapsid structure or replication. The addition of phosphate can be mediated by either cell or virus coded kinases and usually transfers a terminal phosphate from a nucleotide triphosphate to any external serine or threonine residue. The function of this process remains obscure at present, but interest in the process has been stimulated by the discovery that the *sarc* gene product from some retroviruses responsible for oncogenesis has a kinase activity which phosphorylates both itself and a variety of cellular proteins, including several metabolically important enzymes. Moreover, this kinase has a unique action in that it phosphorylates preferentially at tyrosine residues.

Other forms of post-translational modification have been reported including acylation and sulphation, but these have not been studied in depth, and their function, if any, has not been determined.

For additional information, see Lodish *et al.*, 1981 (*30*); Simons *et al.*,

1982 (50); Hanover and Lennarz, 1981 (23) and Varmus and Swanstrom, 1982 (58).

2.6. Assembly and Release of Progeny Virus Particles

Like its counterpart, virus penetration and uncoating, the mechanisms of virus assembly and release are poorly understood. Even in the best studied systems, no attempt is made to explain how what is essentially a macromolecular reversible reaction, virus uncoating $\rightleftharpoons$ virus assembly, proceeds in one direction at the start of infection and in the other at the completion of the infection cycle. In this section, we will concentrate on three virus groups, adenoviruses, picornaviruses and alphaviruses, and where necessary refer to areas of similarity or difference with other virus groups.

The initial stage in adenovirus assembly is the production of empty capsids. But these capsids do not contain all the proteins found in the capsids of mature virions and in addition contain at least five polypeptides which are not found in mature virus particles. These latter polypeptides, however, are not stable and appear to be precursors of other capsid proteins. Nascent DNA, complexed with the two core proteins V and VII, then enters the empty capsids, starting at the left-hand end. The entry of DNA is associated with the disappearance of the 50K and 39K polypeptides from the now full virus capsids. Further processing removes the IVa2 polypeptide and results in the production of several capsid proteins from their respective precursors. Similar mechanisms, i.e. the insertion of DNA into a preformed capsid followed by subsequent modification, appear to occur with other DNA virus groups including the papovaviruses and herpesviruses, although much less is known about their detailed mechanism. Very little is known about the mechanism of release of completed virions from the infected cell.

The assembly of the picornaviruses has many features common to those described above for the adenoviruses. Briefly, five molecules of the NCVP1a polymerize and are cleaved to form an aggregate consisting of VP0, VP1 and VP3. Twelve of these aggregates associate to form the procapsid into which the viral RNA is inserted, and then form the provirion. As discussed in the section above on RNA replication, only those RNA molecules attached to VPg are incorporated into the procapsid.

The provirion undergoes the final cleavage of VP0 to VP2 and VP4 to produce the mature virus particle. Again, very little is known about virus release, but the most likely mechanism would rely on the general disintegration of cellular structures following cell death with the subsequent release of virus particles.

The mechanism of assembly of the enveloped RNA viruses is fundamentally different from that described above for the picornaviruses and DNA viruses. Here, the assembly of the alphavirus SFV is described as a model. As mentioned in the section on RNA replication, the nascent virion RNA is immediately associated along its entire length with a core protein—most vividly demonstrated by the "herringbone" structures seen with the paramyxoviruses. As both virion RNA molecules and their negative strand templates are associated with core proteins, it is not known why mature particles only contain RNA molecules of one polarity. A similar, but inverse, problem occurs with the enveloped negative strand viruses. In contrast, there is some evidence that measles virus may contain RNA molecules of both polarities (J. R. Stephenson, unpublished data) and this may be the true interpretation of the "double-stranded RNA" in the reovirus. The completed nucleocapsid migrates to the membrane, where it attaches to the "cytoplasmic" end of the viral glycoproteins. As these proteins have a higher affinity for the nucleocapsid (and also possibly for each other) than the membrane-associated proteins of the cell, they coalesce around the nucleocapsid. This will eventually cause the membrane to bulge outwards and finally pinch off, resulting in a mature virus particle. Thus, the lipid component of the host cell, but not the protein component, is reflected in that of the progeny virus particle. Similarly, studies on influenza, various paramyxoviruses and rhabdoviruses have demonstrated that very similar mechanisms operate during their assembly process.

For additional information, see Putnak and Phillips, 1981 (*42*); Flint and Broker, 1980 (*18*); Brown, 1980 (*8*) and Simons *et al.*, 1982 (*50*).

2.7. Persistent Infections

Although in the normal course of events, most viruses infect their host in a lytic cycle, members of every well-known virus family have been shown to cause persistent infections, at least in some cell types. A possible exception to this rule is the orthomyxovirus family, although they frequently institute cycles of abortive infection. The importance of persistent virus infection has become increasingly manifest during the past decade. Well-documented examples range from relatively mild diseases such as warts and cold sores to more serious conditions, including certain types of childhood diabetes, recurrent genital herpes and shingles. In addition, recent attention has focussed on the possible association of persistent viral infections with chronic diseases of the central nervous system, e.g. multiple sclerosis and cystic fibrosis. A further intriguing possibility is that of a possible role for persistent virus infection in long-term T-cell memory.

The molecular mechanisms whereby viruses can cause persistent infections are poorly understood, with the possible exception of the retroviruses. Broadly speaking, there are four main mechanisms whereby a virus could possibly cause a persistent infection in the host cell. The virus genome can be integrated into the chromosomal DNA of the host and thus pass unmolested from generation to generation. Only when transcription of the integrated viral DNA is initiated and the resultant gene products appear on the cell surface will the host defences be alerted to the presence of virus. This is the mechanism by which the parvoviruses, papovaviruses, adenoviruses and retroviruses persist, and it is discussed in the above sections on replication. With the former groups, only part of the genome will sometimes be integrated, but with the retroviruses, such a persistent, noncytocidal infection is the normal course of events. As it has been clearly shown that these RNA-containing retroviruses can cause infections in this way, some workers have described experiments to show that many other RNA viruses can cause persistent infections in this way. However, as only the retroviruses have an RNA-dependent DNA polymerase, it is difficult to see how other virus groups can integrate into host DNA unless they utilise similar enzymes from endogenous retroviruses. Furthermore, other workers have shown that at least some RNA viruses can form persistent infections without detectable virus sequences in the DNA of the host cell.

One of the most frequent, but not ubiquitous, characteristics of a cell persistently infected with an RNA virus is the presence of defective interfering (DI) genomes. These, as their name implies, can be of virion sense RNA, complementary viral RNA or both, but with internal deletions of variable size. However, all DI genomes contain the ends of the viral RNA, presumably because they are essential for replicase activity. DNA viruses can also produce DI genomes. Most, but not all, DI genomes can be complexed with viral proteins to form DI particles or, as in the case of the DNA tumour viruses, can integrate into host cell DNA and initiate transformation. Although it is possible that DI particles may affect virus assembly, as with poliovirus, it is generally held that the DI genomes interfere with the replication of viral genetic material, although the mechanism whereby they achieve this is far from clear. The significance of DI genomes in persistently infected cells is difficult to evaluate. Although most DI genomes are not capable of transcription or translation, some have been found which are; but even these DIs, like all others, do not code for the replicase protein. It has been shown that some DI genomes induce interferon, but as this is not true in all cases, it is doubtful whether this can be considered as their mode of operation. Although DI genomes are readily demonstrated in persistently infected tissue cultures and can be induced by sequential passages at high multiplicity of infection, they have only been demonstrated in naturally occurring infections with LCM virus. Thus, it is not clear at present whether DI genomes cause, or can be a by-product of, persistent infections.

Several workers have demonstrated that persistently infected cells give rise to an increase in temperature-sensitive mutants, and recently changes in RNA sequence and amino acid sequence of the structural proteins of VSV have been observed during persistent infections (24, 47). Also, virus derived from a chronic infection with murine coronavirus has been shown to have evolved faster than other viruses from acute infections (61). Moreover, changes in the antigenicity of measles virus envelope glycoprotein from persistently infected cells has also been demonstrated (37). This latter observation is of particular interest, for if a virus can change the antigenicity of its envelope proteins, this could enable it to evade the immune system of the host. Again, however, it is difficult to determine whether mutant viruses are the cause of, or result from, a persistent infection. One feature of RNA genomes should be borne in mind in this context, and that is that as they do not appear to possess genome repair mechanisms, as do the DNA viruses, their mutation rates are several orders of magnitude higher and thus they can change very much faster than their DNA counterparts.

Several host cell factors have been implicated in persistent infection, of which mutant cell populations and interferon (see following section for a discussion of its mode of action) are the most widely studied. Interferon has been demonstrated in several persistent infections, but cannot be implicated in all, as some cell lines (notably Vero cells) are incapable of interferon production, yet readily produce persistent infections.

In conclusion, the cause of persistency may not be ascribed to any one event, but is probably induced by any one of several incidents, which probably differ in each case, thus giving rise to a vast array of different persistent infections, each with its own peculiar characteristics.

For additional information, see Rima and Martin, 1976 (44); Friedman and Ramseur, 1979 (19); Pringle, 1979 (41); ter Meulen and Stephenson, 1980 (36) and Mims, 1982 (34).

3. MOLECULES INVOLVED IN HOST DEFENCE MECHANISMS

3.1. Introduction

In the previous sections in this chapter, the structure and synthesis of eukaryotic viruses have been discussed as a background to the development of new vaccines and chemotherapeutic products. Another approach to the problems of infectious diseases is to augment the natural host defence mechanisms by supplying additional amounts of key components. The most widely known and studied of these are the antibodies, and their structure and synthesis are discussed in Chapter 3; this volume. In this section, those molecules associated with the cellular immune response will be discussed. In particular, the lymphocyte mediators (lymphokines) and the regulatory proteins from mononuclear phagocytes (monokines) will be discussed, along with a closely related family of antivirals—the interferons.

3.2. Lymphokines

All cellular immune reactions are mediated by T lymphocytes and are expressed in a variety of phenomena, including cutaneous delayed-type hypersensitivity, contact allergy, graft rejection, tumour surveillance and resistance to infection by intracellular parasites. These reactions result from complex interactions between T cells, B cells and macrophages and are mediated by the T-cell mediators or lymphokines. These factors are produced in minute quantities, but have dramatic effects in producing inflammatory responses at sites of infection. Because lymphokines are produced in minute quantities, the evaluation of their molecular structure and mode of action has been severely hampered. Moreover, the inability of workers in the field to produce pure cultures of one cell type has prevented the unequivocal association of molecular species to specific biological activities. However, the increasing introduction of monoclonal-lymphokine secreting T-cell lines and hybridomas and the increasing use of highly-specific monoclonal antibodies for both the purification and analysis of lymphokine function should enable significant advances to be made in the next few years.

The most frequently studied lymphokine activities are listed in Table III. It should be noted that these descriptions refer to measurable biological properties and at least in some cases several activities could reside on a single molecule or group of related molecules. Although detailed molecular analysis of most lymphokines has not been possible, most observable biological activities have been assigned to proteins, many of which are probably glycoproteins. The mechanism whereby the synthesis of lymphokines is controlled is at present obscure, although all appear to be made in either T cells or B cells. It must be emphasized, however, that the synthesis of B-cell derived lymphokines is always under the control of T cells. The mode of action of lymphokines is also not well understood, with studies on drug sensitivity revealing no common mechanism of action. The one exception is the effect of raised levels of cellular cyclic AMP, which reduce the activity of all the most widely studied lymphokines. Such a correlation would lend support to the hypothesis that these molecules induce conformational changes in the target cell membrane which then acts via intermediary molecules like cyclic AMP to produce the wide range of effects listed in Table III.

Although most lymphokines have been described *in vitro*, several similar activities have been reported *in vivo* in several body fluids including lymph,

TABLE III Lymphokines

Target cell	Class of lymphokine	Accepted abbreviation
Macrophages	Migration-inhibiting factor	MIF
	Antigen-dependent migration inhibitory factor	—
	Macrophage activating factor	MAF
	Chemotactic factors for macrophages	—
Polymorphonuclear leukocytes	Leukocyte inhibitory factor	LIF
	Eosinophil stimulation factor	ESP
	Chemotactic factors	—
Lymphocytes	Mitogenic factors	—
	Antigen-dependent factors enhancing antibody formation	—
	Antigen-independent factors enhancing antibody formation	—
	Antigen-dependent factors suppressing antibody formation	—
	Antigen-independent factors suppressing antibody formation	—
	Lymphotoxin	LT
Other cells	Growth inhibitory factors	—
	Osteoclastic factors	OAF
	Collagen-producing factor	—
	Colony-stimulating factor	—
	Interferon	IFN
—	Immunoglobulin-binding factor	IBF
	Procoagulant or tissue factor	—

serum and synovial fluid. Abnormal production of various lymphokines has been reported in patients with lymphoproliferative disorders, cancer and infectious mononucleosis, although the exact significance of these observations is not clear at present.

3.3. Monokines

Monocytes and macrophages play a central role in resistance to infection as they not only are actively involved in the removal of micro-organisms, infected cells and cell debris, but they also have evolved a close association with the lymphocytes that are involved in the induction of specific immunity to antigens. Monokines are the molecular mediators whereby it is postulated that these cells are able to perform these functions. The most frequently studied activities are listed in Table IV, but again it must be realized that these are observed biological phenomena and may not represent separate

TABLE IV Monokines

Biological activity	Recognized abbreviation
Plasminogen activator	—
Elastase	—
Collagenase	—
Complement proteins	—
Lymphocyte activating factor	LAF
B-cell differentiation factor	—
B-cell activating factor	BAF
Thymic differentiation factor	TDF
Mononuclear cell factor	MCF
Monocyte pyrogen	EP
Colony-stimulating activity	CSA

molecular species. Indeed, one factor (LAF) has been reported to exhibit several of these biological activities. In general, monokines appear to be smaller than lymphokines, with few, if any, reports of them containing sugar residues or being inserted into membranes. At present, there is little or no information on the control of monokine synthesis or on their mode of action. Although all monocytes and macrophages secrete lysozyme and other hydrolytic enzymes, these are not considered to be lymphokines (although other proteases are; see Table IV) as they are secreted at a constant rate throughout the life-cycle of the cell and thus probably do not exercise a regulatory role.

For additional information, see Altman and Katz, 1982 (4) and Rocklin et al., 1980 (45).

3.4. Interferons

3.4.1. Structure

The interferons (IFNs) are a family of proteins, many sharing considerable sequence homology, which occur in a variety of vertebrates including the bony fishes and man. Broadly speaking, they act as biological regulators of cell function; their ability to induce resistance to viral infection is the property which led to their discovery. The best studied IFNs are those from mice and men. They are among the most potent biological agents known and can operate at 3×10^{-14} M. Human interferons fall into three major antigenically distinct classes, α-IFN (leukocyte), β-IFN (fibroblast) and γ-IFN (type 2 or immune). There are at least eight species of α-IFN, all of which are located on chromosome 9, with six of them in a continuous stretch of DNA. Of the species of β-IFN detected so far, one is located on chromosome

9, one on chromosome 5 and two on chromosome 2. The third type of IFN, γ, can be induced from lymphocytes and can thus be regarded as a lymphokine, but its chromosome location is not known. All interferon genes studied so far are peculiar in that they have no introns. Their messenger RNAs, however, have the normal cap and poly A. tail structures found in other eukaryotic messengers. Mature interferon of all types is a single polypeptide with 166 amino acids, which include a hydrophobic signal peptide of about 20 amino acids. All IFNs, with the exception of human γ-IFNs, are glycoproteins. IFNs of all types from both man and mouse have considerable sequence homology, even though they can be separated antigenically.

3.4.2. Induction and Synthesis

Interferons can be induced in a wide variety of cells by agents including most viruses, some bacteria and protozoa and several natural or synthetic molecules, including double-stranded RNA, endotoxins and polysaccharides. Studies with immobilized inducers indicate that the induction can occur without penetration of the target cell. Thus, it appears that having bound to the target cell membrane, a conformational change is induced that triggers IFN synthesis. Little is known about the molecular mechanism of induction, although it appears to occur at the transcriptional level, as active IFN messenger RNA is only detected in induced cells. This induction is dependent on the continuance of host cell protein synthesis. After an initial period of IFN synthesis, IFN production is shut off. The shut-off can be delayed by treating the cells with inhibitors of RNA synthesis or protein synthesis. This process leads to a greatly increased level of IFN synthesis and is known as superinduction. This phenomenon has led to the hypothesis that translation of IFN messenger RNA may be controlled by a cellular repressor. After a round of IFN synthesis, cells become temporarily refractive to further IFN induction.

3.4.3. The Establishment of the Antiviral State

Most of the studies on IFN action have been performed with virus systems and these studies will be emphasized in this discussion. It should be realized, however, that IFN has many other modes of action, including effects on cell motility, cell proliferation, antibody responses, delayed-type hypersensitivity, graft rejection, macrophage activation, natural killer cell recruitment and as an antitumour agent.

The majority of activities induced in the cell by IFN involve the synthesis and modification of several proteins and the messenger RNAs coding for them. The best studied of these proteins is the $(2'-5')\,(A)_n$ synthetase. This enzyme is activated by double-stranded molecules larger than 30 base pairs and generates $2'-5'$ oligoadenylates from ATP. The unusual linkage in these oligoadenylates makes them very resistant to conventional nucleases. However, these molecules can be degraded by a specific $2'–5'$ phosphodiesterase found in some cells. The presence of $(2'–5')$ $(A)_n$ in a cell activates a latent endoribonuclease, RNase L. This enzyme cleaves single-stranded RNA molecules at the $3'$ side of UA, UG and UU sequences and is probably one of the molecules active in restricting viral replication. This nuclease is, however, not specific for viral RNAs *in vitro* and it is not known how it distinguishes them *in vivo*. Other roles suggested for $(2'–5')$ $(A)_n$ include control of cell division, an activity in accord with some effects of IFN on certain cell types. A protein kinase is also specifically induced by double-stranded RNA in IFN-treated cells. This appears to phosphorylate at least two proteins, one of which has been identified as the translation initiation factor eIF-2. Modification of such a factor could radically alter the binding capacity for viral messengers and thus modulate the production of viral proteins without impairing cellular function.

For additional information, see Lengyel, 1982 (29); Gordon and Minks, 1981 (20) and de Maeyer and de Maeyer-Guignard, 1979 (35).

REFERENCES

1. Acheson, N. H. (1980). Lytic cycle of SV40 and Polyoma virus. In "Molecular Biology of Tumor Viruses" (J. Tooze, ed.), Part 2, pp. 125–204. Cold Spring Harbor Lab., Cold Spring Harbor, New York.
2. Agol, V. I. (1980). Structure, translation and replication of picornaviral genomes. *Symp. Soc. Exp. Biol.* **28**, 419–446.
3. Allison, A. C., and Davies, P. (1974) Mechanisms of endocytosis and exocytosis. *Symp. Exp. Biol.* **28**, 419–446.
4. Altman, A., and Katz, D. H. (1982). The biology of monoclonal lymphokines secreted by T-cell lines and hybridomas. *Adv. Immunol.* **33**, 73–166.
5. Banerjee, A. K., Abraham, G., and Colonno, R. J. (1977). Vesicular stomatitis virus; mode of transcription (a review). *J. Gen. Virol.* **34**, 1–8.
6. Bishop, J. M. (1978). Retroviruses. *Annu. Rev. Biochem.* **47**, 35–88.
7. Bishop, J. M., and Varmus, H. (1982). Functions and origins of retroviral transforming genes. In "Molecular Biology of Tumor Viruses" (R. Weiss, N. Teich, H. Varmus, and J. Coffin, eds.), Part 1, pp. 999–1108. Cold Spring Harbor Lab., Cold Spring Harbor, New York.
8. Brown, D. T. (1980). The assembly of alphaviruses. In "The Togaviruses" (R. W. Schlesinger, ed.), pp. 473–502. Academic Press, New York.
9. Campbell, P. N., and Blobel, G. (1976). The role of organelles in the chemical modification of the primary translation products of secretory organelles. *FEBS Lett.* **72**, 215–226.
10. Carasco, L., and Smith, A. E. (1976). Sodium ions and the shut-off of host cell protein synthesis by picornaviruses. *Nature (London)* **264**, 807–809.
11. Choppin, P. W., and Compans, R. W. (1975). Reproduction of paramyxoviruses. In "Comprehensive Virology" (H. Fraenkel Conrat and R. Wagner, eds.), Vol. 4, pp. 95–178. Plenum, New York.

12. Coffin, J. (1982). Structure of the retroviral genome. *In* "Molecular Biology of Tumour Viruses" (R. Weiss, N. Teich, H. Varmus, and J. Coffin, eds.), Part 1, pp. 261–368. Cold Spring Harbor Lab., Cold Spring Harbor, New York.

13. Dales, S. (1973). Early events in cell–animal virus interactions. *Bacteriol. Rev.* **37,** 103–135.

14. Dickson, C., Eisenman, R., Fan, H., Hunter, E., and Teich, N. (1982). Protein synthesis. *In* "Molecular Biology of Tumor Viruses" (R. Weiss, N. Teich, H. Varmus, and J. Coffin, eds.), Part 1, pp. 513–648. Cold Spring Harbor Lab., Cold Spring Harbor, New York.

15. Dimmock, N. J. (1982). Initial stages in infection with animal viruses. *J. Gen. Virol.* **59,** 1–22.

16. Ehrenfeld, E. (1982). Poliovirus induced inhibition of host-cell protein synthesis. *Cell* **28,** 435–436.

17. Flint, S. J. (1980). Structure and genomic organisation of adenoviruses. *In* "Molecular Biology of Tumor Viruses" (J. Tooze, ed.), Part 2, pp. 383–442. Cold Spring Harbor Lab., Cold Spring Harbor, New York.

18. Flint, S. J., and Broker J. (1980). Lytic infection by adenoviruses. *In* "Molecular Biology of Tumor Viruses" (J. Tooze, ed.), Part 2, pp. 443–546. Cold Spring Harbor Lab., Cold Spring Harbor, New York

19. Friedman, D. D. and Ramseur, D. D. (1979). Mechanisms of persistent infections by cytophatic viruses in tissue culture. *Arch. Virol.* **60,** 83–103.

20. Gordon, J., and Minks, M. A. (1981). The interferon renaissance: Molecular aspects of induction and action. *Microbiol. Rev.* **45,** 244–266.

21. Griffin, B. E. (1980). Structure and genomic organization of SV40 and polyoma virus. *In* "Molecular Biology of Tumor Viruses" (J. Tooze, ed.), Part 2, pp. 61–124. Cold Spring Harbor Lab., Cold Spring Harbor, New York.

22. Halstead, S. B., Nimmannitya, S., and Cohen, S. N. (1970). Observations related to pathogenesis of dengue haemorrhagic fever. IV. Relation of disease severity to antibody response and virus recovered. *Yale J. Biol. Med.* **42,** 311.

23. Hanover, J. A., and Lennarz, W. J. (1981). Transmembrane assembly of membrane and secretory glycoproteins. *Arch. Biochem. Biophys.* **211,** 1–19.

24. Holland, J. J., Graham, E. A., Jones, L. L., and Semler, B. L. (1979). Evolution of multiple genome mutations during long-term persistent infection by vesicular stomatitis virus. *Cell* **16,** 495–504.

25. Huang, R. T. C., Wahn, K., Klenk, H. D., and Rott, R. (1980). Fusion between cell membrane and liposomes containing the glycoproteins of influenza virus. *Virology* **104,** 294–302.

26. Kennedy, S. I. T. (1980). The genome of alphaviruses. *In* "The Togaviruses" (R. W. Schlesinger, ed.), pp. 343–350. Academic Press, New York.

27. Kennedy, S. I. T. (1980). Synthesis of alphavirus RNA. *In* "The Togaviruses" (R. W. Schlesinger, ed.), pp. 351–370. Academic Press, New York.

28. Kolakofsky, D., and Blumberg, B. M. (1982). A model for the control of nonsegmented negative strand virus genome replication. *Symp. Soc. Gen. Microbiol.* **33,** 203–214.

29. Lengyel, P. (1982). Biochemistry of interferons and their actions. *Annu. Rev. Biochem.* **51,** 251–282.

30. Lodish, H. F., Zilberstein, A., and Porter, M. (1981). Synthesis and assembly of vesicular stomatitis virus and Sindbis virus glycoproteins. *Perspect. Virol.* **11,** 31–55.

31. Lonberg-Holm, K., and Philipson, L. (1980). Molecular aspects of virus receptors and cell surfaces. *In* "Cell Membranes and Viral Envelopes" (H. A. Blough and J. M. Tiffany, eds.), Vol. 2, pp. 103–137. Academic Press, London.

32. McCauley, J. W., and Mahy, B. W. J. (1983). Structure and function of the influenza virus genome. *Biochem. J.* **211,** 281–294.

33. Meager, A., and Hughes, R. C. (1977). Virus receptors. In "Receptors and Recognition" (P. Cuatrecasas and M. F. Greaves, eds.), Vol. 4, pp. 141–195. Chapman & Hall, London.

34. Mims, C. A. (1982). Role of persistence in viral pathogenesis. *Symp. Soc. Gen. Microbiol.* **33**, 1–14.

35. de Maeyer, E., and de Maeyer-Guignard, J. (1979). The effect of interferon on the cell-mediated immune system. In "Comprehensive Virology" (H. Fraenkel Conrat and R. Wagner, eds.), Vol. 15, pp. 205–284. Plenum, New York.

36. ter Meulen, V., and Stephenson, J. R. (1980). The possible role of viral infections in MS and other related demyelinating diseases. In "Multiple Sclerosis" (J. Hallpike, C. Adams, and W. Toortellotte, eds.), pp. 241–274. Chapman & Hall, London.

37. ter Meulen, V., Loffler, S., Carter, M. H., and Stephenson, J. R. (1981). Antigenic characterization of measles and SSPE virus, haemagglutinin by monoclonal antibodies. *J. Gen. Virol.* **57**, 357–364.

38. Niemann, H., and Klenk, H. D. (1981). Glycoprotein E1 of coronavirus A59. A new type of viral glycoprotein. In "Biochemistry and Biology of Coronaviruses" (V. ter Meulen, S. Siddell, and H. Wege, eds.), pp. 119–132. Plenum, New York.

39. Peiris, J. S. M., and Porterfield, J. S. (1979). Antibody-mediated enhancement of flavivirus replication in macrophage cell lines. *Nature (London)* **282**, 509–511.

40. Possee, R. D., Schild, G. C., and Dimmock, N. J. (1982). Studies on the mechanism of neutralization of influenza virus by antibody: Evidence that neutralizing antibody (anti-haemagglutinin) inactivates influenza virus *in vivo* by inhibiting virion transcriptase activity. *J. Gen. Virol.* **58**, 373–386.

41. Pringle, C. R. (1979). Virus evolution during persistent infection. *Nature (London)* **280**, 16.

42. Putnak, J. R., and Phillips, B. A. (1981). Picornaviral structure and assembly. *Microbiol. Rev.* **45**, 287–315.

43. Reanney, D. C. (1982). The evolution of RNA viruses. *Annu. Rev. Microbiol.* **36**, 47–73.

44. Rima, B. K., and Martin, S. J. (1976). Persistent infection of tissue culture cells by RNA viruses. *Med. Microbiol. Immunol.* **162**, 89–118.

45. Rocklin, R. E., Bendtzen, K., and Greineder, D. (1980). Mediators of immunity: Lymphokines and monokines. *Adv. Immunol.* **29**, 56–137.

46. Rott, R. (1979). Molecular basis of infectivity and pathogenicity of myxovirus. *Arch. Virol.* **59**, 285–298.

47. Rowland, D., Grabau, E., Spindler, K., Jones, C., Semler, B., and Holland, J. J. (1980). Virus protein changes and RNA termini alterations evolving during persistent infection. *Cell* **19**, 871–880.

48. Schlesinger, R. W., and Kääriäinen, L. (1980). Translation and processing of alphavirus proteins. In "The Togaviruses" (R. W. Schlesinger, ed.), pp. 371–392. Academic Press, New York.

49. Simons, K., Garoff, H., and Helenius, A. (1980). Alphavirus proteins. In "The Togaviruses" (R. W. Schlesinger, ed.), pp. 317–334. Academic Press, New York.

50. Simons, K., Garoff, H., and Helenius, A. (1982). How an animal virus gets into and out of its host cell. *Sci. Am.* **246**, 46–54.

51. Skehel, J. J., and Hay, A. J. (1978). Influenza virus transcription. *J. Gen. Virol.* **39**, 1–8.

52. Spear, P. G., and Roizman, B. (1980). Herpes simplex viruses. In "Molecular Biology of Tumor Viruses" (J. Tooze, ed.), Part 2, pp. 615–746. Cold Spring Harbor Lab., Cold Spring Harbor, New York.

53. Stephenson, J. R., and Dimmock, N. J. (1975). Early events in influenza virus multiplication. I Location and fate of the imput RNA. *Virology* **65**, 77–86.

54. Stephenson, J. R., Hay, A. J., and Skehel, J. (1977), Characterization of virus specific messenger RNAs from avain fibroblasts infected with fowl plague virus. *J. Gen. Virol.* **36**, 237–248.

55. Stephenson, J. R., Hudson, J. B., and Dimmock, N. J. (1978). Early events in influenza virus multiplication. II. Penetration of virus into cells at 4°. *Virology* **86**, 264–271.

56. Temin, H. M. (1971). Mechanism of cell transformation by RNA tumour viruses. *Annu. Rev. Microbiol.* **25**, 609–649,

57. Tooze, J., ed. (1973). "The Molecular Biology of Tumor Viruses." Cold Spring Harbor Lab., Cold Spring Harbor, New York.

58. Varmus, H., and Swanstrom, R. (1982). Replication of retroviruses. *In* "Molecular Biology of Tumor Viruses" (R. Weiss, N. Teich, H. Varmus, and J. Coffin, eds.), Part 1, pp. 369–512. Cold Spring Harbor Lab., Cold Spring Harbor, New York.

59. Wagner, R. A. (1975). Reproduction of rhabdoviruses. *In* "Comprehensive Virology" (H. Fraenkel Conrat and R. Wagner, eds.), Vol. 4, pp. 1–94. Plenum, New York.

60. Walter, P., and Blobel, G. (1982). 7S small cytoplasmic RNA is an integral component of the signal recognition particle. *Nature (London)* **299**, 691.

61. Wege, H., Stephenson, J. R., Koga, M., Wege, H., and ter Meulen, V. (1981). Genetic variation of neurotropic and non-neurotropic murine coronaviruses. *J. Gen. Virol.* **54**, 67–74.

3

Lymphocyte Hybridomas: Production and Use of Monoclonal Antibodies

ALBERT OSTERHAUS
FONS UYTDEHAAG
Rijksinstituut voor Volksgezondheid en Milieuhygiëne
Bilthoven, The Netherlands

1. INTRODUCTION

Transfer of genetic information between cells may be achieved by artificially induced fusion of their cell membranes. The primary objective of fusing somatic cells with appropriate cell lines is the generation of stable

Animal Cell Biotechnology, Vol. 2

hybrid cell lines which immortalize specific functions expressed by normal differentiated cells. The most successful application of somatic cell hybridization at present is the lymphocyte hybridoma technology, introduced by Köhler and Milstein in 1975, which has opened a new area in immunology. As a result of fusion between cells of the B-lymphocyte series and myeloma cells, hybrid cells (hybridomas) are generated which secrete monoclonal antibodies of predefined specificity. The resulting cell lines are cloned and immortal, which assures the monoclonality and permanent availability of their antibody products. In fact the production of any antibody synthesized by the immunized animal can be immortalized by cell fusion. The value of monoclonal antibodies as immunological reagents is being realized as they are used not only in immunological but in virtually the whole field of biomedical research. This increases the demand for monoclonal antibodies with specific characteristics, e.g. immunoglobulin subclass, binding affinity, fine specificity. Also the fusion of activated T lymphocytes with T-cell tumor cell lines has resulted in immortalization of T-cell functions (for review see 24). The present review will mainly describe general principles and different steps involved in the generation of B-cell hybridomas and the production and application of monoclonal antibodies.

2. IMMUNIZATION

The first step in the production of lymphocyte hybridomas is the generation of activated immune B lymphocytes as fusion partners for myeloma cells. It has been shown that after *in vivo* immunization the number of hybridomas resulting from a cell fusion is directly related to the number of freshly stimulated large lymphocytes within the spleen immediately prior to fusion (55), and that immunization procedures should aim at the generation of actively dividing B lymphocytes, rather than cells differentiating into plasma cells. Although mechanisms involved in immunization are not fully understood at present, it is generally appreciated that several factors such as the nature, dose and presentation of the antigen and the route and regimen of immunization are of major importance in obtaining satisfactory cell fusions. The nature of the immunizing antigen—protein, lipid or carbohydrate—and the form in which it is presented influence the response of the immunized animal. Although the purity of the immunogen per se is irrelevant, it may be important if impure material gives weaker responses and if the screening techniques do not distinguish between antibodies to the specific components and antibodies to the impurities. In general, complexed or particulate forms of antigen like surface antigens of cells and microorganisms are highly immunogenic while soluble antigens in aqueous solution are usually poorer immunogens (40). There are different approaches to enhance the antigenicity of such poor antigens. They may be coupled to soluble or particulate carriers or may be administered together with an adjuvant. In particular, protocols using Freund's complete and/or incomplete adjuvant are frequently employed. Also, the way in which the antigen is presented may be of critical importance. It has been shown that monomeric forms of glycoproteins of microorganisms or viruses are very weak or even ineffective as immunogens, whereas multimeric forms of the same proteins may be efficient immunogens. Among these multimeric forms are liposomes (1) and protein micelles (42), and recently a novel highly immunogenic structure has been described for this purpose, the immuno-stimulating complex (ISCOM) (43). Very good immune responses may then be obtained with doses of less than 1 μg of antigen, whereas in conventional immunization protocols between 10 and 100 μg of antigen per inoculation should be given. Immunization is usually carried out in a regimen where the animal is initially primed via a series of intramuscular, subcutaneous or intraperitoneal inoculations and after a rest period of at least 4 weeks is boosted intravenously or intraperitoneally, 3–4 days prior to fusion. Although after short immunization procedures monoclonal antibodies with the desired specificity may be obtained, such protocols lead mainly to antibodies of relatively low affinity. Longer immunization procedures along different routes, more boosters and higher levels of antigen may sometimes be needed. However, repetitive booster immunizations often do not induce the appearance of newly responding clones of lymphoblasts but even progressively reduce the initial response (36). Only after a quiescent period of several weeks or months are clones able to respond again.

The route of inoculation may also influence the level of the immune response. Poor immunogens may e.g. elicit satisfactory responses when administered in a certain regimen in the foot pads of the animals. The strain of mice and rats used for immunization may be of importance since the responsiveness of a certain strain to an antigen may be genetically controlled (6, 8). Therefore Balb/c mice or LOU rats, which are most frequently used, may not always be the best responders to a certain antigen, and if poor responses are obtained the use of other animal strains may be considered. Before the actual fusion is carried out it is always advisable to check the animal for serum antibody against the antigen, both to have an indication of whether the immunization has resulted in the desired response and to have a final check on the serologic screening system to be employed for the selection of hybrid clones.

Recently, different protocols have been described for *in vitro* immunization procedures as a simple and fast method for inducing a blastogenic response in mouse splenocytes, to be used for cell hybridization to produce specific hybridomas (51). These methods offer certain advantages over *in vivo* immunization procedures. With relatively small amounts of permanently available antigen, *in vitro* blast cell induction can be easily established within a few days. It may also be possible to induce *in vitro* antibody responses to auto-antigens or other poor immunogens which are not easily elicited by *in vivo* immunization. Furthermore, in addition to antigens, polyclonal B-cell activators can be used to induce a B-cell blastogenic response *in vitro*. Fusion of these cells with myeloma cells may yield hybridomas reactive with a number of antigens (2, 23). The attraction for the use of *in vitro* methods for the generation of activated *human* B lymphocytes is obvious. Both primary and secondary *in vitro* antibody responses by human B cells after antigen stimulation have been described for a wide range of antigens (4, 9, 10, 14, 15, 21, 25, 35, 38, 39, 44, 54, 58, 60, 61, 63).

Finally, it has been shown that an enrichment in antibody-forming cells of about 80-fold can be achieved by transferring spleen cells of immunized mice into syngeneic irradiated recipients and a subsequent restimulation with the antigen. The most extensive expansion could be obtained by transferring spleen cells of recently re-stimulated mice to irradiated recipients, injecting recipients with antigen on both day 0 and day 4 after transfer and harvesting spleens after 7 days (29).

3. FUSION PARTNERS

In hybridoma technology as it was originally developed by Köhler and Milstein (31), immortalization of immunoglobulin-secreting lymphoblastoid cells is achieved by fusion with mouse myeloma tumor cells and using a metabolic selection system to kill non-hybrid tumor cells. Most of the myeloma cells used for this purpose are deficient in the enzyme HGPRT (hypoxanthine–guanine phosphoribosyltransferase), and the hybridomas resulting from fusion with these cells are selected in the so-called HAT (hypoxanthine–aminopterine–thymidine) selective system. Normal somatic cells can use two pathways for their synthesis of purines and pyrimidines: the principal one, which involves *de novo* synthesis from amino acid and carbohydrate precursors, and the salvage pathways, which synthesize nucleotides from hypoxanthine (purines) or deoxythymidine (pyrimidines). If the principal pathway becomes blocked, e.g. by a folic acid antagonist such as aminopterin, in normal cells the salvage pathways may be used for synthesis of nucleotides. In these pathways the enzymes HGPRT and TK (thymidine kinase) are involved respectively, and since the myeloma cells used are usually selected for a deficiency in the HGPRT enzyme, they only grow if the *de novo* pathway is intact; in the presence of aminopterin they cannot survive. Hybrid lymphoblast myeloma cells resulting from a fusion between immune lymphoblastoid cells and HGPRT-deficient myeloma cells (hybridomas) and lymphoblastoid cells themselves do survive in the presence of aminopterin, provided hypoxanthine and thymidine are present for the salvage pathways. Only the hybrid immortalized cells will eventually be able to proliferate and form continuously growing cell lines.

Myeloma cells deficient in the HGPRT enzyme may be selected from cell populations which are not deficient by culturing them in the presence of a metabolic poison such as 8-azaguanine, which is incorporated into DNA by HGPRT and causes the poisoning of cells that do not show this deficiency. Although the HAT selection system is the most widely used system for the generation of lymphocyte hybridomas, several other methods for the selection of immortalized hybrid cells have been employed: ouabain resistance, irreversible biochemical inhibitors and reversible biosynthetic inhibition (for review, see ref. *11*).

All the myeloma cells initially employed synthesized and/or secreted Ig chains and have gradually been replaced by sublines which do not produce undesired heavy and light chains. Hybridomas constructed with these myeloma cell lines will only produce immunoglobulin chains originating from the lymphoblastoid parent cell and therefore should be preferred over producer lines, especially since lines with good performance are available now. Another important consideration in choosing a myeloma cell line is the stability of the hybridomas formed, which is partly dependent on the cell line. At present, myeloma cell lines from principally three species are available for lymphocyte hybridoma production: mouse, rat and human. Unless there are specific reasons against it, the myeloma cell line should be of the same species as the immunized animal, since this permits tumor development in this species when the hybridomas have been generated. A number of cell lines frequently used to generate hybridomas and some of their properties are shown in Table I. It is extremely important that the myeloma cells at the time of fusion are in the logarithmic growth phase.

The most convenient source of active lymphoblastoid cells from mice or rats to be used for fusion is undoubtedly the spleen of the immunized animal. However, regional lymph nodes, peripheral blood, bone marrow and Peyer's patches have also been used for this purpose. If hybridomas producing monoclonal antibodies of certain immunoglobulin classes are desired, both the immunization protocol and the source of lymphoblastoid cells may be of importance. The majority of monoclonal antibodies derived from fusion with spleen cells are of the IgG and IgM subclasses, although, at very low incidence, IgA-secreting hybridomas have been produced in this way

TABLE I   Some Myeloma Cell Lines Used for Hybridization

Species	Cell line	Ig chains produced/expressed	Reference
Mouse	P_3X63-Ag8 ("P_3")	$\gamma_1 + \kappa$ chain	31
	P_3X63-Ag8.653	—	28
	P_3/NS-1/1-Ag4-1	κ chain	30
	Sp2/0-Ag14	—	53
	F0	—	17
Rat	Y_3-Ag1.2.3	κ chain	20
	210RC43.Ag1	κ chain	12
Human	SK0-007	IgE	47
	GM1500 6TG-A12	IgG	13
	LICR-LON/HMy2	IgG	16

also. It has been demonstrated that IgA-producing hybridomas are more readily derived from fusions with gut-associated lymphoid tissues (32). Monoclonal antibodies of the IgE class have also been described (7).

4. HYBRIDIZATION

Spontaneous cell fusion of differentiated eukaryotic cells is uncommon in nature although it does occur in fertilization, muscle development and the formation of giant cells (18). Fusion of mammalian cells *in vitro* may be induced by many viruses, chemicals, electric fields or the immunizing antigen itself. The mechanisms involved in virus-induced cell fusion have been studied in detail, especially for Sendai and Semliki Forest viruses (3, 26). Since the introduction of polyethylene glycol (PEG) for the production of large numbers of growing somatic cell hybrids, this has become the agent of choice for the generation of hybrid cells including lymphocyte hybridomas, since it is relatively non-toxic and highly effective. The precise mechanism by which PEG induces cell fusion is not entirely understood but seems to involve alterations in the bulk-phase water adjacent to membrane surfaces (57). For successful fusions the optimal concentration, molecular weight and exposure time seem to be crucial. Optimal concentrations of PEG range from 30 to 50% (w/v). Often 5 to 10% (w/v) dimethyl sulphoxide (DMSO) is added, which would improve the fusion efficiency and has a protective effect on the cells. A standard fusion protocol, which we employ routinely in our laboratory and which has been derived from several described methods, is shown in the Addendum (Section 8).

An inherent problem with the PEG-mediated fusion is that PEG treatment of cells results in a non-selective fusion of one cell with another.

Consequently, the final percentage of growing hybrids secreting antibody specific for the immunizing antigen is relatively low. Theoretically one should be able to bind the myeloma cell line selectively to antigen-reactive lymphoblasts by interposing antigen between the two cells as a bridging ligand. When the antigen-coated myeloma cells are incubated with spleen cells from immunized mice, fusion occurs spontaneously. This technique has been described by Bankert *et al.* (5) for the production of antibodies to polysaccharide antigens and to haptenic ligands.

Electrofusion has been used to fuse many different cell types including lymphocytes for monoclonal antibody production (64). In this method cells are aligned by exposure to a low-level non-uniform electric field: "dielectrophoresis". This process generates dipoles in the cells and the attractive forces between the dipoles orient the cells. These oriented cells are subjected to a brief pulse of high voltage. This creates microscopic holes in the cell membranes at the points where they touch by electrical breakdown. Through the expanding holes the contents of the cells are allowed to mix and eventually the cells physically fuse together. Advantages of this process over chemical or biological fusion methods are its efficiency (up to 100% of the cells can be fused) and its gentleness, resulting in high percentages of viable hybridomas.

The potential of these methods as alternatives to PEG-mediated fusion has not fully been evaluated at present.

5. CULTIVATION OF HYBRIDOMAS

After fusion the cells are suspended in HAT medium, which selects for hybridized cells. Some investigators allow the cells to grow in medium without aminopterin for a short period, or increase the aminopterin concentration gradually. Several cell culture media are used for growth of hybridomas *in vitro*, such as Dulbecco's modified Eagle's medium (DMEM) containing the high glucose (4.5 g/litre) formulation, RPMI 1640 and NCTC 109. A number of supplementary tissue culture components are frequently used to facilitate the growth of hybridomas and the production of monoclonal antibodies (see Addendum). Usually cultures are maintained initially in an incubator in a humidified atmosphere of 5–10% CO_2 in air at 37°C. HGPRT-deficient myeloma cells are maintained in tissue culture flasks in suspension culture at densities between 3×10^5 and 3×10^6 cells/ml in the presence of 100 μM 8-azaguanine to select against revertants. Normal culture media, used at the initial stages of hybridoma culture, contain heat-inactivated (30 min at 56°C) fetal bovine serum, which has been screened for supporting the growth of hybrids. Less expensive calf serum may be used for the growth of myeloma cells and hybrids at later stages of cultivation. Recently, several formulations for serum-free media have been developed, and especially for the large-scale *in vitro* production of monoclonal antibodies to be used in purified form, this has obvious advantages. Cloning of hybrid cells should be started as soon as possible either by limiting dilutions or in a semi-solid medium. For limiting dilutions directly after fusion, the cell suspension is seeded in fluid medium in microculture plates or in so-called cloning plates. To enable single hybridoma cell cultures to grow, either feeder cells or conditioned media should be used. Feeder cells are assumed to provide a non-specific and growth-stimulating effect for the hybridomas. For this purpose murine or rat thymocytes, peritoneal macrophages or unfused spleen cells are most frequently used. Media in which these or other types of cells have grown for a certain period of time are used as conditioned media to replace the feeder cell suspension. The presence of macrophages may have the additional advantage of assisting in maintaining the cultures free of contamination. One of the major problems in many laboratories which routinely produce monoclonal antibodies is contamination with mycoplasmas, which have a selective requirement for nucleic acid precursors (purines and pyrimidines) and therefore grow well in hybridoma culture media. Once cultures have been recognised to be contaminated, they should be discarded as soon as possible because of their potential threat to other cultures. Detection methods for mycoplasmas in cell cultures are summarized in Table II.

Valuable hybridoma cultures may be cured of mycoplasma contamination by treatment with specific antisera, by passage through syngeneic animals or by treatment with antibiotics (50, 51). However, the use of antibiotics can lead to antibiotic resistance of mycoplasmas and certain antibiotics such as gentaminin may induce chromosome damage in mouse cells.

The identification of clones secreting the desired antibody should be attempted as early as possible in order to be able to concentrate future efforts on these relevant cells. Sensitive and rapid screening assays should therefore be available. Antibody-producing clones should be cryo-preserved in an early stage in order to minimize the risk of eventually losing a clone, e.g. by

TABLE II Detection Methods for Mycoplasmas in Cell Culture (27,50,51)

Direct culture techniques
Direct culture techniques in presence of animal cells
Radioactive precursor uptake by infected cells
Enzyme assays
Morphological examination (EM)
Fluorescent staining antibodies
Cytochemical fluorescent stains

bacterial contamination or overgrowth by non-producing cells. Standard procedures for freezing animal cells using mixture of bovine serum and DMSO in medium in a programmed or well-defined freezing system in liquid N_2 can be employed. An example is, given in the Addendum.

After positive clones have been subcloned at least twice, frozen in liquid nitrogen and proved stable, they may be grown in larger quantities in cell culture or by inoculation into the peritoneal cavity of syngeneic animals. Both methods have advantages and disadvantages. The main difference between both systems is the yield of antibody per unit volume. In ascites fluid several milligrams per millilitre may be found, whereas in suspension cell culture not more than 0.1 mg/ml can be obtained. The animal serum or ascites fluid is therefore usually between 100 and 1000 times more concentrated. On the other hand, protein impurities in medium can largely be reduced, whereas serum or ascites fluid from tumor-bearing animals always contains other immunoglobulins of the same species. However, in most cases such animals suffer from a severe depletion of their normal immunoglobulins. For the production of monoclonal antibodies from ascites tumors of mice stable clones are injected intraperitoneally into syngeneic animals. In non-syngeneic systems immunosuppressive treatment may be required. Therefore X-ray irradiation (500 rads) or treatment with cyclophosphamide (0.5 mg per mouse) or with anti-thymocyte globulin may be used 24 hr before the tumor is transplanted. Balb/c mice or LOU rats are usually inoculated 14 days prior to inoculation of $1-2 \times 10^6$ cells per animal with 0.5 ml Pristane intraperitoneally. Within about 1–3 weeks ascites fluids may be collected from the distended abdomen. Repeated tappings may be carried out before the animal is sacrificed. Cells are removed by low-speed centrifugation and the oil layer removed, and the resulting fluid may be used as monoclonal antibody-containing ascites fluid. Concentration and partial purification can most easily be achieved by standard ammonium or sodium sulphate precipitation procedures followed by dialysis. Although for most applications culture supernatants or ascites fluids from a small number of mice or rats will suffice, for certain purposes almost unlimited quantities of antibody will be required. Major factors influencing the yields are the intrinsic production potential and the relative stability of the hybridoma clone. Production potentials may vary considerably between clones and also the stability of different clones is not the same. Non-secreting variants tend to finally overgrow the producing cells both in suspension cultures and *in vivo*. This problem of instability seems even more serious in interspecies hybridomas. For large-scale hybridoma production, *in vitro* fermenter systems are already in use, and for both ethical and practical reasons more attention will be focussed on this means of production of monoclonal antibodies. Especially for therapeutic and *in vivo* diagnostic use of monoclonal antibodies in humans, the production in cell culture systems offers certain advantages, e.g. the absence of extraneous viruses introduced by host animals. Also, the availability of serum-free and even protein-free culture media will result in purer starting materials for monoclonal antibody preparations. Some formulations have been made for this purpose, but serum-free DMEM may also be used. Therefore cells in a vigorously growing culture containing 5% fetal calf serum at a density of about 3×10^6 cells/ml are washed and resuspended in the same amount of serum-free medium. Immobilization of these cells, normally grown in suspension, in capsules or beads may give them increased stability and may make such preparations more suitable for continuous operation (37, 45, 46).

6. SCREENING METHODS

It is essential that the producing hybridomas can be identified as early as possible and separated from the clones which do not express the desired functions. Therefore the screening method should be rapid, sensitive and reliable and should select for those antibodies which will eventually be needed. Monoclonal antibody-producing clones may be detected by fundamentally two different methods. The detection of hybrids producing antibody was often carried out by using anti-mouse Ig or protein A preparations in the early days of hybridoma technology. Protein A selectively binds to all mouse IgG's but not to IgM (34). However, since most fusions are carried out to obtain hybridomas of certain specificities, culture supernatants are generally only tested for antibody specific for the immunizing antigen at present. The test to be used will mainly depend on the nature of the antigen involved and the way in which the monoclonal antibody is to be used eventually. Monoclonal antibodies often do not function well in all serological techniques and therefore it is advisable to screen with the method in which the antibody will be used in the end. Before starting with a fusion the proper screening system should be ready. Depending on the nature of the antigen, different techniques may be chosen. Classical tests such as immunoprecipitation in gels or agglutination do not always function well with monoclonal antibodies because these antibodies react with only one antigenic determinant, which may be present in too low a concentration or density on the antigen to permit cross-linking or lattice formation. This is also true because most monoclonal antibodies are of the IgG and not of the IgM class, which is much more likely to cross-link. The sensitivity of the method should allow the detection of antibody in nanogram amounts, which enables the detection of clones consisting of about 100 cells. It should be possible to handle hundreds of samples within 1 or 2 days, but if necessary samples may be pooled in a coordinate system, which offers the opportunity to reduce the number of tests if not too many positive samples are expected. Apart from functional biological assays such as neutralization or agglutination tests for micro-organisms, rapid enzyme-linked immunosorbent assay (ELISA) and radioimmunoassay (RIA) systems are employed for many purposes. Usually the antigen is immobilized on a solid phase and the monoclonal antibody is then detected with an anti-species conjugate. This anti-species conjugate should preferentially react with all classes of antibody.

7. APPLICATIONS OF MONOCLONAL ANTIBODIES

In all fields of science where conventional polyclonal antisera have been used, they may sooner or later be replaced by monoclonal antibodies or mixtures of monoclonal antibodies. A summary of the properties of monoclonal antibodies versus conventional antisera is given in Table III. Mixtures may be needed to overcome some of the theoretical and practical disadvan-

TABLE III   Properties of Monoclonal Antibodies versus Polyclonal Antiserum

	Polyclonal antiserum	Monoclonal antibody
Workload involved in production	Small	Considerable
Content of specific Ig	0.1–1.0 mg/ml	1.0–10 mg/ml (ascites) 0.001–0.025 mg/ml (supernatant)
Other Ig	10 mg/ml	Variable (ascites) Absent (supernatant)
(Sub)class of specific Ig	Mixture	Usually one
Specificity	Low (may be improved by adsorption procedures)	Highest possible (for certain applications even too high)
Reproducibility of specificity and affinity	Varies between batches	Absolute
Homogeneity	Heterogeneous	Homogeneous
Availability	Limited (batchwise)	In principle unlimited
Reacts with	In principle all antigenic determinants in immunizing antigen	*One* antigenic determinant
Cross-reactions (other antigens)	With antigens bearing common determinants	Absent (complete if it reacts with common determinant)
Applicability in serological tests	Good	May not work

tages of monoclonal antibodies. Since the technology has spread into most areas of biological research and clinical medicine within the first 10 years of its existence, it is virtually impossible to summarize all present exploitations of it. A number of already proven successful applications and future applications are discussed below.

Hybridoma technology has definitely transformed the whole field of immunology; almost all immunological laboratories are using monoclonal antibodies in one form or another, whether to identify specific chemical groupings or to assist in the purification of important molecules expressed on cells or present in complex biological mixtures. The generation and use of monoclonal antibodies against human and murine cell surface antigens of lymphocytes has offered the potential to study functional aspects of various subpopulations of lymphocytes on the basis of the presence of different differentiation antigens. These antibodies have already facilitated the understanding of immunoregulation (41) and the detection of immunoregulatory abnormalities in man. Moreover, they have been shown to help in the monitoring of immunoregulatory therapies and are also used as therapeutic immunoregulatory agents themselves. The potential value of anti-human lymphocyte monoclonal antibodies in immunosuppression and bone marrow transplantation is being studied by many groups. In the study of major histocompatibility complex (MHC) antigens monoclonal antibodies have already been used in a variety of serologic, structural and functional studies. It is becoming possible to determine the topographic arrangement of different determinants on these antigens in relation to their various functions.

The use of monoclonal antibodies in the field of infectious agents and their diseases is another example of the enormous potential of the technology. Pioneers in this field were Gerhard and Webster (22), who generated large panels of monoclonal antibodies against the hemagglutinin of influenza virus. They were able to reproduce antigenic drift *in vitro* by propagation of influenza viruses in the presence of selected neutralizing monoclonal antibodies, and demonstrated the presence of variable regions on the hemagglutinin molecule of the virus. Also, Wiktor and Koprowski (62) demonstrated, with the use of monoclonal antibodies, that in rabies virus, which was believed to be antigenically very stable, many different geographically restricted subtypes could be recognised. They showed that these results had major implications for vaccine strategy and the selection of rabies virus vaccine strains. Similar studies have been carried out with a large number of other viruses. We demonstrated that monoclonal antibodies may be used for the intratypic differentiation of poliovirus strains and the differentiation of influenza virus isolates, using a theoretical pattern-fitting computer program (48, 49). Apart from these applications for the analysis of structure and epidemiology of infectious agents including viruses, bacteria and parasites, monoclonal antibodies will play an increasingly important role in diagnosis, therapy and prevention of infectious diseases. Their role in the production of vaccines will not be limited to the analysis of antigenic structures; they will also be employed for the purification of immunogenic structures using affinity chromatography methods, for the selection of recombinant DNA clones producing desired antigenic structures and for the characterization and selection of synthetic peptides to be used as immunogens for vaccination. On the other hand, monoclonal antibodies will also be used for vaccine control purposes, e.g. as controls for identity and potency. In the control of vaccines for the absence of extraneous agents, conventional antisera have been used up to now to neutralize the live vaccine component before the sample is tested in cell culture and/or animals. Of course, the unique specificity of monoclonal antibodies offers the advantage that no unknown agents are neutralized or inhibited, whereas in conventional antisera antibodies against such agents might be present.

An entirely new area for the application of monoclonal antibodies will emerge in the vaccine field since it has been possible to induce specific immunity with anti-idiotypic antibody. Recently it was shown that mice can be immunized against infection with *Trypanosoma rhodesiense* by administration of polyclonal allogeneic anti-idiotypic antibody raised against the idiotypes of protective monoclonal antibody (52). The use of polyclonal xenogeneic or allogeneic anti-idiotypic antibody for this purpose has several disadvantages. It would be very difficult to establish production of polyclonal anti-idiotypic antibody preparations of high consistency and the same identity. Only a small fraction of these anti-idiotypic antibodies will induce idiotype-bearing molecules which will bind antigen. Therefore a monoclonal anti-idiotypic antibody possessing a related epitope would represent the ideal anti-idiotypic vaccine. We demonstrated that it is possible to induce neutralizing antibody against poliovirus type II in mice by vaccination with monoclonal anti-idiotypic antibody (59). Consequently, this monoclonal anti-idiotypic antibody substitutes for antigen in the induction of anti-viral neutralizing antibody.

Another area in which monoclonal antibodies have been applied successfully is the investigation of many aspects of cancer. The main option has invariably been the generation of monoclonal antibodies which distinguish between normal and malignant cells; surface antigens have been sought which unequivocally identify cells as being malignant. Although the search for these tumor-specific antigenic determinants is still going on, most of the determinants which were supposed to be truly tumor specific have also been found on embryonic ("onco fetal antigens") and certain normal cells. They can, however, still be used to identify tumor cells under certain circumstances. Another approach is the use of monoclonal antibodies against products of differentiated cells of a particular tissue to determine the functional and differentiation status of the cells in a tumor, e.g. as practiced with monoclonal antibodies against the membranes of human milk fat globules (19). Apart from the use of these monoclonal antibodies for the diagnosis of malignancies with tissue samples, another possibility is the use of monoclonal antibodies conjugated with radioactive agents to image tumors and metastases. The potential of immunotherapy of humans with monoclonal antibodies directed against these tumor antigens is a subject of study at present. Perhaps the best example of this approach at present is the use of the Ig idiotype on neoplastic B lymphocytes as a tumor antigen, in that the elimination of all cells bearing it by using anti-idiotypic monoclonal antibody will ablate or substantially diminish the tumor in a specific way. Surface Ig could serve as a prototype tumor antigen since a great deal is known about its structure, antigenic make-up, extracellular analogues, metabolism and responses to union with antibody (56). In many of the studies using monoclonal antibodies against tumor antigens, cytotoxic agents such as ricin or diphtheria toxin have been attached to the antibodies, to determine whether these agents may be targeted selectively to tumor cells: the "magic bullet" concept. One of the problems in these studies is that they have to be carried out with mouse or rat monoclonal antibodies in man. For both *in vivo* diagnosis and therapeutic trials in humans, the use of human monoclonal antibodies should be preferred so that repeated treatment will not result in allergic reactions.

The importance of anti-idiotypic antibodies generated during the course of malignant disease, or during this treatment with monoclonal antibodies, is being emphasized at present (33). As has been shown for the prevention of infectious diseases, exogenously administered anti-idiotypic antibody may also become important for the prevention or treatment of malignant disease. It was shown that inoculation of human subjects with mouse monoclonal antibody directed against certain cancer antigens resulted in some patients in the development of anti-idiotypic antibody, and these patients improved clinically and had long remission from their disease. The possible presence of the internal image of the cancer antigen on the human immunoglobulin molecule may change the conditions under which the immune system reacts to the tumor antigen and may open new approaches to the control of tumor growth.

8. ADDENDUM: PREPARATION OF MOUSE–MOUSE HYBRIDOMAS

The next sections review the procedure for preparation of mouse–mouse hybridomas routinely used in our laboratory (Fig. 1).

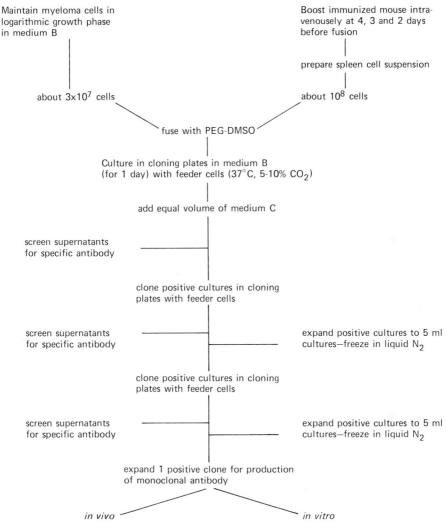

Fig. 1. Preparation of mouse–mouse hybridomas.

8.1. Media and Solutions

Medium A: DMEM with 4.5 g/litre glucose to 1000 ml
10 ml penicillin–streptomycin (5000 IU/ml—5000 μg/ml)
20 ml L-glutamine (0.02 M)
0.3 ml β-mercaptoethanol (0.13 M)
20 ml hypoxanthine–thymidine (1.0 × 10^{-4} M; 1.6 × 10^{-5} M)

Medium B: medium A + 20% fetal calf serum[1]
Medium C: medium B + 0.4 mM aminopterin
PEG–DMSO: 2.5 g PEG-4000
                         3 ml medium A
                         0.5 ml DMSO
                         Add medium A and DMSO to PEG-4000 after melting.
NH$_4$Cl solution for lysis of erythrocytes: 0.83 g NH$_4$Cl
                                                             H$_2$O to 100 ml
                                                             adjust pH to 7.4
All serum is heat-inactivated by placing the bottle in a 56°C water bath for 30–45 min.

8.2. Fusion and Cultivation of Hybridomas

1. Place all media and PEG–DMSO in 37°C water bath.
2. Kill immunized mouse by placing in CO$_2$ atmosphere for 2 min.
3. Remove spleen under sterile conditions.
4. Suspend spleen cells in 10 ml medium A.
5. Spin cells at room temperature for 20 sec at 400 g to remove tissue clumps.
6. Spin supernatant suspension at room temperature through a 1.5-ml FCS cushion for 10 min at 400 g.
7. Remove supernatant.
8. Suspend pellet in 1 ml NH$_4$Cl solution at room temperature.
9. Add 9 ml medium B after 1 min and spin this suspension at room temperature through a 1.5-ml FCS cushion for 10 min at 400 g.
10. Suspend pellet in 10 ml medium A containing 2.5 × 10^7 myeloma cells.
11. Wash suspension three times with medium A.
12. Remove last supernatant carefully and completely.
13. Gently tap tube to disrupt pellet.
14. While holding tube in hand, slowly, over a period of 1 min, add 1 ml of PEG–DMSO.
15. Mix during this period by gently tapping. Do *not* draw up into pipette.
16. Add 1 ml medium A while gently tapping.
17. Keep tube in hand for 2 min, gently tapping now and then.
18. Add 2 ml medium A while gently tapping.
19. Keep tube in hand for 3 min, gently tapping now and then.

---

[1]Fetal calf serum (FCS): Batches of FCS are screened for supporting the growth of hybrids at the initial stages of hybridoma culture.

20. Add 5 ml medium A while gently tapping.

21. Keep tube in hand for 2 min while gently tapping.

22. Add 5 ml medium B and mix by turning tube upside down.

23. Centrifuge for 8 min at 400 g.

24. Suspend pellet carefully in 200 ml medium B in which 2×10^8 freshly prepared rat thymocytes have also been suspended.

25. Dispense this suspension in 192 cups of cloning plates (Greiner).

26. Incubate for 24 hr at 37°C in a humidified 10% CO_2 atmosphere.

27. Add an equal volume of medium C to each cup.

28. Incubate plates for 48 hr.

29. Change medium for medium C.

30. Incubate further under the same conditions for about 2 weeks, changing the medium every 4 days, until clones of about 100 cells can be visualized.

31. Screen culture supernatants for presence of desired antibody.

32. Expand positive cell clones in 25-ml tissue culture flasks to be frozen in liquid N_2 and reclone cells directly.

8.3. Cloning in Fluid Medium

Positive cell cultures are cloned in cloning plates (Greiner) at least twice by limiting dilutions or by micromanipulation. For this purpose rat thymocytes are added in a concentration of 1×10^6/ml at the start of the cloning procedure. After 14–21 days supernatants are screened again and positive clones are transferred to 25-ml tissue culture flasks for expansion. Positive subclones are frozen in liquid N_2.

8.4. Freezing and Thawing of Cells

Only freeze cells in full logarithmic growth. At least 10^5 cells are frozen per ampule.

1. Suspend 10^5–10^6 cells in 1 ml medium B containing 7.5% DMSO at a temperature of 0°C in freezing ampule.

2. Keep cells for 20 min at 0°C.

3. Place ampules in a programmed freezing device for programmed temperature decrease, or wrap them in cotton and place them in a styrofoam box in −70°C freezer.

4. Place ampules in a liquid N_2 freezer for storage.

For thawing place the ampules as quickly as possible in a 37°C water bath. When thawing is nearly complete, transfer the suspension to a 10-ml tube placed in an ice bath. Add dropwise 10 ml of ice-cold medium B, during about 3 min, mixing very carefully. Spin at 4°C for 10 min at 400 g and resuspend the cells in fresh medium B.

8.5. Expansion of Hybridomas—Production of Monoclonal Antibodies

Upon expansion of hybridoma cultures *in vitro* cells should be maintained at concentrations between 3×10^5 and 3×10^6 cells/ml. If cells are grown at lower concentrations either feeder cells or conditioned media should be used. Usually expansion *in vitro* can best be performed by splitting the cultures at a 1 : 2 or 1 : 3 ratio every 4–6 days. At regular intervals the culture supernatants should be tested for the presence of specific antibody.

If monoclonal antibodies are produced in the form of ascites fluid in Balb/c mice, the following procedure may be used:

1. Prime mice with 0.5 ml Pristane (2,6,10,14-tetramethylpentadecane) at least 5 days before injecting the hybridomas.
2. Inject $1–2 \times 10^6$ hybridomas in logarithmic growth phase intraperitoneally in 0.5 ml saline.
3. Tap the mice when abdomen is distended (1–3 weeks after inoculation of hybridomas). Repeated tappings may be carried out.
4. Collect ascites fluid by low-speed centrifugation.
5. Precipitate the gamma globulin fraction with ammonium sulphate (45% saturated) and resuspend and dialyze the antibody.

REFERENCES

1. Allison, A. C., and Gregoriadis, G. (1974). Liposomes as immunological adjuvants. *Nature (London)* **252**, 252.
2. Andersson, J., and Melchers, F. (1978). The antibody repertoire of hybrid cell lines obtained by fusion of X63-Ag8 myeloma cells with mitogen activated B-cell blasts. *Curr. Top. Microbiol. Immunol.* **81**, 130–139.
3. Bächi, T., Deas, J. E., and Howe, C. (1977). Virus–erythrocyte membrane interactions. *Cell Surf. Rev.* **2**, 83–127.
4. Ballieux, R. E., Heijnen, C. J., UytdeHaag, F., and Zeegers, B. J. M. (1979). Regulation of B cell activity in man: Role of T cells. *Immunol. Rev.* **45**, 3.
5. Bankert, R. B., Des Soye, D., and Powers, L. (1980). Antigen-promoted cell fusion: Antigen-coated myeloma cells fuse with antigen-reactive spleen cells. *Transplant. Proc.* **12**, 443–446.
6. Biozzi, G. C., Stiffel, D., Mouton, D., and Bouthillier, Y. (1975). *In* "Immunogenetics and Immunodeficiency" (B. Benacerraf, ed.), p. 179. MTP Press, Ltd., Lancaster, England.
7. Böttcher, I., Hämmerling, G., and Kapp, J. F. (1978). Continuous production of mouse IgE antibodies with known allergenic specificity by a hybrid cell line. *Nature (London)* **275**, 761–762.

8. Boumsell, L., and Bernard, A. (1980). High efficiency of Biozzi's high responder mouse strain in the generation of antibody secreting hybridomas. *J. Immunol. Methods* **38**, 225–229.

9. Brenner, M. K., and Munro, A. J. (1981). Human anti-tetanus antibody response in vitro: Autologous and allogeneic T cells provide help by different routes. *Clin. Exp. Immunol.* **46**, 171.

10. Callard, R. E. (1979). Specific in vitro antibody response to influenza virus by human blood lymphocytes. *Nature (London)* **282**, 734.

11. Clements, G. B. (1975). Selection of biochemically variant in some cases mutant, mammalian cells in culture. *Adv. Cancer Res.* **21**, 273–390.

12. Cotton, R. G. H., and Milstein, C. (1973). Fusion of two immunoglobulin-producing myeloma cells. *Nature (London)* **244**, 42–43.

13. Croce, G. M., Linnenbach, A., Hall, W., Steplewski, Z., and Koprowski, H. (1980). Production of human hybridomas secreting antibodies to measles virus. *Nature (London)* **288**, 488–489.

14. Delfraissy, J. F., Galanaud, D., Dormot, J., and Wallon, C. (1977). Primary in vitro antibody response from human peripheral blood lymphocytes. *J. Immunol.* **118**, 630.

15. Dosch, H. M., and Gelfand, E. W. (1976). In vitro induction of hemolytic plaqueforming cells in man. *J. Immunol. Methods* **11**, 107.

16. Edwards, P. A. W., Smith, C. M., Neville, A. M., and O'Hare, M. J. (1982). A human–human hybridoma system based on a fast-growing mutant of the ARH-77 plasma cell leukaemia-derived line. *Eur. J. Immunol.* **8**(8), 641–648.

17. Fazekas de St. Groth, S., and Scheidegger, D. (1980). Production of monoclonal antibodies: Strategy and tactics. *J. Immunol. Methods* **35**, 1–21.

18. Foster, C. S. (1982). Lymphocyte hybridomas. *Cancer Treat. Rev.* **9**, 59–84.

19. Foster, C. S., Edwards, P. A. W., Dinsdale, E. A., and Neville, A. M. (1982). Moncolonal antibodies to the human mammary gland. I. Distribution of determinants in non-neoplastic mammary and extra-mammary tissues. *Virchows Arch. A: Pathol. Anat. Histol.* **394**, 279–293.

20. Galfrè, G., Milstein, C., and Wright, B. (1979). Rat × rat hybrid myelomas and a monoclonal anti-Fd portion of mouse IgG. *Nature (London)* **277**, 131–133.

21. Geha, R. S., Mudawar, F., and Schneeberger, E. (1977). The specificity of T cell helper factor in man. *J. Exp. Med.* **145**, 1436.

22. Gerhard, W., and Webster, R. G. (1978). Antigenic drift in influenza A viruses. I. Selection and characterization of antigenic variants of A/PR/8/34 (HONI) influenza virus with monoclonal antibodies. *J. Exp. Med.* **148**, 383–392.

23. Goldsby, R. A., Osborne, B. A., Surt, D., Mandel, A., Williams, J., Gronowicz, E., and Herzenberg, L. A. (1978). Production of specific antibody without specific immunization. *Curr. Top. Microbiol. Immunol.* **81**, 149–151.

24. Hämmerling, G. J., Hämmerling, U., and Kearney, J. F., eds. (1981). Monoclonal antibodies and T-cell hybridomas, perspectives and technical advances. *Res. Monogr. Immunol.* **3**, 587.

25. Heijnen, C. J., UytdeHaag, F., Gmelig-Meyling, H. J., and Ballieux, R. E. (1979). Localization of human antigenic-specific helper and suppressor function in distinct T-cell subpopulations. *Cell. Immunol.* **43**, 282.

26. Hosaka, Y., and Shimizu, K. (1977). Cell fusion by Sendai virus. *Cell Surf. Rev.* **2**, 129–144.

27. Kaplan, D. R., Henkel, T. J., Braciale, V., and Braciale, F. J. (1984). Mycoplasma infection of cell cultures: Thymidine incorporation of culture supernatants as a screening test. *J. Immunol.* **132**(1), 9–11.

28. Kearney, J. F., Radbruch, A., Liesegang, B., and Rajewsky, K. (1979). A new mouse myeloma cell line that has lost immunoglobulin expression but permits the construction of antibody-secreting hybrid cell lines. *J. Immunol.* **123**, 1548–1550.

29. Kenny, P. A., McCaskill, A. C., and Boyle, W. (1981). Enrichment and expansion of specific antibody-forming cells by adoptive transfer and clustering, and their use in hybridoma production. *AJEBAK* **59**, Part 4, 427–437.

30. Köhler, G., and Milstein, C. (1976). Derivation of specific antibody-producing tissue culture and tumour lines by cell fusion. *Eur. J. Immunol.* **6**, 511–519.

31. Köhler, G., and Milstein, C. (1975). Continuous cultures of fused cells secreting antibody of predefined specificity. *Nature (London)* **256**, 495–497.

32. Komisar, J. L., Fuhrman, J. A., and Cebra, J. J. (1982). IgA producing hybridomas are readily derived from gut-associated lymphoid tissue. *J. Immunol.* **128**, 2376–2378.

33. Koprowski, H. *et al.* (1984). Human anti-idiotype antibodies in cancer patients: Is the modulation of the immune response beneficial for the patient? *Proc. Natl. Acad. Sci. U.S.A.* **81**, 216.

34. Kronvall, G., and Williams, R. C. (1969). Differences in anti-protein A activity among IgG subgroups. *J. Immunol.* **103**, 828–833.

35. Lane, H. C., Volkman, D. J., Whalen, G., and Fauci, A. S. (1981). In vitro antigen-induced antigen-specific antibody production in man. *J. Exp. Med.* **154**, 1043.

36. Lee, W., and Köhler, H. (1974). Decline and spontaneous recovery of the monoclonal response to phosphorylcholine during repeated immunization. *J. Immunol.* **113**, 1644–1646.

37. Lim, F. (1980). Micro-incapsulated islet as bio-artificial endocrine pancreas. *Science* **210**, 908–910.

38. Luzzati, A. L., Taassing, M. J., Meo, T., and Pernis, B. (1976). Induction of an antibody response in cultures of human peripheral blood lymphocytes. *J. Exp. Med.* **144**, 573.

39. Misiti, J., and Waldmann, T. A. (1981). In vitro generation of antigen-specific hemolytic plaque-forming cells from human peripheral blood mononuclear cells. *J. Exp. Med.* **154**, 1069.

40. Mitchison, N. A. (1969). The immunogenic capacity of antigen taken up by peritoneal exudate cells. *Immunology* **16**, 1–14.

41. Möller, G., ed. (1983). Functional T cell subsets defined by monoclonal antibodies. *Immunol. Rev.* **74**.

42. Morein, B., Sharp, M., Sundquist, B., and Simons, K. (1983). Protein subunit vaccines of parainfluenza type 3 virus: Immunogenic effect in lambs and mice. *J. Gen. Virol.* **64**, 1557–1569.

43. Morein, B., Sundquist, B., Höglund, S., Dalsgaard, K., and Osterhaus, A. (1984). ISCOM, a novel structure for antigenic presentation of membrane proteins from enveloped viruses. *Nature (London)* **308**, 457.

44. Morimoto, C., Reinherz, E. L., and Schlossman, S. F. (1981). Primary in vitro anti-KLH antibody formation by peripheral blood lymphocytes in man: Detection with a radioimmunoassay. *J. Immunol.* **127**, 514.

45. Nilsson, K., and Mosbach, K. (1980). Preparation of immobilised animal cells. *FEBS Lett.* **118**, 145–150.

46. Nilsson, K., Scheirer, W., Merten, O. W., Ostberg, L., Liehl, E., Katinger, H. W. D., and Mosbach, K. (1983). Entrapment of animal cells for production of monoclonal antibodies and other biomolecules. *Nature (London)* **302**, 629–630.

47. Olsson, L., and Kaplan, H. S. (1980). Human–human hybridomas producing monoclonal antibodies of predefined antigenic specificity. *Proc. Natl. Acad. Sci. U.S.A.* **77**, 5429–5431.

48. Osterhaus, A. D. M. E., Weijers, T. F., Bijlsma, K., de Ronde-Verloop, F. M., van Asten,

J. A. A. M. and de Jong, J. C. (1984). Comparison of monoclonal antibodies with ferret sera for the characterization of influenza A (H_3N_2) virus strains in a computer system. *Dev. Biol. Stand.* **57**, 245.

49. Osterhaus, A. D. M. E., van Wezel, A. L., Hazendonk, T. G., UytdeHaag, F. G. C. M., Asten, J. A. A. M., and Van Steenis, B. (1983). Monoclonal antibodies to polioviruses: Comparison of intratypic strain differentiation of poliovirus type 1 using monoclonal antibodies versus cross-absorbed antisera. *Intervirology* **20**, 129–136.

50. Polak-Vogelzang, A. A. (1983). Mycoplasmatales in cell cultures. Detection and control. Thesis, Amsterdam.

51. Reading, C. L. (1982). Theory and methods for immunization in culture and monoclonal antibody production. *J. Immunol. Methods* **53**, 261–291.

52. Sacks, D. L., Esser, K. M., and Sher, A. (1982). Immunization of mice against African trypanosomiasis using anti-idiotypic antibodies. *J. Exp. Med.* **155**, 1108.

53. Shulman, M., Wilde, C. D., and Köhler, G. (1978). A better cell line for making hybridomas secreting specific antibodies. *Nature (London)* **276**, 269–270.

54. Souhami, R. L., Babbage, J., and Callard, R. E. (1981). Specific in vitro antibody response to varicella zoster. *Clin. Exp. Immunol.* **46**, 98.

55. Stähli, C., Staehelin, T., Miggiano, V., Schmidt, J., and Haring, P. (1980). High frequencies of antigen specific hybridomas: Dependence on immunization parameters and prediction. *J. Immunol. Methods* **32**, 297–304.

56. Stevenson, G. T., and Stevenson, F. K. (1983). Treatment of lymphoid tumors with anti-idiotype antibodies. *Springer Semin. Immunopathol.* **6**, 99–115.

57. Tanford, C. (1961). "Physical Chemistry of Macromolecules," pp. 114–200. Wiley, New York.

58. UytdeHaag, F., Heijnen, C. J., Pot, K. H., and Ballieux, R. E. (1981). Antigen-specific human T cell factors. II. T cell suppressor factor: Biologic properties. *J. Immunol.* **126**, 503.

59. UytdeHaag, F. G. C. M., and Osterhaus, A. D. M. E. (1985). Induction of neutralizing antibody in mice against poliovirus type II with monoclonal anti-idiotypic antibody. *J. Immunol.* **134**, 1225.

60. UytdeHaag, F. G. C. M., Osterhaus, A. D. M. E., Loggen, H. G., Bakker, R. H. J., Van Asten, J. A. A. M., Kreeftenberg, J. G., van der Marel, P., and Van Steenis, B. (1983). Induction of antigen-specific antibody response in human peripheral blood lymphocytes in vitro by a dog kidney cell vaccine against rabies virus (DKCV). *J. Immunol.* **131**(3), 1234–1239.

61. Volkman, D. J., Allyn, S. P., and Fauci, A. S. (1982). Antigen-induced in vitro antibody production in humans: Tetanus toxoid-specific antibody synthesis. *J. Immunol.* **129**, 107.

62. Wiktor, T. J., and Koprowski, H. (1980). Antigenic variants of rabies virus. *J. Exp. Med.* **152**, 99–112.

63. Yarchoan, R., Murphy, B. R., Strober, W., Schneider, H. S., and Nelson, D. L. (1981). Specific anti-influenza virus antibody production in vitro by hyman peripheral blood mononuclear cells. *J. Immunol.* **127**, 2588.

64. Zimmermann, U., and Scheurich, P. (1981). High frequency fusion of plant protoplasts by electric fields. *Planta* **151**, 26–32.

4

Production of Insect-pathogenic Viruses in Cell Culture

R. A. J. PRISTON

Shell Research Laboratories
Sittingbourne, Kent, United Kingdom

1. INTRODUCTION

The rapid advances made in invertebrate cell culture techniques since the early 1960s are primarily the result of pressures that came from a variety of sources to develop alternatives to chemical pesticides. At that time there were growing public concerns over the effects of these products on

Animal Cell Biotechnology, Vol. 2

wildlife and the presence of toxic residues in food crops. These concerns led to the introduction of strict regulations controlling their safety evaluation and added appreciably to final costs. The same period saw a sharp rise in the cost of raw materials for chemical products, and as a result the search for less expensive alternative ways to control insect pest species became more attractive.

Viruses from several taxonomic groups have been used in pest control programmes. However, several of these groups contain morphologically similar viruses (picorna, parvo and pox viruses) that infect vertebrates, and for this reason those viruses normally considered for insect control programmes are predominantly of the occluded type. This large category includes the baculoviruses [nuclear polyhedrosis (NPV) and granulosis (GV)] and the cytoplasmic polyhedrosis viruses (CPV). Another group, containing the iridescent viruses (IV), has also been used against *Tipula* spp. larvae, though with disappointing results (2). Insect viruses have been used and are in use in forestry, horticulture and agriculture. Entwhistle (6) listed 31 lepidopteran, 6 hymenopteran and 1 coleopteran pest species for which baculovirus control has been considered likely.

Those viruses that are used in pest control programmes are produced exclusively in whole insects reared in the laboratory or purpose-built breeding facilities. Six NPV and one CPV have been adequately evaluated and have been granted licences for use, mainly in the United States and Canada (3). Several patents on the production or formulation of insect viruses have been filed (47).

At present there is no facility for the large-scale production of insect viruses in invertebrate tissue cultures. However, this approach offers several advantages over using the whole animal. Tissue culture allows the pursuit of basic information on host cell and virus metabolism free from complications imposed by the specialisation of the whole animal. Tissue culture must be operated under sterile conditions and a far cleaner product is obtained. Production is less labour-intensive than mass rearing and, especially where fermenters are to be used, it offers greater flexibility; i.e. a fermenter can be diverted to production of another virus within a few days. These advantages, on the other hand, must be balanced against disadvantages such as high capital costs, requirement for maintenance of sterile conditions and media costs.

In this chapter I shall consider some of the recent advances made in large-scale production of insect tissue cultures and viruses. Several comprehensive reviews on replication of baculoviruses (27), large-scale production of insect viruses (49) and microbial pesticides (3, 47) are available for further reference.

2. BASIC *IN VITRO* METHODS

2.1. Selection of Viruses

Unlike traditional chemical pesticides and biological agents that act through production of toxins (e.g. *Bacillus thuringiensis*), viruses have a limited spectrum for target pests. Some, such as NPV of *Spodoptera* species, have an extremely narrow spectrum, while others, for example *Autographa californica* (AC) NPV, will infect several pest species. In addition, viruses isolated from different localities can show different degrees of virulence (*44*). It is, therefore, crucial to take account of the intended target species when considering the use of viruses. If a virus is to be considered as a pest control agent a great deal of information on its structure, its interaction with the host and its infectivity spectrum *in vivo* and *in vitro* must be obtained.

Historically, occluded viruses were chosen because of their characteristic pathology in the diseased animal and because they were believed to be more stable than non-occluded viruses. Baculoviruses were thought to be exclusively insect pathogens but they are now known in other groups (*4, 8*). Most occluded viruses handled *in vitro* are NPV (*19*), although CPV (*15*) and entomopox (*16*) also grow *in vitro*. Serial cultivation of GV *in vitro* has yet to be demonstrated. Among non-occluded viruses IV (*1*), Oryctes (*26*) and cricket paralysis virus (*42*) have been established in cell culture.

A brief introduction, only, to the structure and replication of baculoviruses is given here. For further details the reviews by Kelly (*27*), Granados (*14*) and Carter (*3*) are recommended.

Baculoviruses are DNA-containing viruses that replicate in the nucleus of infected cells to produce rod-shaped virions. Once formed, the virions are wrapped, individually or in groups, in lipid-containing envelopes, and many of these become embedded in a proteinaceous matrix to form storage bodies, the polyhedral inclusion bodies (IB). The protein matrix is produced in large quantities in the nucleus of infected cells.

The IB are of various shapes, depending upon the virus strain. More than 100 may be produced in each infected cell before it eventually bursts, releasing both IB and surplus virions to begin the infectious cycle once again.

The infectious unit is the free virion. *In vitro* either haemolymph from infected insects or media from infected cell cultures can be used to infect healthy cell cultures, but in the conditions prevailing in the field the virion is very unstable and it is the IB that is responsible for spread of infection between susceptible hosts. The IB is first ingested; then the protein matrix is digested by the enzymes in the gut and the free virions are released to infect cells of the gut wall.

Because of the protective nature of the protein matrix and the number of encapsulated virions, inclusion bodies are highly stable. Indeed, dry suspensions of IB can be stored for several years at $-20°C$ without detectable loss of activity.

Inclusion bodies are formed by CPV and pox viruses of insects as well as the baculoviruses, and their survival under field conditions is far superior to that of viruses that do not produce them.

2.2. Selection of Cells

A large number of insect cell lines are available (20) and many of these are amenable to virus replication. Most of the cell lines are from lepidopteran and dipteran insects and none have so far been established from the order Hymenoptera. In attempts to establish cell lines, workers have generally chosen material that is potentially able to multiply rapidly and is easily handled under sterile conditions. Consequently, embryonic tissue has been widely employed, but larval and adult tissues have also been used [see review by Stockdale and Priston, (49)]. One of the most widely used cell lines was derived by Hink (18) from the ovaries of adult *Trichoplusia ni* (*T. ni*) and has been designated TN-368. Tissues not immediately considered as suitable have also been used. At least one line has been established from pupal fat body of *T. ni* (12) and lines have been derived from larval and pupal haemocytes from several insect species.

Many of the cell lines appear to grow equally well in suspension and as monolayers on the walls of culture vessels or other support media. When attached to a substrate they may be detached for subculture by vigorous shaking and scraping with a rubber policeman.

Not all cell lines support the growth of viruses and in others IB may not be formed. The latter does not always indicate a failure to grow because replication may be incomplete. Ignoffo *et al.* (25), for example, were unable to demonstrate formation of IB by *Heliothis zea* (*H. zea*) NPV in an homologous cell line designated IMC-HZ-1, although they did succeed in raising the tissue culture infectivity titre to 10^8 by serial passage. On the other hand, Goodwin *et al.* (12) grew the same virus in another cell line of *H. zea* (IPLB-1079) and reported that only cells growing attached to a substrate produced IB; those in suspension did not.

Procedures for measuring virus titres have been developed in tissue culture, using end-point dilution and plaque assays. These have been reviewed by Knudson and Buckley (28). They offer an excellent mechanism by which virus preparations may be cloned to obtain "pure" virus cultures.

Invertebrate cell cultures, like those of vertebrate origin, are vulnerable to contaminants. Apart from bacterial and fungal contamination resulting from poor aseptic techniques, intrinsic contamination by microsporidia, mycoplasma and viruses has been described. For a review of the literature in this area see Knudson and Buckley (28) and Steiner and McGarrity (46).

Invertebrate cell lines are generally amenable to preservation by storage in liquid nitrogen, using 10% dimethyl sulphoxide as a protective agent in the culture medium. General procedures for storing cells are described by Shannon and Macy (43).

2.3. Culture Media

The composition of the body fluids of Insecta varies widely throughout the group and during development of a particular insect. For this reason it seems unlikely that a single medium will be suitable for the culture of cells from the different orders.

Animal cells have simple nutritional needs [a usable carbon and energy source, amino acids and vitamins; see review by Stockdale and Priston, (49)] but they are unable to concentrate small molecules to any extent or, unlike vertebrate cells, synthesise sterols. Consequently, if cell growth is to be established the balance between extra- and intracellular concentrations of these molecules must be maintained. In practice, this is achieved either by retaining a certain amount of "conditioned" medium at the time of sub-culturing or by adding a large proportion of nutrients of biological origin. Foetal bovine serum (FBS) is most commonly added, but chicken egg ultra-filtrate (CEU), bovine plasma albumin (BPA), tryptose phosphate broth and tissue culture yeastolate are also used.

Development of chemically defined media for large-scale production of insect cells and viruses would be advantageous in terms of handling, cost and quality control and current progress in this area will be discussed later. At present, however, high cell concentrations are obtained by using partially defined media in which known quantities of amino acids and vitamins are supplemented with biological additives (13, 35) or in which extracts of biological origin are present.

3. LARGE-SCALE PRODUCTION

Several laboratories are attempting to produce reliable and economic cell culture systems for the mass production of insect viruses. There are, unfortunately, a number of technical and economic obstacles that must first be overcome and a large-scale facility does not, as yet, exist. Several estimates of comparative *in vivo/in vitro* costs have been made (49). Recently, from studies with ACNPV in *T. ni* 368 cells, Hink (21) obtained yields of 10^8

IB /ml, but it was concluded that a 20-fold increase would be necessary before the process became economic. In the following paragraphs I shall examine what progress has been made towards the problem of producing insect viruses in large-scale culture systems.

3.1. Design of Equipment

There have been a number of different approaches to large-scale production of cells and viruses. Roller bottles have been used on several occasions. Weiss et al. (51) optimised a procedure using Spodoptera frugiperda cells growing in 430-cm² roller bottles. Cells were infected with ACNPV 6 days after seeding and IB were collected after a further 8 days' incubation. They reported yields of 9 × 10⁹ IB per bottle (equivalent to the yield from 1.5 infected larvae) but concluded that production costs made the system uneconomic. Similar studies have been conducted by Hilwig and Alapatt (17) and by Vaughn and Dougherty (50), who also used a perfusion culture system. However, because of the bulky nature of this type of system and the high labour costs, it seems unlikely to be attractive to industry.

Fermenter vessels are being used in several laboratories. Röder and Groner (41) described the growth of cells of S. frugiperda, S. littoralis and Mamestra brassicae in Biostat (Braun Melsungen) and microferm (New Brunswick) fermenters of up to 10-litre capacity. Cultures were maintained for periods of up to 200 days by regular replenishment of the cell suspension with fresh growth medium. They reduced medium costs by replacing, probably, the most expensive ingredient (foetal bovine serum) with egg yolk emulsion (40) and obviated the problem of sparging by passing air through a silicone rubber tube coiled inside the vessel as described by Miltenburger and David (34).

Difficulties with aeration were also encountered by Hink and Strauss (23) when they studied the growth of ACNPV in T. ni 368 cells with fermenters of 2- to 3-litre capacity. Vigorous aeration was required at these volumes and this caused excessive foaming and cell death. To overcome this, Hink and Strauss added antifoam and increased the concentration of methylcellulose in the growth medium.

At present, the only publication that deals specifically with the problems of large-scale production is that by Pollard and Khosrovi (37). They concluded that scaled-up laboratory batch procedures are unlikely to be economic. Taking into account oxygen requirements and the fragility of insect cells, they proposed that the most suitable design of vessel will be a vertical tubular flow reactor in which cells are maintained in suspension by a medium current upstream of the inoculation point. By using a small proportion of the product cells or inoculum, such a system could be maintained in continuous operation, and by providing aeration across a silicone membrane in a way similar to that used by Röder and Groner (*41*) the maximum shear stress present would be several orders of magnitude lower than that encountered in a mechanically agitated vessel. Pollard and Khosrovi presented data for the operation of a 183-m^3 vessel and pointed out that production on this scale would be equivalent to that of 1500 roller cultures of 5-m length and 2-m diameter or that of 400 separate stirred-tank fermenters with similar shear forces. The fixed costs of such a vessel would be 21.4% and capital costs would be 9.1% over a 10-year depreciation period. Variable costs would form a major part of the total costs (68.1% for materials and 1.4% for utilities).

3.2. Choice of Viruses

Once a target species has been identified and a suitable virus has been shown to replicate *in vitro*, there are other characteristics to be taken into account. For example, yields of IB and virions are affected by incubation temperature (*10*), the growth phase of cells (*48*) and cell density (*53*).

It may be an advantage to isolate virus clones with more rapid growth rates (*31*), higher temperature stability (unpublished), increased yields of IB or resistance to ultraviolet light.

Perhaps one of the most significant, and as yet unexplained, observations on the replication of baculoviruses in cell culture is that serial passage causes a second form of virus to appear. Several studies have shown that IB produced during the first few passages in cell culture are as infective as IB produced *in vivo*. Serial passage, especially at high multiplicities of infection, leads to reduced yields of IB. Faulkner and Henderson (*7*) observed changes in the quality of IB of *T. ni* NPV during repeated passage in *T. ni* cells. Hirumi *et al.* (*24*) made similar observations with ACNPV in *T. ni* cells. McKinnon *et al.* (*32*) found that the yield of *T. ni* NPV IB fell from 28 to 2.5 per cell after 50 passages and similar findings have been described for *S. frugiperda* NPV in *S. frugiperda* cells (*29*) and *H. zea* NPV in *H. zea* cells (*54*).

This change is caused by the mergence and increasing dominance of a new form of virus during subculture. Hink and Strauss (*22*) isolated two variants after passage of ACNPV *in vitro*. These yielded 81–352 and 2–13 IB per cell and were called MP (many polyhedra) and FP (few polyhedra) respectively. Similar observations have been reported for *T. ni* NPV (*38*) and *Galleria mellonella* NPV (*9*).

Compared with MP, the inclusion bodies of FP virus contain fewer, single nucleocapsid virions (*39*) and incomplete virions, which implies lower infectivity and reduced stability.

Both MP and FP are infectious *in vivo* and give rise to MP to the virtual exclusion of FP (9). Plaque-purified MP continues to give rise to FP on passage *in vitro*, whereas FP is genetically stable.

If large-scale production is to be contemplated this problem will need to be overcome or at least minimised. Enough infectious haemolymph is unlikely to be available to inoculate large fermenters and several virus replication cycles will be necessary for maximal infection.

3.3. Choice of Cells

Population doubling times of insect cells are similar to those of vertebrate cells with the fastest growth recorded for *T. ni* (TN-368) cells at 17 hr (*18*). A fast-growing cell line may offer reduced overall incubation times with a consequent reduction in maintenance requirements and a higher yield of cells per unit of a given nutrient.

Method of cell culture, i.e. in suspension or attached to a substrate, can influence the yield of IB (see Section 2.2). Cell morphology may also be important. For example, the spheroidal cell of *S. frugiperda* may be more resilient to mechanical damage than the fusiform type typified by TN-368.

One additional factor is the origin of the cell line. Lynn and Hink (*30*) compared the growth of ACNPV in cells from the lipidopterans *Estigmene acrea, Lymantria dispar, M. brassicae, S. frugiperda* and *T. ni*. They observed significant differences in their sensitivities to infection and in yields of virons and IB.

3.4. Medium Modifications

Media are generally difficult to handle and store and their costs are high because of a requirement for expensive nutrients such as animal serum. Dougherty *et al.* (5) recently calculated that for *G. mellonella* NPV production *in vitro* the cost of serum was half the total costs—including labour.

Considerable progress has been made in developing more defined and cheaper media [see review by Stockdale and Priston (*49*)] and serum-free media are now available for some cell lines. Wilkie *et al.* (52) developed two chemically-defined media that supported the growth of *S. frugiperda* cells, two mosquito cell lines and ACNPV in *S. frugiperda* cells. These media were supplemented with increased concentrations of vitamins and amino acids such as cystine, hypoxanthine and tyrosine. Röder (*40*) substituted FBS with 1% egg yolk emulsion and grew cells of *S. frugiperda, S. littoralis* and *M. brassicae. Autographa californica* NPV also replicated in *S. frugiperda* cells to yield IB with unaltered virulence and morphology. Goodwin and

Adams (*11*) and Mitsuhashi (*36*) have also developed serum-free media. Weiss *et al.* (*51*) incorporated FBS in medium to grow *S. frugiperda* cells in large volume but reduced costs by omitting antibiotics.

3.5. Physicochemical Conditions

Both cells and viruses will replicate over a wide range of physicochemical conditions, i.e. osmotic pressure pH and temperature, and this area is un-likely to present problems in large-scale production. Because of the fragility of the cells, however, supply of gases may be difficult in large volumes. Some efforts have already been made to address this problem (see Section 3.1) and the subject is adequately reviewed by Stockdale and Priston (*49*).

4. VIRUS PURIFICATION

Preparations of IB produced *in vivo* are inevitably contaminated with insect fragments, gut contents and micro-organisms and there is a strong case for some degree of purification before use to minimise possible health hazards posed by these contaminants. This purification, however, generally involves sequential filtration and centrifugation and can make the final prod-uct prohibitively expensive. Moreover, it is well known that proteins can protect viruses from inactivation, so their removal may lead to lower infec-tivity and reduced environmental persistence. It is anticipated that viruses produced in cell culture will be free from contaminating micro-organisms and minimal purification will be necessary. The hazards posed by cellular fragments and by intrinsic contaminants such as viruses are as yet unknown.

5. PROSPECTS

The future of large-scale *in vitro* cell culture and virus production depends upon overcoming major technical problems and also upon the attitudes of commercial and regulatory organisations. Viruses are not expected to re-place traditional pesticides but to supplement or to support their use.

Considerable progress has been made towards their efficient and econom-mic production and current technical obstacles such as development of inex-pensive media and provision of adequate aeration with minimal stress should be solved, given time. A major obstacle is the reduction in IB yield that is seen in continuous culture and at high cell densities. There is, as yet, no clear understanding of how this may be overcome.

One possible answer may lie in new virus isolates, for it seems probable that those available to date represent only a fraction of those present in nature.

Another approach is to clump together or wrap up the enormous quantities of free virions produced in infected cells to form synthetic IB.

Discovery of new viruses may itself provide new strains with desirable characters. In the meantime, desirable properties such as increased virulence, resistance to UV light and increased temperature stability will continue to be addressed using classical genetic approaches.

Genetic engineering of ACNPV is already under way. Future possibilities include the introduction of an insect-specific toxin gene into the virus genome, using recombinant DNA technology (33). Such engineering could hasten the rate at which virus kills the host and possibly broaden the host range. In this way viruses may become more attractive to industrial producers.

It seems likely that *in vitro* production will initially be achieved by a fermentation organisation with spare fermenter capacity. In this context NPV and insect cells may play an important role in future recombinant DNA research. Human β-interferon has already been synthesised in insect cells, using ACNPV as vector (45), and other compounds such as insect pheromones may be produced in a similar way.

ACKNOWLEDGMENTS

I wish to thank Mrs. Brenda Whitehead for typing this manuscript and for her patience. I am also indebted to H. Stockdale for his helpful comments.

REFERENCES

1. Bellet, A. J. D., and Mercer, E. H. (1964). The multiplication of *Sericesthis* iridescent virus in cell cultures from *Antheraea eucalypti* Scott. 1. Qualitative experiment. *Virology* 24, 645–653.
2. Carter, J. B. (1978). Field trials with *Tipula* iridescent virus against *Tipula* spp. larvae in grassland. *Entomophaga* 23, 169–174.
3. Carter, J. B. (1984). Viruses as pest control agents. *Biotechnol. Genet. Eng. Rev.* 1 (in press).
4. Couch, J. A., Summers, M. D., and Courtney, L. (1975). Environmental significance of baculovirus infections. *Ann. N.Y. Acad. Sci.* 266, 528–536.
5. Dougherty, E. M., Cantwell, G. E., and Kuchinski, M. (1982). Biological control of the greater wax moth (Lepidoptera: Pyralidae), utilising *in vivo* and *in vitro*–propagated baculovirus. *J. Econ. Entomol.* 75, 675–679.
6. Entwhistle, P. F. (1983). Viruses for insect pest control. *Span* 26, 59–62.

7. Faulkner, P., and Henderson, J. F. (1972). Serial passage of a nuclear polyhedrosis disease virus of the cabbage looper (*Trichoplusia ni*) in a continuous tissue culture cell line. *Virology* **50**, 920–924.

8. Federici, B. A., and Humber, R. A. (1977). A possible baculovirus in the insect-parasitic fungus *Strongwellsea magna*. *J. Gen. Virol.* **35**, 387–392.

9. Fraser, M. J., and Hink, W. F. (1982). The isolation and characterisation of the MP and FP plaque variants of *Galleria mellonella* nuclear polyhedrosis virus. *Virology* **117**, 336–378.

10. Gardiner, G. R., Priston, R. A. J., and Stockdale, H. (1976). Studies on the production of baculoviruses in insect tissue culture. *Proc. Int. Colloq. Invertebr. Pathol., 1st, pp.* 99–103.

11. Goodwin, R. H., and Adams, J. R. (1980). Liposome incorporation of factors permitting serial passage of insect viruses in lepidopteran cells grown in serum-free medium. *In Vitro* **16**, 222.

12. Goodwin, R. H., Vaughn, J. L., Adams, J. R., and Louloudes, S. J. (1974). Replication of a nuclear polyhedrosis virus in an established insect cell line. *Misc. Publ. Entomol. Soc. Am.* **9**, 66–72.

13. Grace, T. D. C. (1962). Establishment of four strains of cells from insect tissue grown *in vitro*. *Nature (London)* **195**, 788–789.

14. Granados, R. R. (1980). Infectivity and mode of action of baculoviruses. *Biotechnol. Bioeng.* **22**, 1377–1405.

15. Granados, R. R., McCarthy, R. J., and Naughton, M. (1974). Replication of a cytoplasmic polyhedrosis virus in an established cell line of *Trichoplusia ni* cells. *Virology* **59**, 584–586.

16. Granados, R. R., and Naughton, M. (1975). Development of *Amsacta moorei* entomopoxvirus in ovarian and hemocyte cultures from *Estigmene acrae* larvae. *Intervirology* **5**, 62–68.

17. Hilwig, I., and Alapatt, F. (1981). Insect cell lines in suspension cultivated in roller bottles. *Z. Angew. Entomol.* **91**, 1–7.

18. Hink, W. F. (1970). Established insect cell line from the cabbage looper, *Trichoplusia ni*. *Nature (London)* **226**, 466–467.

19. Hink, W. F. (1976). A compilation of invertebrate cell lines and culture media. *In* "Invertebrate Tissue Culture Research Applications" (K. Maramorosch, ed.), pp. 319–369. Academic Press, New York.

20. Hink, W. F. (1980). The 1979 compilation of invertebrate cell lines and culture media. *In* "Invertebrate Systems *In Vitro*" (E. Kurstak, K. Maramorosch, and A. Dubendorfer, eds.), pp. 553–578. Elsevier/North-Holland, Amsterdam.

21. Hink, W. F. (1982). Production of *Autographa californica* nuclear polyhedrosis virus in cells from large-scale suspension cultures. *In* "Microbial and Viral Pesticides" (E. Kurstak, ed.), pp. 493–506. Dekker, New York.

22. Hink, W. F., and Strauss, E. (1976). Replication and passage of alfalfa looper nuclear polyhedrosis virus plaque variants in cloned cell cultures and larval stages of four host species. *J. Invertebr. Pathol.* **27**, 49–55.

23. Hink, W. F., and Strauss, E. M. (1980). Semi-continuous culture of the TN-368 cell line in fermenters with virus production in harvested cells. *In* "Invertebrate Systems *In Vitro*" (E. Kurstak, K. Maramorosch, and A. Dubendorfer, eds), pp. 27–33. Elsevier/North-Holland, Amsterdam.

24. Hirumi, H., Hirumi, K., and McIntosh, A. H. (1975). Morphogenesis of a nuclear polyhedrosis virus of the alfalfa looper in a continuous cabbage looper cell line. *Ann. N.Y. Acad. Sci.* **266**, 302–326.

25. Ignoffo, C. M., Shapiro, M., and Hink, W. F. (1971). Replication and serial passage of

infectious *Heliothis* nucleopolyhedrosis virus in an established line of *Heliothis zea* cells. *J. Invertebr. Pathol.* **18**, 131–134.

26. Kelly, D. C. (1975). 'Oryctes' virus replication: Electron microscopic observations on infected broth and mosquito cells. *Virology* **69**, 596–606.

27. Kelly, D. C. (1982). Baculovirus replication. *J. Gen. Virol.* **63**, 1–13.

28. Knudson, D. L., and Buckley, S. M. (1977). Invertebrate cell culture: Methods for the study of invertebrate-associated animal viruses. *In* "Methods in Virology" (K. Maramorosch and H. Koprowski, eds.), Vol. 6, pp. 323–391. Academic Press, New York.

29. Knudson, D. L., and Harrap, K. A. (1976). Replication of a nuclear polyhedrosis virus in a continuous cell culture of *Spodoptera fnugiperda:* Microscopy study of the sequnce of events of the virus infection. *J. Virol.* **17**, 254–268.

30. Lynn, D. E., and Hink, W. F. (1980). Comparison of nuclear polyhedrosis virus replication in five lepidopteran cell lines. *J. Invertebr. Pathol.* **35**, 234–240.

31. McIntosh, A. H., and Rechtoris, C. (1974). Insect cells: Colony formation and cloning in agar medium. *In Vitro* **10**, 1–5.

32. McKinnon, E. A., Henderson, J. F., Stoltz, D. D., and Faulkner, P. (1974). Morphogenesis of nuclear polyhedrosis virus under conditions of prolonged passage in vitro. *J. Ultrastruct. Res.* **49**, 419–435.

33. Miller, L. K., Luigg, A. J., and Bulla., L. A., Jr. (1983). Bacterial, viral and fungal insecticides. *Science* **219**, 715–721.

34. Miltenburger, H. G., and David, P. (1980). Mass production of insect cells in suspension. *Dev. Biol. Stand.* **46**, 183–186.

35. Mitsuhashi, J. (1973). Establishment of cell lines from the pupal ovaries of the swallow tail papilio-zuthus lepidoptera papilionidae. *Appl. Entomol. Zool.* **8**, 64–72.

36. Mitsuhashi, J. (1982). Media for insect cell cultures. *Adv. Cell Cult.* **2**, 133–196.

37. Pollard, R., and Khosrovi, B. (1978). Reactor design for fermentation of fragile tissue cells. *Process Biochem.* **13**, 31–37.

38. Potter, K. N., Faulkner, P., and Mackinnon, E. A. (1976). Strain selection during serial passage of *Tricholusia ni J. Virol.* **18**, 1040–1050.

39. Ramoska, W. A., and Hink, W. F. (1974). Electron microscope examination of two plaque variants from a nuclear polyhedrosis virus of the alfalfa looper, *Autographa californica. J. Invertebr. Pathol.* **23**, 197–201.

40. Röder, A. (1982). Development of a serum-free medium for cultivation of insect cells. *Naturwissenschaften* **69**, 92–93.

41. Röder, A., and Groner, A. (1982). Growth conditions of insect cells and NPV in suspension. *Proc. Int. Colloq. Invertebr. Pathol.* 3rd, 82, p. 166.

42. Scotti, P. D. (1976). Cricket paralysis virus replicates in cultured *Drosophila* cells. *Intervirology* **6**, 333–342.

43. Shannon, J. E., and Macy, M. L. (1973). Freezing, storage and recovery of cell stocks. *In* "Tissue Culture: Methods and Applications" (P. F. Kruse, Jr. and M. K. Patterson, eds.), pp. 712–718. Academic Press, New York.

44. Shapiro, M., and Ignoffo, C. M. (1970). Nucleo polyhedrosis of *Heliothis*: Activity of isolates from *Heliothis zea. J. Invertebr. Pathol.* **16**, 107–111.

45. Smith, G. E., Summers, M. D., and Fraser, M. J. (1983). Production of human beta interferon in insect cells infected with a baculovirus expression vector. *Mol. Cell. Biol.* **3**, 2156–2165.

46. Steiner, T., and McGarrity, G. (1983). Mycoplasmal infection of insect cell cultures. *In Vitro* **19**, 672–682.

47. Stockdale, H. (1984). Microbial insecticides. *In* "Comprehensive Biotechnology," Vol. 2. Pergamon, Oxford (in press).

48. Stockdale, H., and Gardiner, R. (1977). The influence of the condition of cells and medium on production of polyhedra of *Autographa californica* nuclear polyhedrosis virus *in vitro*. *J. Invertebr. Pathol* **30**, 330–336.

49. Stockdale, H., and Priston, R. A. J. (1981). Production of insect viruses in cell culture. *In* "Microbial Control of Pests and Plant Diseases 1970–80" (H. D. Burges, ed.), pp. 313–328. Academic Press, London.

50. Vaughn, J. L., and Dougherty, E. M. (1981). Recent progress in *in vitro* studies of baculoviruses. *Beltsville Symp. Agric. Res.* **5**, 249–258.

51. Weiss, S. A., Smith, G. C., Kalter, S. S., and Vaughn, J. L. (1981). Improved method for the production of insect cell cultures in large volume. *In Vitro* **17**, 495–502.

52. Wilkie, G. E. I., Stockdale, H., and Pirt, S. V. (1980). Chemically-defined media for production of insect cells and viruses *in vitro*. *Biol. Stand.* **46**, 29–37.

53. Wood, H. A., Johnston, L. B., and Burand, J. P. (1982). Inhibition of *Autographa californica* nuclear polyhedrosis virus replication in high density *Trichoplusia ni* cell cultures. *Virology* **119**, 245–254.

Yamada, K. I., Sherman, K. E., and Maramorosch, K. (1982). Serial passage of *Heliothis zea* singly embedded nuclear polyhedrosis virus in a homologous cell line. *J. Invertebr.*

54. *Pathol.* **39**, 185–191.

PART II

DOWNSTREAM PROCESSING

5

Clarification and Sterilisation

G. D. BALL

The Wellcome Research Laboratories
Beckenham, Kent, United Kingdom

1. INTRODUCTION

The separation techniques discussed are those of sedimentation and filtration. Sedimentation techniques are dependent upon gravitational or centrifugal forces to achieve the separation of the solid from the liquid phase. To achieve the removal of particulate matter to a given standard, e.g. 0.2 μm, clarification must be followed by "sterile" or final filtration.

Filtration mechanisms, systems and equipment are discussed, together with disease security considerations. Apparatus and filtration media de-

Animal Cell Biotechnology, Vol. 2

scribed are those most commonly used in biotechnology processes. Manufacturers named in this chapter do not represent a definitive list.

2. SEDIMENTATION

When the stirring or agitation energy is withheld from anchorage-inde-
pendent cell systems, cells and associated debris sediment.

If a culture is arranged to allow for decantation, by the positioning of a
syphon tube or vessel port, the supernatant fluid and the settled cells can be
independently removed. It is advantageous to allow cells to settle at a controlled temperature, e.g. 4° or 25°C. The removal of cell aggregates, complete cells or cell sub-units in a compact volume will obviously improve the
economy of subsequent separation processing. If the objective is to harvest
the cells, the unwanted supernatant fluid is independently removed. If the
objective is to harvest supernatant fluid for further processing, the lower the
unwanted solids content, the easier the separating purification process
becomes.

The volume of the culture system will determine the time required for
particulate matter to sediment. Usually an overnight time interval is acceptable. The viability of the sediment itself may be reduced. Wang *et al.* (55)
demonstrated that Burkitt lymphoma cells sedimenting in 35-ml centrifuge
tubes at 4° and 25°C maintained >90% viability for 50 hr, while at 37°C
viability was reduced to 14% in the same period.

3. CENTRIFUGATION

The use of centrifugal force to sediment whole cells and to remove coarse
particles, as quoted by Paul (38), is an elementary technique. The factors
influencing sedimentation include the proportional difference between the
densities of the particles and the suspending liquid (equal densities result in
zero sedimentation), the size and shape of the particles and the viscosity of
the fluid. The ratio of centrifugal to gravitational acceleration, or relative
centrifugal force (R.C.F.), is an expression of the separating power of a
centrifuge, where

$$\text{R.C.F.} = 1118 \times r \times \text{rpm}^2 \times 10^{-8}$$

r being the radial distance in centimetres from the tip of the bucket to the
centre of the shaft.

Manufacturers quote K factors as a means of comparison of rotors; the
lower the value the more efficient is the performance. The factor produces

an estimate of the time required to sediment a particle of a given sedimentation coefficient. The performance index (P_i) is the expression of relative performance of a rotor under ideal conditions. The sedimentation time, termed precipitation time (T_s), is dependent upon rotor characteristics and the physical properties of biological material; it can be calculated from data published for each rotor.

The volume of fluid to be centrifuged together with the density and size of the suspended particles will influence the choice between batch centrifugation, i.e. the distribution into containers of pre-determined capacity for sedimentation, and continuous flow centrifugation. In the latter case the suspension is passed through a sealed rotor whilst retaining the sediment and discharging the separated fluid.

3.1. Batch Centrifugation

A centrifuge rotor assembly with a capacity of 6 × 1.0 litres can be used to sediment tissue culture cells or large particles, e.g. bacteria or cellular debris. Such a rotor, equipped with swinging buckets, has a maximum velocity of 4200 rpm and R.C.F. = 5010 (M.S.E. Coolspin). A fixed angle head operates at a higher velocity, e.g. 9000 rpm and 13,600 R.C.F. (Sorvall G.F.3. rotor), but with a capacity reduced to 6 × 0.5 litres. Tissue culture cells can be sedimented readily with such equipment.

For the recovery of viable tissue culture cells the centrifugal force applied should be as low as is consistent with the time available and the degree of separation required. Wang et al. (55) demonstrated the effect of time, temperature and centrifugal force. They centrifuged cells at 25°C and at 42,200 g force for 80 min, where viability was reduced to 57%, but at 24,800 g 77% viability was recorded. In practice a 30-min duration is often adequate.

If the recovery of viable cells is not a consideration, the higher the centrifugal force applied the more efficient will be the sedimentation, to the benefit of any subsequent separation process.

3.2. Continuous Flow Centrifugation

The continuous flow method is designed for the centrifugation of larger volumes than is possible with the batch method. The system can be used to remove debris or to recover viable cells. Continuous flow rotors can be fitted to conventional centrifuges to operate at a flow rate of up to 100 litres/hr, whereas industrial centrifuges operate at flow rates in the order of tens of litres per minute.

The technologist can manipulate two factors to control the separation process: (1) the relative centrifugal force applied and (2) the flow rate

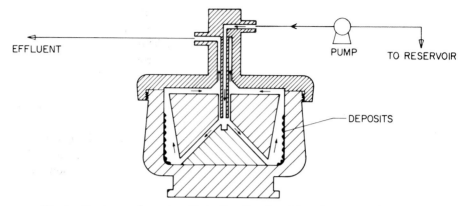

Fig. 1.   Continuous flow rotor, illustrating direction of flow during centrifugation.

through the rotor. For a centrifuge rotor of 660-ml capacity with a flow rate
of 30 litres/hr a dwell time of 1.3 min will result. To improve the sedimentation efficiency the flow can be decreased and/or the speed of centrifugation
increased.

The advantage offered by continuous flow systems is that the basic equipment can be used for the coarse removal of debris prior to the further
processing, by a sedimentary technique, of the remaining particles. A virus
suspension containing cell debris and some tissue culture cells can be clarified by applying a (comparatively) low g force, but at a high flow rate, e.g. 45
litres/hr at 6000 rpm. To concentrate or pellet the virus, the fluid can be re-
processed through the same machine at 20,000 rpm but at 3.0 litres/hr,
when a more efficient rotor would be preferred. The rate of flow can be
controlled by the use of a peristaltic feed pump, a diaphragm pump in the
larger systems (see Fig. 1) or sometimes by a throttle applied to a gravity
feed system.

If the objective is to collect the deposit, the recovery can be achieved by
disassembling the rotor. The Sorvall system collects the deposit into tubes
for recovery outside the equipment.

Industrial separators, constructed of a revolving 100-mm-diameter basket
contained within a long cylinder, generate 50,000 g from a very high rotation
speed. By the introduction of conical discs into a bowl, as illustrated by Fig.
2, the flow is divided into thin layers, which decreases the distance the
particles have to sediment. Consequently, the rotor speed and bowl diameter are reduced in comparison to those of conventional machines. The systems are described by Coulsen and Richardson (9) and both Alfa-Lavell and
Westfalia Separators, Ltd. manufacture such machines.

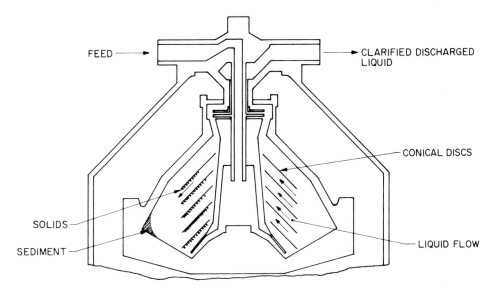

FEED ⟶   ⟶ CLARIFIED DISCHARGED
                                            LIQUID

                                            CONICAL DISCS

SOLIDS

SEDIMENT   LIQUID FLOW

Fig. 2.   Industrial separating centrifuge bowl, illustrating conical discs. Reproduced by permission of Westfalia Separation, Ltd.

Zwerner *et al.* (59) reported that the Sharples Laboratory Super centrifuge could process 50 litres/hr but with a 25–50% cell loss. They successfully used a basket-type centrifuge to operate at 90–120 litres/hr with 95% recovery of cells. The basket centrifuge required the scraping of cells from the rotor to be followed by batch centrifugation for final recovery; the cells were reported to metabolise actively.

Continuous flow rotors can be sterilised by autoclaving and assembling with associated reservoirs, using techniques to ensure the maintenance of sterile conditions. The sterilisation of large-scale commercial machines may, however, impose difficulties. Zwerner *et al.* (59) sterilised their machine by a "combination of steam, autoclaving and washing with disinfectant".

The capacity of the centrifuge bowl to retain debris is important. In small-volume systems there will be sufficient capacity to contain cells, or cell debris, accumulated within a run. With larger volumes, in a commercial system, it is possible to automatically discharge the sedimented contents during operation (steam sterilisable, Alfa-Laval Co., model AX213, and Westfalia Separator, model SB14).

Machines can be operated at a constant temperature, 0–25°C, depending on the objective. Wang *et al.* (55) demonstrated with Burkitt lymphoma cells that viability remained constant when they were centrifuged at 0°C with 42,200 g. With the flow rate indicated, for cell recovery, the dwell time in the machine is unlikely to have a significant effect, provided the system is cooled to 25°C or less.

For thermolabile viruses centrifugation at 0°C is essential.

3.3. Disease Security

The centrifugation by batch, continuous flow or zonal systems of biologically hazardous materials is a potentially dangerous procedure, as described by Baldwin (1), Baldwin et al. (2) and Dimmick et al. (11). Even when innocuous material is being processed the safety precautions for centrifuge operation must be observed, for example:

1. Correct operation of the equipment
2. Regular inspection and maintenance
3. Close attention to rotors
4. Accurate balancing of opposing buckets where necessary
5. Avoidance of blockage and froth formation of continuous flow systems
6. Adherence to good laboratory practices to maintain a clean environment

Batch Centrifugation.　The main centrifugation hazards result from (1) aerosol escape through inadequate sealing of the containers, (2) outside contamination of the container and (3) operational breakage. The use of polycarbonate or polypropylene centrifuge tubes or bottles overcomes breakages, but repeated autoclaving combined with centrifuge stress weakens the plastic and breakage cannot be excluded.

Aerosol escape is countered by containment. The most reliable solution is the confinement of the machine within its own "safety" environment, as described by Evans et al. (14), Webb et al. (56), and Godders et al. (18). The loading of hazardous fluids to bottles or tubes must be performed in a disease security cabinet with decontamination of all external surfaces prior to transfer to the rotor.

The protection offered by bottle or centrifuge bucket closure and rotor shields is limited, being dependent upon operator efficiency.

Continuous Centrifugation.　Continuous flow and zonal rotor centrifuges permit the release of significant amounts of aerosol from leaking seals (2, 11). Total confinement offers a permanent and safe solution. The centrifuge is located in an elaborate box, constructed from polycarbonate, with ventilation and exhaust protection by absolute air filters. The cabinet should be leak-proof and held under a negative pressure of at least $\frac{1}{2}$ in. water gauge.

The mechanisms of modern centrifuges are complicated, containing refrigeration and oil circulation systems. In the event of a failure or when serviced, a standard operating procedure should be provided which includes the drainage of refrigeration and oil systems and the decontamination of the recovered fluids. The emptied systems require sterilisation by, e.g. ethylene oxide (18) or heat and ethylene oxide. The de-contamination procedure should be validated with an innocuous contaminant (e.g. bacteria or bacteriophage) to prove efficiency.

At initial fitting the satisfactory operation of the air filter must be proved by a test which measures the retention of either an aerosol generated from sodium chloride solution (7) or dioctyl phthalate (DOP) particles. Both tests are briefly discussed by Denyer *et al.* (10). Thereafter tests should be repeated at, for example, 4-week intervals, to ensure routine safety operating standards. Also, the integrity of the construction should be measured by performing the operations with a suspension of an innocuous bacterium, e.g. *Serratia marcescens*, and establishing the degree of confinement (14).

4. FILTRATION

The filtration techniques to be considered are intended to either recover cells for further processing or produce a clarified or cell-free filtrate.

To achieve these objectives the mechanisms of filtration—depth adsorption, zeta potential and sieve retention—should be understood. The use of more than one mechanism simultaneously or in sequence will often occur. The technologist today has available a large variety of filter materials held in a variety of configurations (Table I).

4.1. Mechanisms of Filtration

4.1.1. Depth Adsorption

The retention of particles within the tortuous passages of a filter medium is defined as depth filtration by adsorption. Other mechanisms (see the following section) may simultaneously occur. The filter material may be constructed of asbestos fibre, glass fibre, polyester or polypropylene and may be in sheet form, for filter press use, or as a disc or cartridge (see Tables I,II). The system will have a high dirt removal capacity compared to a screen-type filter (see Fig. 3).

A filter aid, e.g. diatomite, can be used to form a porous cake maintained on a supporting screen, in which situation the structure performs as a depth filter. Filter binding is also reduced by the addition of a filter aid, and Telling *et al.* (47) used this system for BHK cell removal from culture fluids (see Fig. 4).

TABLE I Common Filtration Media, Trade Names, Suppliers, Sterilisation Methods, Mechanism and Process Applications

Filter media	Trade name	Manufacturer[a]	Presentation	Pore rating (μm)	Sterilisation methods[b]	Filtration mechanisms[c]	Process application
Asbestos	—	3, 11	Sheet Disc	Not applicable	S.A.	Depth, Z.P.	Clarification and final filtration
Asbestos glass epoxy resin	Cox M-780	4	Sheet Disc	0.45–2.0	S.A. DH.	Depth Z.P.	Clarification and final filtration
Acrylic co-polymer	Versaflow	5	Cart.	0.45–0.8	ETO	S.R.	Clarification
Cellulose + inorganic filter aids	Versapor Zeta Plus	1	Disc Cart. Disc Sheet	1.0–1.5 0.2–10	S.A.	S.R. S.R. Z.P.	Clarification
Cellulose acetate	Sartoban	10	Disc Cart.	0.2–1.2 0.2–0.8	A. DH A. S.	S.R.	Clarification and final filtration
Cellulose nitrate	—	8	Disc	0.1–5.0	A.	S.R.	Clarification and final filtration
Cellulose tri-acetate	GA Metricel	10 5	Disc, sheet Disc	0.01–8.0 0.2–5.0	A. A. ETO.	S.R.	Clarification and final filtration Final filtration
Cellulose ester mixed	GN Metricel MF - Millipore	5 7	Disc Disc	0.45–0.8 0.025–8.0	A.ETO. A.	S.R.	Clarification Final filtration
Cellulose ester and polypropylene	Membra-Fil MilliStak Milligard	8 7 7	Disc Capsule Cart.	0.1–5.0 0.22 Not applicable	A. Pre-sterilised A.	S.R.	Final filtration Final filtration Clarification

Cellulose ester + glass fibre	Polysep	7	Cart.	0.2–1.0	A.	Depth, S.R.	Clarification
Glass fibre	—	2	Tube	0.3–25	S.A.	Depth	
	AP. 15-25 Lifeguard.	7	Disc, Cart.	Not applicable	A.	Depth	
	—	8	Disc	0.65–12.0	S.A.	Depth	Clarification
	Ultipore GF	9	Cart.	0.65–10.0	S.A.	Depth, Z.P.	
	Ultipore GF Plus	9	Cart.	1.0–40.0 absolute	S.A.	Depth, Z.P.	
Nylon	—	5	Disc	1.0	A.	Depth	Final filtration
	Ultipore N66	9	Cart., Disc	0.1–0.45	S.A.	S.R.	Clarification
			Cart.	Bacterial pre-filter	S.A.	Depth	
	Nylaflo	6	Cart.	0.2–0.4	A	S.R.	Final filtration
	Posidyne	9	Cart., Disc	0.1–8.0	S.A.	Z.P., S.R.	Clarification and final filtration
Polycarbonate polyester	Zeta Por	1	Disc	0.1–0.45	A.	Z.P., S.R.	Final filtration
	Nuclepore track etched	8	Disc	0.05–12.0	S.A.	S.R.	
Polyester	Tortuous pore	8	Cart.	0.2–0.4	S.A.	S.R.	Final filtration
Polymer aromatic	HT - Tyffryn	5	Disc	0.1–0.64	S.A.	S.R.	Final filtration
Polypropylene	H.D.C.	9	Cart.	0.6–70	S.A.	Depth	Clarification
	—	5	Capsule	0.22	Pre-sterilised	S.R.	Final filtration
Polyvinylidene fluoride	Durapore		Disc	0.2–0.45	S.A.	S.R.	Final filtration
	Millipak	7	Multi-disc Capsule	0.2	A.	S.R.	

(continued)

TABLE I (*Continued*)

Filter media	Trade name	Manufacturer[a]	Presentation	Pore rating (μm)	Sterilisation methods[b]	Filtration mechanisms[c]	Process application
Potassium titanate	Ultipore	9	Cart.	0.2–3.0	S.A.	S.R., depth	Clarification and final filtration

[a] 1. A.M.F. Cuno Division, 2. Balston, Ltd., 3. Carlson Ford, Ltd., 4. Cox Instrument Division Lynch Corporation, 5. Gellman Sciences, Ltd., 6. Membrana GmbH, 7. Millipore Corporation, 8. Nuclepore Corporation, 9. Pall Process Filtration, Ltd., 10. Sartorius Instruments, Ltd., 11. Seitz.
[b] A (autoclave cycle; consult manufacturer's technical information for specification), DH (dry heat), S (in-line steam; consult manufacturer's technical information for specification), ETO (ethylene oxide).
[c] S.R. (sieve retention), Z.P. (zeta potential, electrokinetic).

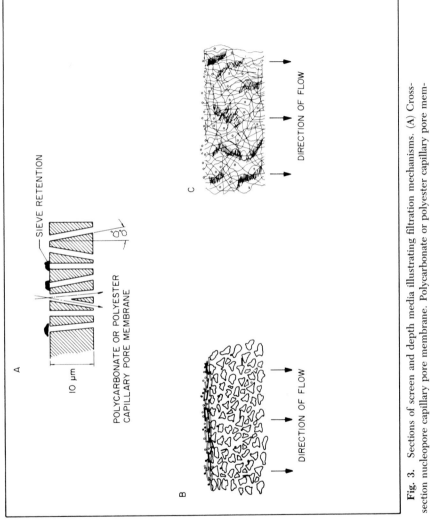

Fig. 3. Sections of screen and depth media illustrating filtration mechanisms. (A) Cross-section nucleopore capillary pore membrane. Polycarbonate or polyester capillary pore membrane. Open area: ~15%; 20–30 million pores/cm². Reproduced by permission of Nucleopore Ltd. (B) Retention by filter-cake formation. (C) Depth-type retention.

97

TABLE II Characteristics of Common Filtration Media

	Asbestos (3)	Glass fibre (2)	Glass asbestos (4)	Polypropylene (9)	Charged			
					Cellulose diatomite (1)	Nylon membrane (9)	Track etched membrane (8)	Cast membrane
Coarse clarifying grade	S9–S10	B–C	AA1000–100	H.050	1–10	NCZ–NPZ	N.A.	[a]
Fine clarifying grade	EK–EKS	A B	AA100–30	H012 H2-15	30–50	NBZ–NNZ	N.A.	[a]
Sterilising grade	EKS2	No	No	No	No	NFZ–NTZ	Yes	[a]
Mechanisms operating[b]	Depth, Z.P.	Depth, Z.P.	Depth, S.R.	Depth	Depth, Z.P.	Z.P., S.R.	S.R.	S.R. (Z.P.)
Spectrum of particles retained	Wide	Wide	Grade-dependent	Wide	Wide	Wide	Grade-dependent	Moderate
Cut-off performance	Poor	Poor	Moderate	Good	Moderate	Good	Very good	Good
Particle migration	Yes	Yes	Yes	Possible	Yes	Yes	No	Yes
Fibre release	Yes	No	Some	No	No	No	No	No
Fluid retention	High	Small	Small	Small	Moderate	Small	Very small	Small

98

Effect of pH	On Z.P.	c	c	No	c	d	No	Material-dependent
Retention affected by flow rate and Δp	Yes	Yes	Yes	Yes	Yes	Yes	No	Yes
Pre-wash necessary	Yes	No	Yes	No	Yes	No	No	No
Virus adsorption	Possible	e	e	Not known	e	Probable	Possible	Material-dependent
Dirt retention	High	Good	Moderate	Good	Good	Very good	Poor	Limited
Integrity tests	No	No	Possible	No	No	Yes	Yes	Yes
Notes	Low operating costs	Low operating costs	Low operating costs	Negative zeta potential	Pressure limitations		Strong; can be back-washed	a
	(3) Carlson Ford, Ltd.	(2) Balston Ltd. [a]	(4) Cox Instrument Div., Lynch Corp.	(9) Pall Filtration Ltd.	(1) AMF Cuno Div.	(9) Pall Process Filtration Ltd.	(8) Nuclepore Corp.	

[a] See Table I.
[b] Z.P. (zeta potential), S.R. (sieve retention).
[c] See Fig. 5.
[d] See Fig. 6.
[e] See Fig. 7.

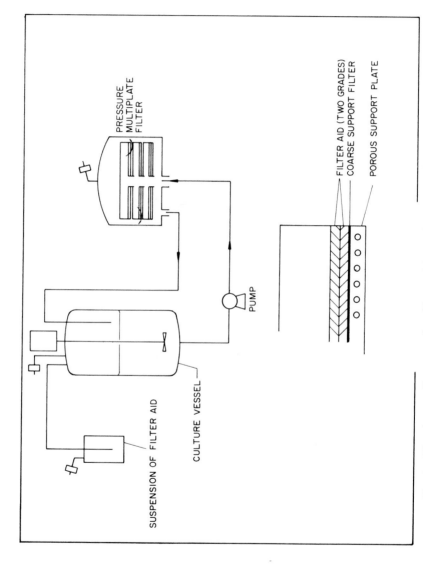

Fig. 4. Flow diagram illustrating the addition of filter aid to a suspended culture system.

Depth filters have disadvantages. The characteristics of flow rate and differential pressures, Δp, are not independent. Also, particle migration through the filter can occur from an increase in the pressure applied to a system to maintain a desired flow rate, particularly as the filter binds. Transen (50) recommends a low Δp to reduce this problem.

Fibre release of asbestos-type filters led to the banning of the use of such material for sterilization of fluids for human injection. It has been replaced by synthetic and other materials. Depth media, generally, have high dirt retention properties. Few depth filters are satisfactory as a sterilising medium, yet "sterilising grades" of Seitz asbestos filters have been successfully used (48).

4.1.2. Zeta Potential (ζ)

Zeta or *electrokinetic potential*, also referred to as *electrostatic attraction*, is defined by the Filtration Dictionary (15) as the electric potential within a fluid where a liquid is in contact with an ionogenic solid. Particles smaller than the pores of a filter medium are removed if the zeta potentials of the pore walls and the suspended particles are of opposing signs. If the charge difference is small or nil, the attraction will be small or nil, according to Pall *et al.* (35). Cells, bacteria and viruses are negatively charged (5, 19, 44). Zierdt (58) demonstrated the retention of bacteria, erythrocytes, lymphocytes, platelets and bacteriophage by different filter materials with a wide range of pore sizes, 1.0–14.0 μm. Separations were explained by electrostatic properties of filter materials.

Charged filters can carry a positive or negative charge, resulting either from the manufacturing process, or the natural structure of the fibre may result in a strongly positive charge, e.g. magnesium hydroxide functional groups of chrysotile asbestos fibres, which have an isoelectric point quoted by Sobsey and Jones (45) of about pH 12. Figures 5 and 6 present the variation of zeta potential with pH for various materials.

In practice, the choice of filter material will depend upon the purpose of the operation and the zeta potential of the suspended material. To use a filter medium with a high positive zeta potential to clarify tissue culture harvest fluid from which a virus is to be recovered would also result in removal of the virus [see Fig. 7, reproduced from Sobsey and Jones (45)]. Gerba and Hou (17) showed that a 0.22-μm pore rated, positive zeta potential, nylon filter removed 99% of piliovirus type 2 at a concentration of 10^4 plaque-forming units (PFU) per millilitre from distilled water at pH 6.5, where removal by negatively charged 0.22-μm cellulose nitrate was <10%.

Cliver (8) and Mix (28) referred to the use of solutions of proteins, e.g. serum or gelatine, as a means of blocking the action of the functional groups, to prevent the adsorption of viruses to a filter. In addition, there are other

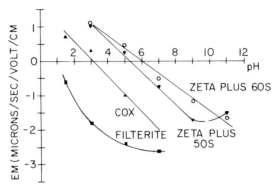

Fig. 5. Electrophoretic mobility of filter media as a function of pH. Reproduced by permission of Sobsey and Jones (45).

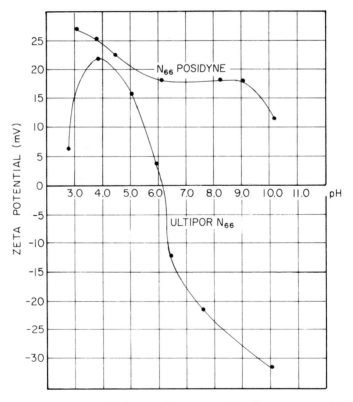

Fig. 6. Zeta potential of Pall nylon membranes. Reproduced by permission of Pall Process Filtration, Ltd.

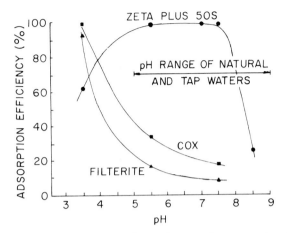

Fig. 7.   Poliovirus adsorption of filter media as a function of pH. Reproduced by permission of Sobsey and Jones (45).

ways of manipulating the zeta potential. Keswick and Wagner (23) reviewed the effect of salts, pH and ionic strength upon zeta potential and the influence upon the filtration performance with respect to virus and bacterial retention.

If the zeta potential is similar to that of the suspended particle or is intentionally destroyed, other filtration mechanisms dependent on the structure of the filtration medium still operate. See Fig. 8, reproduced from Hou et al. (19).

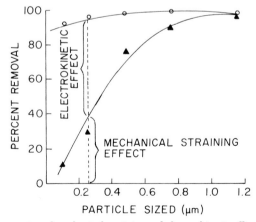

Fig. 8.   Demonstration of mechanical straining and electro-kinetic effect by 90S filter media by using polystyrene beads. Symbols: (O) 90S; (▲) 90S alkaline-treated to destroy positive charge. Reproduced by permission of Hou et al. (19).

The zeta potential can cease to operate or the efficiency decrease to an unacceptable level during an operation. Particles initially removed will be detectable in the filtrate (4). The phenomena will not be detected by pressure drop (Δp) increase across the filter unless other filtration mechanism changes are coincident.

The properties of charged media are summarized in Table II.

4.1.3. Sieve Retention

A true sieve can be visualized as a series of uniform pores set in an otherwise unbroken membrane. The capillary pore membranes produced by a patented irradiation etching process of polycarbonate or polyester composition are the only examples of such a material available. See Figs 3a and 9 and Tables I and II.

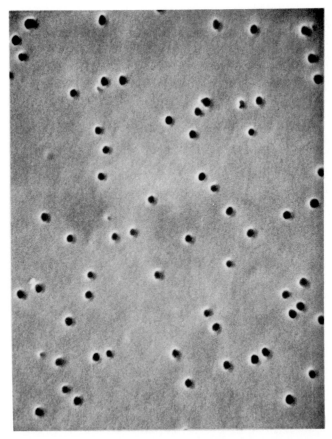

Fig. 9. Scanning electron micrograph, membrane manufactured by track-etch process. Reproduced by permission of Nuclepore, Ltd.

It is obvious that as non-deformable particles larger than the pore diameter will be retained, cake formation on the surface of the filter can result in filter blinding, Fig. 3a,b. Such filters can also perform as a charged filter expressing a zeta potential. Zierdt (58) reported that 3.0-μm pore size filters removed 99% *Escherichia coli* challenge, 68% *Staphyloccous aureus* and 94% latex spheres of 0.5-μm diameter.

There is, sometimes, confusion between true capillary pore membranes and "tortuous pore" membranes, also referred to as depth membranes. Such filters are manufactured from a great variety of materials (see Table I). Sieve retention and depth mechanisms can occur simultaneously, depending on the pore size of the depth membrane and the diameter of the particle.

The pore size of the microporous membrane filter assumes a greater significance than that of a depth filter, because of its "absolute" filtration characteristics. Yet Lukeszewicz *et al.* (25) commented that pore size measurements are rarely reported; manufacturers refer to "narrow distribution limits" without substantiation. However, Millipore Corporation publish tolerances for selected membranes, e.g. mean pore size 0.2 ± 0.02 μm, MF-Millipore Membrane. Nuclepore claim diameter variations of $+0\%$ to -20% at a pore density of up to $6 \times 10^8/cm^2$; Pall (33) quoted 6×10^{10} pores/cm² of 0.125 to 0.22 μm. From the appearance of Fig. 10, pore variation is obviously present. Pall *et al.* (35) demonstrated the passage of 0.5-μm latex particles through such a membrane and Wallhausser (52, 53) demonstrated failures in terms of bacterial breakthrough.

The success of filtration will depend on whether a particle meets a hole larger than its diameter. Despite integrity test data (Section 4.4), latex tests show such holes to be present in membranes. Duberstein (12) postulates the probability to be 10^{-10} for a bacterium to pass through a 1-ft², 0.2-μm membrane filter.

Lukeszewicz *et al.* (25), amongst others, have demonstrated improved bacterial retention on a 0.45-μm membrane at 5 psi compared to 30 psi operating pressures. Biological particles are not rigid structures. Reti *et al.* (43) speculate on whether distortion might occur under such filtration conditions. It is the opinion of the author that cells can be deformed to pass through holes smaller than their diameter.

4.2. Equipment and Filtration Systems

The filtration system of choice is dictated by the scale of the system and the mechanism of filtration required. Filters are manufactured as discs, pleated cartridges or sheets, each requiring a holder commonly constructed of stainless steel. A limited, but increasing, range of disposable units is available. Re-usable membranes in the form of multiple layers or capillary tubes have a particular application for tangential flow systems.

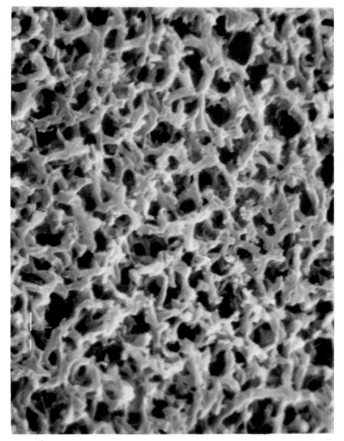

Fig. 10.   Scanning electron micrograph of Nylon 66 0.2-μm filter, magnification × 5000. Reproduced by permission of Pall Process Filtration, Ltd.

Membrane holders retain discs of diameter 13, 25, 47, 142 and 293 mm, representing filtration areas of 0.7–468 cm². The filter membrane disc is usually supported on a screen and sealed by O rings between filter housing surfaces.

To obtain filter areas greater than 468 cm², two 293-mm units can be connected in parallel to give a maximum area of 0.1 m²; however, the hardware is expensive (about £1000 per filter).

For 0.1-m² areas and larger, a cartridge containing a pleated sheet of filter medium provides an economical answer. A common cartridge form is the 10-in. or 254-mm unit containing at least 0.45 m² (or larger, depending on the material used) of filter medium; multiple units are achieved by the fusing of modules to a maximum of 40 in. and by using housings containing two or more cartridges arranged on a common base. Smaller cartridges (Sealklean or Sartobarn-mini) and disposable units of approximately 0.1 m² are satisfactory for small-scale systems.

Figure 11 illustrates a typical construction of a 10-in. cartridge housing. Cartridges are located by spring pressure or tie-rod and the seals effected by O rings or combined O rings and locating flanges. The seal of the cartridge to the housing base must be faultless.

The direction of flow is usually from the outside to the inside of the cartridge and the construction allows for maximum differential pressure, Δp, in that direction. Figures 12 and 13 illustrate various fittings used to connect a filter housing to a pressure reservoir or pump.

The multi-plate sheet filter unit illustrated by Fig. 14 is sterilised in place by steam. Single plate disc units have been a long-established feature of laboratories, pre-dating membrane filters. Multi-plate units are assembled

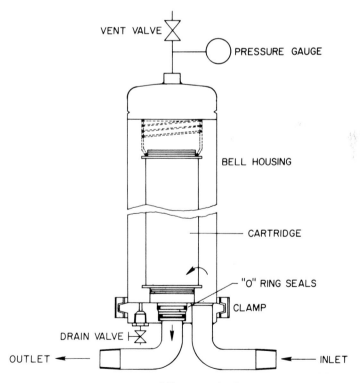

Fig. 11.   Typical filter cartridge housing.

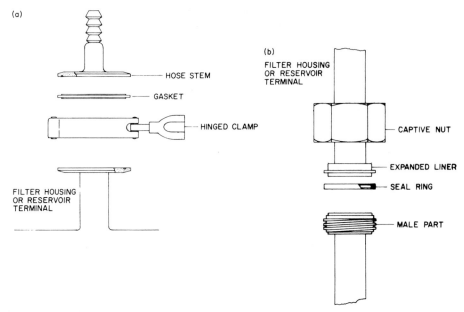

Fig. 12. (a) Typical Tri-Clover or Tri-Clamp sanitary fitting. (b) Typical industrial dairy fitting.

with attention to the correct direction of flow through the filter (rough side top). The sheets are clamped between hollow plates, to channel the flow uniformly through the filter. Change-over plates are possible to allow two grades of filter in a single press.

4.2.1. Depth Filters

These filters are composed of materials forming a labyrinth of tortuous pathways (Fig. 3c) such as chrysotile asbestos, glass fibres, cellulose and diatomaceous earth and, more recently, polypropylene. Table II lists their properties.

Asbestos. The valuable depth-dirt retaining capacity of the material can be considered in many applications. Asbestos grades are notated:

HP S 10		HP 7A	
HP EK	Sterilising,	HP 7	Fine
HP EKS	5–20 μm	HP 8	clarifying,
HP EK2	mean pore	HP 9	20–30 μm

Coarser clarifying grades exist. Filters treated with hydrochloric acid to reduce extractable metal are coded HP.

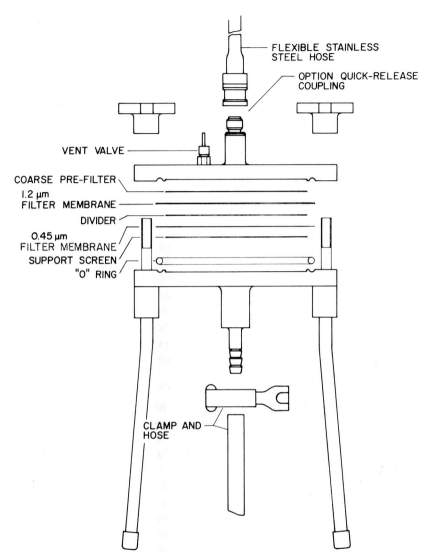

Fig. 13. Typical filter disc holder assembly with membranes of increasing retentive properties divided by nylon mesh.

Asbestos pads can be successfully used (not for human parenteral preparations) as a sterilising filter [Telling and Radlett (48)], although the performance is dependent on factors which are difficult to control. Before use in sterilising animal cell culture media and in other applications, a pre-filtration wash of saline, buffer or water is required to remove particles and fibres released from the system by steam sterilisation.

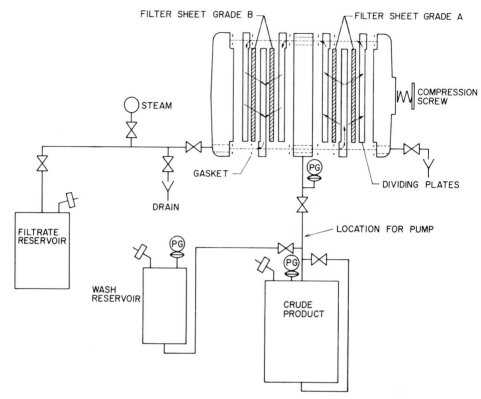

Fig. 14. Flow diagram of multi-plate filter press with change-over plate.

The filter materials exhibit a high zeta potential, but have been quoted by Osgood (32) to be satisfactory for the filtration of vaccines on an industrial scale. By contrast, >99% removal of poliovirus from tap water has been demonstrated by Sobsey and Jones (45) through grade S Seitz filters. To utilize only the depth properties of the material, the zeta potential sites can be neutralised to some extent by pre-washing the filter with 2% gelatine or a protein solution, e.g. calf serum (54).

Glass Fibre. A long-established depth filter medium is that of glass fibre. Filters can be constructed from borosilicate glass or microfibre, with bonding of an epoxy resin. Material is available in the form of discs, cartridges or tubes. Performance will not be absolute, e.g. Balston state 98% removal efficiency of a specific particle size. All presentations can be autoclaved and are stable with in-line steam. The permitted period of exposure reflects the thermal stability of components of the filter.

The zeta potential of the pure material is generally negative and it will have little virus adsorption activity at neutral or alkaline pH ranges, as demonstrated by the removal of viruses from water in experiments of Sobsey and Jones (45) and Hou et al. (19), but positively charged glass-based materials can be made (Pall Ultipore G.F. plus).

The material, having good dirt retention capacity, is often used as a clarification pre-filter.

Glass–Asbestos Combination.   A further variation is a combination of glass and asbestos microfibres, permanently bonded with an epoxy resin to form the filter medium. Known by the code M780 series, manufactured by Cox Instruments Division of Lynch Corporation, the filter is described by the manufacturer as "not a screen and not a depth filter". The construction, in disc form, is too thick to be a membrane at 0.78 or 0.50 mm; it is claimed to contain "a maze of ever narrowing vertical and horizontal flow paths". The filter has better dirt retention capacity than membranes. Graded porosities from 20 to 0.45 μm allow for a wide spectrum of duty, but values are not absolute and represent 99.999% efficiency.

Despite the asbestos content the grade AA (0.2 μm) was reported by Sobsey and Jones (45) to possess an isoelectric point of pH 3.0–4.0 and to have a negative zeta potential at higher pH values (see Fig. 5). This is consistent with the material being suitable for the filtration of virus-containing fluids with little loss of activity due to zeta potential adsorption (see Fig. 7).

The system is useful for virus suspensions containing cell debris of 10 to 100 litres volume, where graded filtration will precede final sterilising filtration.

Polypropylene.   This is an example of a depth filter medium manufactured from synthetic materials specifically designed to have a high dirt removal capacity. Available in disc form (Gelman Sciences, Ltd.) at 10-μm pore rating, or in pore ratings of 0.6–70 μm at 100% removal efficiency (Pall Filtration, Ltd.) in cartridge form, with filtration areas from 0.1 m^2 (66 mm length) to 0.5–0.79 m^2 (dependent on removal rating) for 254-mm-long cartridges.

The material is claimed not to be subject to fibre release or to the unloading of an entrapped particle during the filter life. The zeta potential of the material is quoted by Blosse (4) to be negative at neutral pH.

Cellulose–Diatomaceous Earth Media.   Depth filters are constructed from cellulose and diatomaceous earth with a pre-determined zeta potential and exhibit the combined benefits of depth filtration and the presence of a positively charged filter. Particle retention capacity is indicated by eight grades ranging from 4 to 10 μm (grade 01) and 0.1 to 0.25 μm (grade 90). The latter is not claimed to be a sterilising filter; however, it will give excellent service upstream of a (sterilising grade) membrane filter.

The zeta potential value is plotted as a function of pH in Fig. 5, and the efficiency of the electrokinetic effect and mechanical sieving is illustrated in Fig. 8. The removal of bacteria, virus and latex particles has been demonstrated by such filters expressing a positive zeta potential (19, 24, 45).

Migration of particles will occur when the charged sites become saturated, resulting in the appearance of particulate matter in the filtrate. Considerable differential pressures can be applied to a maximum of 2.5 kg/cm². During use, filtration pressures should be gradually increased within this system as the Δp increases; this only illustrates the performance of the depth filtration function without providing information on saturation of the charged sites.

4.2.2. Membrane Filters

Table I lists 10 materials from which membrane filters of 0.1- to 5.0-μm pore size are manufactured, together with their recommended sterilisation methods. Membranes are manufactured by two methods: (1) the track etched process of Nuclepore Corporation, using polycarbonate or polyester, and (2) the casting of a solution to form a fine layer upon a moving belt, which results in an open but tortuous pore structure (see Fig. 10); both methods are described by Denyer et al. (10). The resulting membranes are produced with controllable porosity ratings, 8.0 to 0.2 μm or smaller. The thickness of the track etched membrane is stated to be between 5.0 and 10 μm with less than 5% variance; that of the cast membrane is 80–150 μm. The pore structures of opposing surfaces may differ, and the direction of flow, therefore, influences the performance of the disc or sheet filter, whilst nylon filter discs are not isotropic.

No fibres can be released, except those residual from the packaging, so membrane filters can be used to process materials destined for human injection. The 0.22-μm size is generally accepted as a sterilising grade. However, *Pseudomonas diminuta* cultured on deprived media by Wallhausser (53) and *Corynebacterium aquaticum* by Phelps et al. (39) have both been demonstrated to pass 0.2-μm rated membranes.

Disc filters of decreasing pore size can be combined in a holder (see Fig. 13), but the dirt retention capacity is limited. Graded membranes are produced with two or more layers, e.g. Sartorius Sartoban or Pall Ultipore nylon, to combine some of the advantages of both depth and membrane filtration and also to extend the life of the cartridge or improve its protection. Double-layer membranes improve the theoretical removal efficiency, as stated by Wallhausser (52, 53), Duberstein (12) and Reti et al. (43). Pall and Kirnbauer (34) expressed the relationship between membrane thickness and titre reduction by the equation $T_R = T_{R1}t$, where

T_R = ratio of influent to effluent bacteria
T_{R1} = T_R for a membrane of "unit" thickness
t = thickness of membrane as multiples of "unit" thickness

A typical cartridge system consists of:

1. Depth pre-filter
2. Combined layer (at least 0.45 μm) cartridge
3. Final filter, 0.2 μm

each in series.

For the harvesting of virus suspensions the choice of membrane or the use of filter wash to minimize filtration loss is paramount (see section on "Zeta Potential" and Fig. 6). The performance of the recently introduced nylon media with pre-determined positive zeta potential (37) is yet to be independently reported. The manufacturers claim 100% removal efficiency of *Mycoplasma* by membranes of 0.65-μm pore rating or less (see Table III).

The properties of membrane filters are summarized in Table II.

4.2.3. Filter Aids

The use of a filter aid is defined by the Filtration Dictionary (*15*) as the addition of powders to the suspension being filtered in order to improve the operational life or the efficiency of the filter. The action is to increase the porosity of the cake of retained solids upstream of the filter surface and so to maintain the filter permeability.

Activated carbon, diatomite, pearlite (natural volcanic mineral) and cellulose fibre preparations are commonly used. The aid can be added to the

TABLE III Performance of Positive Zeta Potential Nylon Filter, Removal Rating 0.65 μm[a]

Test contaminant	Diameter (μm)	Quantity per 254 mm cartridge	Removal efficiency (%)	Pressure drop at end of test (bars)
Spherical silica	0.21	8.9 g	>99.99	
	0.021	18.7 g	90.8	<0.07
Spherical polystyrene	0.038	1.26 g	>99.99	
		2.48 g	73.7	<0.07
Mycoplasma[b]	0.1–0.4	1.5×10^{12}	100	<0.07
Ps. diminuta	0.3	10^{13}	100	2.67
E. coli endotoxin	0.001	0.016 g	>99.997	<0.07

[a] Reproduced by permission of Pall Process Filtration, Ltd. Abstracted from Technical Data Sheet, SD 905.
[b] *Acholeplasma laidlawii*.

unfiltered fluid and maintained in suspension, or deposited on a surface prior to the beginning of filtration, as described by Telling *et al.* (*47*). The use of multi-layers of differing particle retention properties is also described by the same authors. Hyflo Supercel, retention 0.5 μm; Superaid, retention 0.2 μm; additional Hyflo Supercel; and Dicolite grade 4200, retention 1.2 μm, are sequentially deposited. Telling *et al.* preferred a bed thickness of 10–14 mm, resulting in 0.8–1.6 kg/m^2 in the example quoted. The usual quantity of filter aid added is in the order of 1% (w/v).

The deposition system will depend upon the scale of the operation and the filter area employed. The use of multi-plate filters is illustrated by Fig. 4, where the filter aid is supported on suitable grade of Whatman filter paper. However, other filters can be substituted, e.g. a fibrous pad. Ballew (*3*) briefly described a similar system.

The deposited cake produced by a filter aid will exhibit the properties of a depth system with additional characteristics of the agent used.

4.2.4. Tangential Flow System

Sections 4.2.1 to 4.2.3 refer to so-called dead-end flow systems in which the flux becomes uneconomical or separation efficiency is lost as cake formation develops. Tangential or cross-flow filtration (see Fig. 15) attempts to prevent cake formation by the continuous circulation of the suspension across the surface of the membrane. To be effective a suitable velocity of flow is required, at an adequate differential pressure, to maintain the desired flow rate through the membrane.

A constant-volume circulatory system can be maintained by automatically restoring the filtered volume with fresh medium or buffer, as indicated in Fig. 15. By this means an automatic harvesting system or a means of modifying the conditions of a cell culture is simply achieved.

The system has been applied to:

1. The separation of F 49 lymphoma cells with 1.2-μm membrane by Howlett *et al.* (*20*).
2. The separation of virus suspensions with molecular grade membranes by Gangemi *et al.* (*16*) and Mathes *et al.* (*26*).
3. The recovery of feline leukemia virus in a continuous culture system by Ratner *et al.* (*40*).

For low-volume systems a holder to house a 90-mm disc is available (Nuclepore); the unit is autoclavable provided the selected membrane is compatible. Large-scale units with areas of 0.5–25 ft^2 or 170–2550 cm^2 are manufactured by Millipore Corporation or Sartorious Instruments, Ltd. Autoclavable assemblies or plastic, chemically sterilised, assemblies are manufactured. Gangemi (*16*) used 2.0% formalin in isotonic saline solution as a sterilising agent.

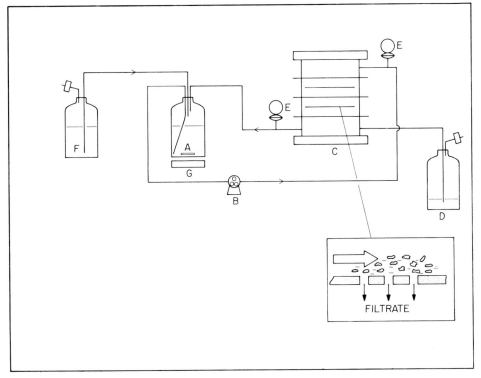

Fig. 15. Tangential cross-flow system. A, sealed suspension cells or virus; B, peristaltic or membrane pump; C, multi-plate tangential flow filter unit; D, vented filtrate vessels; E, pressure gauge; F, vented vessel closed circuit to A. For concentration system, F is omitted and A is vented.

Equivalent systems consisting of synthetic microfibres and known as hollow fibres have been much used for reverse osmosis and ultrafiltration applications and are available in both molecular and the microporous screen ranges.

4.3. Filtration Conditions

The performance of filters is measured in terms of the sieve retention of latex particles (35). Sterilising grades are evaluated by retention of bacterial test strains (42, 46). See also the manufacturer's technical literature. Information on the removal of cell debris or other particles is useful for the processing of animal cell culture fluids, but little has been published relevant to such processes. Practically, the efficiency of a system is important in terms of (1) its capacity, (2) retention performance, (3) flow rates and (4) operating costs.

The factors affecting performance include filtration pressure, flux (i.e.

volume per unit area) and fluid velocity; these affect (1) the dwell time of a particle within the filter matrix and (2) the exposure of particles to surface contact. Olson (29) suggested a flow rate of 2 ml/min/cm² to be optimum. Tanny et al. (46) investigated 0.45-μm cellulose tri-acetate membranes by bacterial challenge and concluded that adsorptive retention becomes progressively compromised as pressure increases; titre reduction is inversely related to Δp.

As blockage occurs the flow rate decreases. If the applied pressure is increased the effect will be to increase the chance of release of particles downstream (filter migration); the compacting of surface cake will also increase with excessive pressure. Similar effects occur when pumps generate flows which pulsate.

The quantity of particles present does not affect the efficiency of removal by sieve retention mechanisms, but will limit the filterable volume. The nature of the particle, e.g. size and charge, influences the efficiency of depth filtration where frictional and electrostatic forces are operating (35); gelatinous particles tend to increase filter binding.

4.4. Integrity Tests

These tests serve to show that the filter conforms to a standard and demonstrates the integrity of the system. These tests are particularly significant when it is intended to produce a sterile filtrate. It is important that the relationship of the test measurements to the theoretical bacterial retention of the filter is understood.

The information the technologist requires includes confirmation that (1) the correct membrane is fitted, (2) its performance is as specified and (3) the filter is still intact after use.

4.4.1. Bubble Point Test

The filter membrane is a collection of capillaries (see Fig. 10) which will retain liquid by surface tension. The minimum pressure required to displace the liquid from capillaries is a measure of tube diameter and will indicate the largest pore detectable. The equation below was published by Jacobs (21) and others:

$$D = K4\gamma\cos\theta/P$$

where D = diameter, K = experimental constant, γ = surface tension, P = pressure and θ = angle of contact. The test is performed by exposing the wetted membrane to increasing pressure and observing the first appearance of a bubble downstream of the filter; the pressure is then noted. Standard values are published by manufacturers.

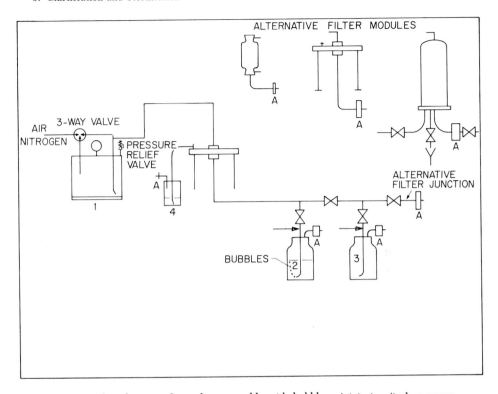

Fig. 16. Flow diagram of autoclave assembly with bubble point test unit. 1, pressure reservoir; 2, reservoir to receive bubble point test fluid; 3, reservoir to receive filtrate; 4, disease security container; ⋈, valves or clamps; →, junctions, sterile connections; A, air filters.

There are difficulties as the pores are not uniform capillaries (see Fig. 10). Meltzer and Leukaszewicz (27) and Leukaszewicz *et al.* (24) calculate the largest pore of a 0.45-μm cellulose ester membrane to be 0.94 μm. The anomaly is explained by the shape of the pore and a corrected relationship is suggested:

$$P = \gamma 4 \cos \theta \ (L)/A$$

where L = perimeter of pore and A = area.

Test accuracy is at its best ±10%, which has serious implications. Operational difficulties occur in the recognition of downstream bubbles, which increases with the large areas found in multi-plate housings and pleated cartridges (22, 31, 49). The test manual published by Millipore Corporation suggests the application of 80% of the published bubble point pressure value [Reti (41)]. Figure 16 is a diagrammatic representation of the filter test system based upon recommendations by Millipore Corporation.

The illustrated assembly shows a membrane disc holder complete with (glass) collecting vessel to receive the "bubble test" fluid, a harvest vessel and a pressure container. The lengths of connecting tubes should be as short as is convenient. A three-way valve can be fitted to a stainless steel pressure reservoir to allow for both a filtration pathway and a bubble test pathway. To operate, water (or the fluid to be filtered) is passed through the pre-sterilised filter to be collected in reservoir 2. Nitrogen or air pressure is applied upstream of the filter, and the pressure is slowly increased to the maximum of 80% of the stated bubble point of the membrane. If no bubbles are observed, the pressure can be further increased to that of the expected bubble point value. Greater increases in pressure will result in a major flow of gas.

If the membrane is broken a major flow of gas will occur immediately and at a pressure below the manufacturer's stated pressure limit. If the intact membrane fails to meet specification a detectable and persistent bubble stream will be found at some value less than the specified pressure. If no bubbles appear at 80% of the manufacturer's stated pressure, then the membrane may be used.

4.4.2. Forward Flow Test

Fluid retained by the matrix of the membrane will allow the diffusion of gas below the bubble point pressure, which forms the basis of measurement of the larger pores. The diffusive flow can be determined particularly with membranes of larger area, i.e. cartridges. The technique can be performed *in situ* by collecting gas downstream of the filter in a graduated burette or by measuring the pressure decay of the gas, held in the housing, upstream of the filter.

The pressure hold test, which is described in this section, is simple to perform, but it requires special equipment, though not necessarily of commercial origin. A sensitive and accurate pressure gauge is the minimum requirement. Figure 17 is a flow diagram assuming in-line sterilisation of a cartridge filter with the pressure hold equipment attached. The system is so arranged as to allow the sterilisation of the cartridge housing complete with filter and its subsequent *in situ* testing, before and after use.

The basis of the test consists of thoroughly wetting the filter to be tested and then applying a suitable pressure of air or nitrogen to the upstream side of the filter membrane. Following the initial displacement of water, the rate at which the upstream pressure decreases is a measure of the pore size of the filter and confirms the integrity of the membrane system. Failure to wet a portion of the membrane results in erroneous readings.

It is important to realise that the pressures applied for these tests are considerable, 40–70 psi. Operators should ensure that the equipment is

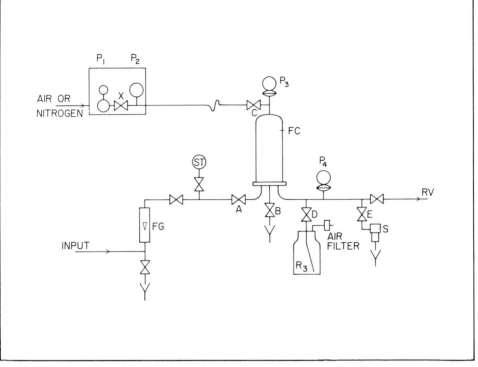

Fig. 17. Flow diagram for *in situ* testing of filter cartridge sterilised by in-line steam. P_1, small pressure gauge and regulating valve; P_2, test pressure gauge; R_3 test fluid reservoir; S, steam trap; FG, sterilizable flow gauge; P_3–P_4, process pressure gauges with chemical seals, A–F, valves or tubing clamps; RV, sterile fluid receiving vessel; Y, drain symbol; ST, steam source; X, isolating valve.

certified to withstand test pressures and take account of the possible hazard which could result from misuse of the system. Test values from (Pall) Nylon 66 medium are included in Table IV.

The bubble point procedure has been criticised and compared to the forward flow test by Pall (*33*), Pall and Kirnbauer (*34*), Leukaszewicz *et al.* (*24*), Olson *et al.* (*31*) and Olson (*30*). Filters marketed by Pall Filtration, Ltd. are validated by forward flow procedures; the test values and, therefore, the performances of the filters are related to a given bacterial challenge as stated in Manufacturer's Technical Data. The forward flow test is criticised for inaccuracies by Transen (*49*) and Johnston *et al.* (*22*), who hold that the procedure has not been widely reviewed.

Pall and Kirnbauer (*34*) and Pall *et al.* (*35*) characterised the capillary displacement of membrane filter media by a K_L value and compared this

TABLE IV   Standard Values for Pressure Hold Test for Ultipore
N66 Nylon Filter Medium[a]

Cartridge grade	Test fluid	Test pressure (psi)	Permitted psi decay, 254-mm cartridge, specified housing[b] 10 min
0.2 μm NR (P)	Water	40	0.85
	ETOH	12	1.71
	I.P.A.	11	1.52
0.45 μm NXP	Water	25	1.07
	ETOH	8	1.91
	I.P.A.	7	1.71

[a] Values supplied by Pall Process Filtration, Ltd.
[b] Housing MSS or MAS 500I.

parameter with the removal of bacteria and latex beads. The K_L value is defined as a pressure below which air flow through a water-wet filter is purely diffusional; above this pressure, air flow increases rapidly due to expulsion of water from the largest pores in the filter. The forward flow test pressure recommended for Pall nylon media is 85–90% of the K_L value (36). The filter media are validated by given bacterial challenges. It is claimed that 10^{13} *Ps. diminuta* can be retained without passage by a 254-mm cartridge. However, Wallhausser (52) reports failures under strictly controlled conditions with nylon media. Pall *et al.* (35) also quote work by Howard and Duberstein in which filters with K_L values of 50–53 psi pass small bacteria and filters with a K_L of 90 psi are required to prevent all forms of bacterial passage. Such filters imply small pores or, alternatively, thick membranes, where a depth mechanism is also contributory.

4.4.3. Electronic Test Equipment

This equipment is a recent addition to integrity test technology. Bubble point tests and forward flow tests require reliable pressure gauges as the minimum tools to perform the work. As stated above (Section 4.4.1), the bubble point test is dependent upon the visibility of bubbles. The electronic test equipment will perform both the bubble point and the forward flow tests automatically and deliver a graphic display or a digital print of the performance data, thereby reducing human error. Sartorious Instruments, Ltd., Pall Filtration, Ltd. and Millipore Corporation each offer an automatic diffusion tester. Olson (30) reviews both the bubble point test and the forward flow procedure with reference to the electronic test equipment. He indicates the necessity of calibrating the equipment in terms of the filtration system used and concludes that each machine functions as claimed.

The high capital cost (about £5500, 1982) limits the application of the equipment; yet it is probable that it will be favoured in pharmaceutical applications, where the Good Manufacturing Practice Code requires documentation of integrity tests.

4.5. Sterilisation of Filtration Equipment

Despite the ability of some material to withstand high air temperatures (cellulose acetate at 180°C), steam sterilisation by autoclave or direct line steam is the method of choice.

4.5.1. Autoclave Sterilisation

It is essential that the filter holder and all associated downstream equipment is sterilised. It may be convenient or necessary to sterilise items as separate units, then to unite them to form an assembly, preferably making the connection under the protection of a laminar air cabinet with aseptic precautions.

The filter medium, disc or cartridge, must be assembled in its housing and then autoclaved by following the manufacturer's instructions exactly. The steam pathway through the system is ensured by the inclusion of (removable) air filters. Pressure differences above and below the membrane must be avoided, especially for certain materials (cellulose esters).

Figure 16 illustrates a filter unit complete with reservoirs. Depending on the scale of operation, it can be autoclaved assembled or as sub-assemblies. The connectors must be protected with a cover of aluminium foil or craft paper until assembled. Valves, if fitted, are open during sterilisation to ensure a free pathway for steam and/or air. Membrane disc holders, small cartridges and simple forms of "sheet" holders, i.e. single disc, can be autoclaved. Systems involving filtration of volumes up to 100 litres can be autoclaved provided the steriliser capacity is available to house the reservoir(s). Pharmaceutical and industrial scale installations generally exceed these volumes. With an increase in capacity of systems larger cartridges or large sheet filters are sterilised by in-line steam. Such systems require auxiliary equipment including pressure gauges, flow meters, large-volume vessels and a steam source.

Figure 17 is a flow diagram of a filter cartridge system complete with pressure gauges, flow meter and reservoir to receive the filter test fluid.

Valves are often regulated to maintain the minimum sterilising pressure of 1.0 kg/cm², 100 kN/m² (kilonewtons per square metre) or 15 psi, although a higher pressure of 1.3 kg/cm² or 140 kN/m² is preferable (1.0 kg/cm² saturated steam pressure = 121°C). It is important that the filter side of the valve F in Fig. 17 (if not previously sterilised with a reservoir) is sterilised. The drain valves are adjusted during the sterilisation cycle to ensure the escape of "condensate". After the steam pressure is maintained for a suitable time (e.g. 60 min) drain valves are closed and filter-sterilised compressed air is introduced to prevent formation of a vacuum as the system cools.

4.6. Disease Security

The disease security aspects of filtration are principally those of containment, i.e. prevention of leakage of process fluids. The factors discussed here are those appropriate to systems which may be operated in general or production laboratories, engaged in processing pathological virus-containing fluids. The nature of the infectious agent must be considered; the degree of containment necessary for processing an attenuated virus strain by staff with demonstrated immunity will be less than that necessary for operations with foot-and-mouth disease virus in England, where work is permitted only in a specialised licensed institution.

Room-scale containment is achieved by maintaining a negative pressure with respect to the external environment by means of air locks and transfer hatches, with tested absolute air filters (HEPA, 99.999% efficiency) fitted to the exhaust air system (see Section 3.3). At a smaller scale a disease security cabinet will provide an enclosure at a negative pressure with respect to the surroundings. The capacity and type of filter are further factors. Many filter systems can be made secure. The multi-plate press is subject to leakage; however, the external gasket type of British Filter overcomes the problem when correctly used.

The transfer of infective suspensions to a reservoir by pouring, on the open bench or within a safety cabinet, will result in considerable environmental contamination by aerosols. Fluids can be retrieved from small containers, e.g. tissue culture flasks, by pipetting, using a vacuum applied to a reservoir. To prevent aerosol transfer to the vacuum line an absolute filter must be inserted between the reservoir and the source. Secured transfer lines should be used where possible, e.g. for large-scale monolayer culture systems. Before use, the assembled system or parts of the system should be pressure tested above the pressure to be used during the operation. The sealed system can be held under pressure and observed. (Note: small-scale leaks may take hours to be detected, so a sensitive pressure gauge is required.)

Glass and plastic containers should not be subjected to pressure. All pressure vessels must be metal and stainless steel is used universally. The operation of the pressure relief valve will result in aerosol generation; this is overcome by the use of an additional valve with an attached absolute air filter to vent the vessel. Vessel lids should not be removed until de-contamination has been performed.

The connection of the vessel or pump to the filter holder is of major concern. The system will be subjected to a considerable pressure when effecting an integrity test, especially when a post-filtration integrity test is used. Connector options include:

1. Rubber or silicone hose attached to hose adaptors. The seal must be secured; nylon or wire ties applied with a tension gun (57) and hose clamps are equally effective.

2. Tri-Clamp or Tri-Clover fittings. These consist of a flange which is integral to a vessel or filter; a corresponding flange, complete with gasket, is secured with a hinged clamp. A hose adaptor or a permanent pipe flange can be fitted (Fig. 12a).

3. Quick Connect couplings offer more security than the rubber hose system, especially when used with flexible stainless steel hose (Fig. 13). A sliding collar of the female part is moved to allow the male adaptor to be engaged or released. The seal is effected by pressure against an O ring. The system can be refined by the use of integral shut-off valves, which are open only when the couplings are engaged. Whilst preventing accidental parting of equipment, leakage is still possible when the O ring seal fails.

4. Sanitary fittings are common (Fig. 12b) for semi-permanent assemblies, e.g. commercial scale. Here a captive nut will secure a corresponding male thread with an interposed gasket. The system is less open to disturbance than that in 2 above. The system is not necessarily leakproof; gasket failure can result in fluid escape.

5. Permanent welded pipe or tube systems offer the greatest security. Threaded joints are also used to fabricate a system; polytetrafluoroethylene tape is necessary to effect a seal and such joints can leak under repeated heating–cooling cycles, but massive escape under pressure is almost impossible. A variant to the above is the use of O ring flange joints to ensure pressure or vacuum security.

Membrane disc holders rely on O rings to effect seals (see Fig. 13). Careful cleaning, handling and fitting are necessary. The O rings are damaged by overtightening, especially during sterilisation, where distortion results in subsequent leakage. Figure 16, detail 4, illustrates the disease security fitting assembled to a disc holder vent valve.

Cartridge holders are readily connected to vessels by one of the systems described above. However, drain and vent valves present a potential hazard. Both should be connected to removable containers equipped with absolute air filters to allow the collection of fluids or aerosols. The *in situ* decontamination of an in-line steam system (see Fig. 17) may present initial problems. The first displacement of fluid through the drain or vent valves and the steam trap will not be at sterilising temperature. These fluids should be collected for decontamination into a vented container, as previously described. To detach the filtrate-containing vessel from the system, the reservoir isolating valve requires decontamination, or the output side must be sealed. All associated equipment should be decontaminated by autoclaving before it is discarded or cleaned for re-use.

Preventive maintenance of equipment is important; O rings, connecting systems, valves and joints must be regularly inspected and replaced. The practice of preparing full and detailed operating procedures, as recommended by codes of Good Manufacturing Practice or Good Laboratory Practice, provides the basis of safe and consistent operation.

5. STERILITY CONTROL

To ensure sterile products from the separation techniques described in this chapter, the following controls can be applied to the systems:

1. Validation of the chosen system prior to its routine use
2. Use of aseptic working methods within a clean environment
3. Proof of sterility by the testing of process samples

A separation system should not be assumed to operate in a sterile manner unless prior physical and microbiological tests have been conducted. The approach to such validation is illustrated by Olson (29). It is the opinion of the author that a system should be tested, as detailed below, prior to its introduction and thereafter, if in regular use, at annual intervals.

The tests will include:

1. Use of temperature-recording apparatus and/or spore strips, implanted within equipment during the steam sterilisation cycle
2. Validation of the accuracy of equipment measuring sterilising conditions, e.g. temperature recorders and pressure gauges
3. Performing the process with a representative volume of bacteriological culture media, which should remain sterile upon subsequent incubation

The success of a sterilising filtration system is not ensured by the procedures listed above. Dependable performances are obtained from filtration media for which the bacterial removal capability is published and to which integrity tests can be applied.

The laminar air flow cabinet provides the cleanest environment within a laboratory or production unit. The discipline of work in an area when sterile operations are performed is paramount, including the use of sterilised clothing, the containment and removal of used equipment and the avoidance of spillage. Microbiological monitoring of contamination levels within the working area, by exposure of settle plates and the use of contact plates applied to work surfaces, indicates the approach of conditions requiring remedial action, such as special cleaning or room fumigation.

The techniques for sterility testing of process samples are detailed by the British Pharmacopoeia (6), the United States Pharmacopoeia (51) and the European Pharmacopoeia (13). Two methods are suggested: (1) culturing of 0.45-μm analytical membranes, through which the test sample is filtered, and (2) dilution of the sample with the culture medium, e.g. 1 : 100. Both overcome the suppression of microbiological growth, in the culture media, by the presence of antibacterial agents. Mycoplasmas will not be detected by standard culture media used for bacterial and fungal growth.

The Pharmacopoeia methods are oriented to the testing of products filled into containers; however, they are adaptable to the testing of process samples. It is important that the samples submitted to any test procedure be representative of the bulk.

REFERENCES

1. Baldwin, C. L. (1973). Biological safety recommendations for large scale zonal centrifugation on oncogenic virus. In "Proceedings of the National Cancer Institute Symposium on Centrifuge Biohazards," pp. 61–71. Cancer Research Safety Monograph Ser., Vol. 1. U.S. Dep. Health, Educ., Welfare, Washington, D.C.
2. Baldwin, C. L., Lemp, J. F., and Barbeito, M. S. (1975). Biohazards assessment in large scale centrifugation. Appl. Microbiol. 29, 484–490.
3. Bellew, H. W. (1978). In "Basics of Filtration and Separation," pp. 9–12. Nucleopore Corporation, California.
4. Blosse, P. (1982). The new attraction in membrane filters. Process Eng. (London) 63, 34–37.
5. Brinton, C., Jr., and Lauffer, M. A. (1959). The electrophoresis of viruses, bacteria, and cells, and the microscope method of electrophoresis. In "Electrophoresis" (M. Bier, ed.), Vol. 1, pp. 427–487. Academic Press, New York.
6. British Pharmacopoeis (1980). Vol. II, Appendix XVI, pp. 186–190. H. M. Stationery Office, London.
7. British Standards Institution (1969). "Method for Sodium Flame (other than for air supply to I.C. engines and compressors)," BS3928. Br. Stand. Inst., London.
8. Cliver, D. D. (1968). Virus interactions with membrane filters. Biotechnol. Bioeng. 10, 877–889.
9. Coulson, J. M., and Richardson, J. F., with Buckhurst, J. R., and Harker, J. H. (1978). Sedimentation centrifuge separation. In "Chemical Engineering," Vol. 2, pp. 203–219. Pergamon, Oxford.
10. Denyer, S. P., Russell, A. D., and Hugo, W. B. (1982). In "Principles and Practice of Disinfection, Preservation and Sterilization" (A. D. Russell, W. B. Hugo, and G. A. J. Ayliffe, eds.), pp. 569–609. Blackwell, Oxford.
11. Dimmick, R. L., Vogl, W. R., and Chatigny, M. A. (1973). Potential for accidental microbial aerosol transmission in the biological laboratory. In "Proceedings of Conference, Biohazards in Biological Research," (A. Hellman, M. N. Oxman, and R. Pollack, eds.), pp. 246–260. Cold Spring Harbor Lab., Cold Spring Harbor, New York.

12. Duberstein, R. (1979). Mechanism of bacterial removal by filtration. *J. Parenter. Drug Assoc.* **33**, 250–256.
13. European Pharmacopoeia (1980). 2nd ed., Sect. V.2.11. Maisonneuve.
14. Evans, C. G. T., Harris-Smith, R., Stratton, J. E. D., and Melling, J. (1974). Design and construction of a ventilated cabinet for a continuous flow centrifuge. *Biotechnol. Bioeng.* **16**, 1681–1687.
15. Filtration Dictionary (1975). "Filtration Society." Uplands Press, London.
16. Gangemi, J. D., Connell, E. V., Mahlandt, B. G., and Eddy, G. A. (1977). Arena virus concentration by molecular filtration. *Appl. Environ. Microbiol.* **34**, 330–332.
17. Gerba, C. P., and Hou, K. C. (1982). Enhanced containment removal by charged-modified membrane filters (personal communication).
18. Gooders, A. P., Webb, N. L., Allen, G. J., and Richards, B. M. (1981). Total containment of a continuous-flow zonal centrifuge. *Biotechnol. Bioeng.* **23**, 1–18.
19. Hou, K. C., Gerba, C. P., Goyal, S. M., and Zerda, K. S. (1980). Capture of latex beads, bacteria, endotoxin and viruses by charge-modified filters. *Appl. Environ. Microbiol.* **40**, 892–896.
20. Howlett, A. C., Sternweiss, P. C., Macik, B. A., Van Arsdale, B. M., and Gilman, A. (1979). Reconstitution of catecholamine-sensitive adenylate cyclase. *J. Biol. Chem.* **254**, 2287–2295.
21. Jacobs, S. (1972). The distribution of pore diameters in graded ultra-filter membranes. *Filtr. Sep.* **9**, 525–530.
22. Johnston, P. R., Leukeszewicz, R. C., and Meltzer, T. H. (1981). Certain impression in the bubble point measurement. *J. Parenter. Sci. Technol.* **35**, 36–39.
23. Keswick, M. A., and Wagner, R. A. (1978). Electrophoretic mobilities of virus adsorbing filter material. *Water Res.* **12**, 263–268.
24. Logan, K. B., Rees, G. E., Seely, N. D., and Primrose, S. D. (1980). Rapid concentration of bacteriophage from large volumes of fresh water. Evaluation of positively charged microporous filters. *J. Virol. Methods* **1**, 87–97.
25. Lukeszewicz, R. C., Tanny, G. B., and Meltzer, T. H. (1978). Membrane-filter characterisations and their implications for particulate retention. *Pharm. Technol.* **2**, 77–83.
26. Mathes, L. E., Yohn, D. S., and Olson, R. G. (1977). Purification of infectious feline leukemia virus from large scale volumes of tissue culture fluids. *J. Clin. Microbiol.* **5**, 372–374.
27. Meltzer, T. H., and Leukaszewicz, R. C. (1979). *In* "Quality Control in the Pharmaceutical Industry" (M. S. Cooper, ed.), Vol. 3, pp. 145–212. Academic Press, New York.
28. Mix, T. W. (1974). The physical chemistry of membrane–virus interaction. *Dev. Ind, Microbiol.* **15**, 136–142.
29. Olson, W. P. (1979). Validation and qualification of filtration systems for bacterial removal. *Pharm. Technol.* **3**, 85–90.
30. Olson, W. P. (1982). A system for integrity testing of disc and cartridge membrane filters. *Pharm. Technol.* **6**, 42–60.
31. Olson, W. P., Martinez, E. D., and Kern, C. R. (1981). Diffusion and bubble point testing of microporous cartridge filter: Preliminary results at production facilities. *J. Parenter. Sci. Technol.* **35**, 215–222.
32. Osgood, G. (1967). Filter sheets and sheet filtration. *Filtr. Sep.* **4** (July/Aug.) 1–11.
33. Pall, D. B. (1975). Quality control of absolute bacteria removal filters. *Bull. Parenter. Drug Assoc.* **29**, 129–204.
34. Pall, D. B., and Kirnbauer, E. A. (1978). Bacteria removal prediction in membrane filters. *Colloid Interface Sci. [Proc. Int. Conf.]*, 52nd, 1978.
35. Pall, D. B., Kirnbauer, E. A., and Allen, B. T. (1980). Particle retention by bacteria retentive membrane filters. *Colloid Surf.* **1**, 235–256.

36. Pall Process Filtration, Ltd. "The Pall Ultipore N66 Membrane Filter Guide," Tech. Bull. SD 872b. P.P.F.
37. Pall Process Filtration, Ltd. "N66 Posidyne Filter Guide," Tech. Bull. SD 905. P.P.F.
38. Paul, J. (1970). "Cell and Tissue Culture." Livingstone, Edinburgh and London.
39. Phelps, W., Baughn, R., and Black, H. S. (1980). Passage of *Corynebacterium aquaticum* through membrane filters. *In vitro* 16, 751–753.
40. Ratner, P. L., Cleary, M. L., and James, E. (1978). The production of 'rapid-harvest' Moloney murine leukaemia virus by continuous cell culture on synthetic capillaries. *J. Virol.* 26, 536–539.
41. Reti, A. R. (1977). An assessment of test criteria for evaluating the performance and integrity of sterilising filters. *Bull. Parenter. Drug Assoc.* 31, 187–194.
42. Reti, A. R., and Leahy, T. S. (1979). Validation of bacterially retentive filters by bacterial passage testing. *J. Parenter. Drug Assoc.* 33, 257–272.
43. Reti, A. R., Leahy, T. S., and Meier, P. M. (1979). The retention mechanism of sterilising and other sub-micron high efficiency filter structures. *Proc. World Filtr. Congr. 2nd,* pp. 427–435. Eur. Fed. Chem. Eng. and Am. Inst. Chem. Eng., New York.
44. Sherbet, G. V. (1978). "The Biophysical Characterisation of the Cell Surface," p. 55. Academic Press, London.
45. Sobsey, M. D., and Jones, B. L. (1979). Concentration of polio virus from tap water using positively charged microporous filters. *Appl. Environ. Microbiol.* 37, 588–595.
46. Tanny, G. B., Strong, D. K., Presswood, W. G., and Meltzer, T. H. (1979). Absorptive retention of *Pseudomonas diminuta* by membrane filter. *J. Parenter. Drug. Assoc.* 33, 40–51.
47. Telling, R. C., Passingham, R. J., Kitchener, P. L., and Hopkinson, D. G. (1972). Improvements in cell and virus culture systems. British Patent 1,436,323.
48. Telling, R. C., and Radlett, P. (1970). Large-scale cultivation of mammalian cells. *Adv. Appl. Microbiol.* 13, 91–119.
49. Transen, B. (1979). Non-destructive tests for bacterial filters. *J. Parenter. Drug Assoc.* 33, 273–279.
50. Transen, B. (1981). Designing process microfiltration systems. *Pharm. Technol.* (Nov.) 62–69.
51. United States Pharmacopoeia (1980). 20th revision, pp. 878–880. U. S. Pharmacopoeia Convention Inc. Mack Publ. Co., Easton, Pennsylvania.
52. Wallhausser, K. H. (1979). Is the removal of micro-organisms by filtration really a sterilising method? *J. Parenter. Drug. Assoc.* 33, 156–170.
53. Wallhausser, K. H. (1979). Neue Untersuchunger zur Sterilfiltration. *Pharm. Ind.* 41, 475–481.
54. Wallis, C., and Melnick, J. L. (1967). Concentration of enteroviruses on membrane filters. *J. Virol.* 1, 472–477.
55. Wang, D. I. C., Sinskey, J. J., Gerner, R. E., and De Filippi, R. P. (1968). Effect of centrifugation on the viability of Burkitt lymphoma cells. *Biotechnol. Bioeng.* 10, 641–649.
56. Webb, N. L., Richards, B. M., and Gooders, A. P. (1975). The total containments of a batch-type zonal centrifuge. *Biotechnol. Bioeng.* 17, 1313–1322.
57. Whiteside, J. P., Rayner, R. W., and Spier, R. E. (1983). The improvement and standardisation of the simplified process for the production of F.M.D. virus from B.H.K. suspension cells. *J. Biol. Stand.* 11, 145–155.
58. Zierdt, C. H. (1979). Adherence of bacteria, yeasts, blood cells and latex spheres to large porosity membrane filters. *Appl. Environ. Microbiol.* 38, 1162–1172.
59. Zwerner, R. K., Cox, R. M., Lynn, J. D., and Acton, R. T. (1981). Five-year perspective of the large-scale growth of mammalian cells in suspension culture. *Biotechnol. Bioeng.* 23, 2717–2735.

6

Inactivation of Viruses Produced in Animal Cell Cultures

T. R. DOEL

Vaccine Research Department
Animal Virus Research Institute
Pirbright, Woking
Surrey, United Kingdom

1. INTRODUCTION

Since the pioneering studies of Louis Pasteur almost a century ago, inactivated vaccines have been developed against many of the most serious virus diseases of mankind and his livestock. Pasteur was able to demonstrate that the infectivity of rabies virus in rabbit spinal cord was destroyed by drying the tissue over potassium hydroxide. He subsequently tested the inactivated materials in dogs and later used a 14-day-old dried preparation to vaccinate and save the life of an 8-year-old boy who had been bitten a few days previously by a rabid dog [see review by Turner (70)]. Further progress in the early part of the 20th century led to the development of rabies vaccines either inactivated or partially inactivated with phenol [Semple and Fermi vaccines, respectively (70)] or ether [Hempt vaccine (70)].

This progress was not confined to rabies virus. Vaccines against foot-and-mouth disease (71), yellow fever [Hindle, 1929, cited by Spier (67)], New-castle disease [cited by Lancaster (45)], hog cholera (52), canine distemper (17) and smallpox [Janson, 1891, cited by Kaplan (43)] were among the first to be developed. Not all of these were successful; for example Janson's heat-inactivated smallpox preparations gave equivocal results. Indeed the quest for high-potency vaccines led many workers from killed to attenuated virus vaccines, of which smallpox is one of the best known examples. Other factors influenced this trend, for example the tragic circumstances associated with the use of some of the early killed polio vaccines which had not been inactivated adequately. However, recent years have seen a partial reversal of this trend. A considerable body of evidence, particularly from the European countries that use killed poliovirus vaccine, indicates that safe and potent inactivated preparations may be made. Furthermore, there are a number of advantages associated with the use of killed poliovirus vaccine. In the first place it would appear to be more effective than the attenuated type when used in under-developed regions of the world (63). Secondly, there no longer appears to be any risk of vaccine-associated disease when the killed poliovirus vaccine is used. This latter point owes much to the very considerable effort made to improve both inactivation and innocuity testing procedures (8).

The continued use of inactivated virus vaccines is probably assured for at least the next 10 years. The justifications for this statement include the fact that not all attempts to produce attenuated vaccines have been successful, or alternatively, the level of reversion to the wild-type virus has proved unacceptable. A notable example here is the very considerable effort made by workers at Pirbright in the early 1960s to develop attenuated foot-and-mouth disease (FMD) vaccines. It was subsequently demonstrated that the host response of different animals to laboratory-attenuated viruses was highly variable and, therefore, totally unsatisfactory for safe, effective, disease control (G. N. Mowat, personal communication).

Another factor favouring reliably inactivated vaccines is that while attenuated vaccine-associated disease may be considered tolerable against the background of serious endemic disease, this is not the case when the control programme is highly effective. Indeed, even inactivated vaccines come under close scrutiny in disease-free situations. The European experience with FMD is an excellent example. Over the last decade a large proportion of outbreaks have been attributed either to improperly inactivated vaccine or the escape of virus from vaccine production plants, rather than the introduction of new strains from other parts of the world (3). Although the proportion may have been overstated by some authorities, there is little doubt that some outbreaks have been associated with the use or production of vaccine.

Some recent developments bear on the future of inactivated vaccines, namely the advances made with the production of genetically engineered antigenic fragments of virus and, even more recently, the organic synthesis of peptide sequences corresponding to the active site of antigenic proteins (9, 44). While it is impossible to predict the eventual outcome of these preliminary experiments, there is no doubt in the writer's mind that the newer approaches to vaccine production will usurp some of the currently used vaccines.

In this chapter I shall consider the mode of action of various inactivating agents and their effects, if known, on the antigenicity of viruses. The increasingly important area of innocuity testing will also be reviewed, for although many of us would wish to achieve complete inactivation of all virus particles in a given preparation, there will always remain the statistical probability that some infectious particles persist. Furthermore, the risk of very small quantities of residual infectious virus causing disease in a population will increase with the numbers of doses of a vaccine administered. It perhaps would be useful to record that approximately 500 million doses of FMD vaccine are used annually in South America alone.

Finally, it should be said that the author's failure to cite all relevant literature and his preoccupation with FMD are purely a function of the limitations of space and his particular interest in virus inactivation.

2. KINETICS OF INACTIVATION

Ideally, inactivation of virus preparations should follow an exponential relationship, i.e. the survival rate should be a decreasing exponential function of time (e^{-kt}). It therefore follows that the logarithm of the survival ratio is a straight line when plotted against time. Unfortunately, many workers were soon to recognise that significant departures from the exponential law were relatively commonplace. This was particularly the case with early attempts to inactivate poliovirus with formaldehyde. Explanations for the so-called tailing effect observed by many workers with formalin-inactivated preparations of foot-and-mouth disease virus (FMDV) and poliovirus included aggregation of virus particles into inactivant-resistant clumps, adsorption of virus to walls of vessels and the presence of aerosol droplets above the liquid surface. However, Gard's theory of reduced permeability due to formaldehyde interaction with the coat protein found widespread acceptance and fitted the empirical equation and experimental data exceptionally well (33).

It is worthwhile to discuss inactivation briefly in terms of single- and multiple-hit target theory. With the single-hit mechanism, as the name implies, modification of a single base within the nucleic acid would be sufficient to inactivate the whole molecule. Multiple-hit mechanisms require the accumulation of a number of sublethal hits before the organism is killed. Which mechanism applies to a given virus/inactivant mixture will depend on a number of factors including the nature of the modification to the nucleic acid, for example predisposition to hydrolysis, and the presence or absence of genetically inactive regions in the viral genome.

Single-hit mechanisms are characterised by linear survival 'curves', whereas multiple-hit curves show shoulders of varying breadth depending on the number of sublethal hits which can be tolerated by the organism (41). However, in Hiatt's view it is erroneous to equate single-hit survival curves with first-order kinetic reactions. This, he points out, implies an understanding of the mechanism, which may not be justifiable by the facts available. By the same token, the presence of a shoulder on a survival curve does not necessarily indicate a multiple-hit process and may, in fact, be due to a mechanistic effect (41). A good example of this is given by the photodynamic inactivation of toluidine blue-treated S-13 coliphage. The presence of a shoulder in this system would appear to indicate a multiple-hit mechanism whereas, in fact, the shoulder can be attributed to the rate of binding of the dye to the RNA (41).

It may be concluded that if the logarithm of the survival ratio plotted against time gives a straight line from the origin, then the process is exponential as far as the data extend. Thus, the virus population is homogenous and inactivation does not require cumulative damage. Appearance of an initial shoulder on the semilog survival curve indicates either a complex reaction or a mechanism based on cumulative damage. If the period of induction is constant with differing experimental conditions, it is likely to be a cumulative damage effect.

3. INACTIVANTS

3.1. Chemical

3.1.1. Aziridines

Chemistry and Mode of Action. The aziridines owe much to the development of mustard gases in World Wars I and II and, in particular, the β-chloroethylamines or nitrogen mustards. Indeed, ethyleneimine (EI) is com-

TABLE I   Properties of the Aziridines[a]

i)	Miscible with water and most organic liquids
ii)	Ring structure quantitatively opened by thiosulphate
iii)	Substitution of alkyl groups at one of ring carbon atoms increases rate of ring opening
iv)	Substitution at ring nitrogen decreases rate of ring opening unless alkyl group contains an electronegative group (e.g., acetyl) when rate is increased.
v)	Common derivatives have low boiling point, e.g. 56.7°C for EI. Vapour pressure high enough to be an inhalation hazard
vi)	Objectionable ammonia-like odour
vii)	Ethyleneimine is known to react with α and ε-amino, imidazole, carboxyl, sulphydryl and phenolic groups of proteins, inorganic phosphate, glycero and hexose phosphates and amino groups of adenine and thiamine.

[a] Sources: Dermer and Ham (18); Gilman and Philips (34).

monly prepared by cyclization of bromoethylamine hydrobromide under alkaline conditions

$$BrCH_2CH_2NH_2 \xrightarrow{\text{NaOH}} CH_2\!\!-\!\!CH_2$$
$$\underset{NH.HBr}{\diagdown\diagup}$$

[After Bahnemann (6)]

The relevant properties of the aziridines are summarised in Table I.

Clearly the aziridines are a highly reactive group of substances and they have been used to mutagenise a wide range of organisms. Although undoubtedly toxic, there is, to the author's knowledge, no published evidence of tumour induction in man. Furthermore, Fellowes (25) failed to produce tumours in rats given 0.5 mg of acetylethyleneimine and kept for 515 days.

Applications.   The current world-wide use of the aziridines for the production of inactivated FMD vaccines owes much to the work of Brown and his colleagues at Pirbright (12, 13). One stimulus for their work was the often-observed 'tail' of infectious virus with formalin-inactivated FMD preparations. Thus 0.05% acetylethyleneimine (AEI) for 12 hr at 37° was effective, whereas 0.05% formaldehyde for 144 hr at 26°C, pH 8.9, gave incomplete inactivation (13).

Other viruses which have been inactivated with aziridines include rabies (16) (0.05% AEI, 6.5hr, 37°C), pseudorabies (37) (0.05% AEI, 4 hr, 37°C) and porcine parvovirus (46) (EI—5 mM, 48–72 hr, 26°C). We have also used AEI to inactivate swine vesicular disease and African swine fever viruses. Nitrogen mustards have been used to inactivate hog cholera and influenza A viruses, possibly by the spontaneous formation of an aziridine (6). There is some evidence that AEI is less effective with double-stranded RNA viruses. J. Parker (unpublished results) was unable to produce complete inactivation of African horse sickness and bluetongue viruses.

In addition to the commonly used AEI and EI, Bahnemann (5) and Wittman et al. (78) have reported that propyleneimine and ethylethyleneimine respectively may be used to inactivate FMDV.

Operational Considerations. Both AEI and EI are thought to inactivate FMDV by a first-order reaction. Whereas AEI is traditionally used directly from a pure preparation of the chemical (usually one to two doses of 0.05% for 24–48 hr at 26–37°C), EI is commonly manufactured either by *in situ* cyclization of bromoethylamine hydrobromide in NaOH-supplemented virus suspensions or *in vitro* cyclization immediately prior to addition to the culture vessel (6). Freshly prepared EI is considered to be more active than AEI with the inactivation of FMDV (5) and porcine parvovirus (46). This may be due in part to the fact that the half-life of EI is greater than that of AEI. Bahnemann (5) reported that pure EI stored for 14 to 20 months at 20–40°C did not lose potency, whereas Graves and Arlinghaus (36) observed a 69% reduction in potency of AEI stored for 42 days at 23°C. Storage of AEI at −20°C prevents deterioration over a considerable period of time, in our experience at least several years. At room temperature and above, all of the aziridines are prone to polymerisation.

Although both AEI and EI are, clearly, highly reactive, a considerable body of evidence points to a selective activity towards the nucleic acid of viruses. Thus, the capsids of the majority of FMDV strains are not disrupted by AEI or EI treatment (11, 21, 36, 76) and the presence of extraneous protein such as bovine serum albumin does not reduce the efficacy of AEI towards the virus (36). Nevertheless, there is evidence that some strains of FMDV are extremely labile to AEI [e.g. SAT 2, Kenya 227/66 (62)] and that inactivation of all strains of FMDV results in a lowering of thermal and pH stabilities (21; T. R. Doel and R. F. Staple, unpublished results). The thermal stability of AEI-treated empty particles (i.e. RNA-free) of FMDV is the same as that of the untreated controls (21). There is no evidence to suggest that the immunogenicity of the more stable virus strains is significantly impaired or altered by aziridine treatment.

While *in situ* cyclization to produce EI presents less of a potential health hazard to production personnel, the efficiency of the inactivation procedure is lower than that with the externally prepared EI. Bahnemann's work (6) indicated that the presence of bicarbonate in the medium had an inhibitory effect on the cyclization reaction.

Both AEI and EI may be neutralized in the culture medium by the addition of sodium thiosulphate.

3.1.2. Formaldehyde and Other Aldehydes

Chemistry and Mode of Action. Formaldehyde has been used more widely for the inactivation of viruses than any other single chemical. Despite the widespread use of aldehydes, and formaldehyde in particular, the chemistry of commercial solutions remains somewhat obscure. Most aldehydes, including formaldehyde and glutaraldehyde, readily polymerise to a whole series of derivatives and it becomes difficult to propose specific mechanisms of inactivation against this background of aldehyde-related substances. It is clear, however, that formaldehyde reacts with both nucleic acids and proteins primarily through exposed amino groups. Fraenkel–Conrat (29) suggested that besides the reversible addition of formaldehyde to amino groups, there are slower and more stable cross-linking reactions of the resultant amino methylols through condensation with other amino acid side chains yielding methylene bridges. Similar reactions occur with the amino groups of nucleic acids (particularly single-stranded molecules) and probably give rise to cross-links within the nucleic acid and between the nucleic acid and any adjacent protein.

During the inactivation of poliovirus there is a considerable shift in the surface charge of the virus particle with the probable formation of bonds between adjacent protein molecules (33). In Gard's view, the particle becomes increasingly impermeable to formaldehyde during inactivation. The kinetics of inactivation of FMDV and poliovirus by formaldehyde is consistent with this model of an initially permeable virus particle in which the nucleic acid is available for inactivation. Thus 'locking up' of the protein coat would prevent further ingress of formaldehyde and explain the residue of infectious virus observed with both vaccines (33, 35).

Despite the ability of aldehydes to cross-link adjacent proteins, we have not observed significant levels of cross-linking between particles of FMDV or between particles of FMDV and BHK cell proteins (21). These experiments were conducted with glutaraldehyde and clearly demonstrated the capability of this reagent to cross-link within virus particles and stabilise.

Glycidaldehyde has been used to inactivate FMDV, Newcastle disease virus and T1 bacteriophage (50, 51). However, its mode of action probably does not relate solely to its aldehyde properties. It also carries an epoxide group and it has been shown that epoxy compounds react readily with proteins and nucleic acids (50).

Applications. It would almost be easier to list those viruses which have not been inactivated at some time or other with formaldehyde. It was first used to inactivate FMDV by Vallée et al. (71) and, indeed, is still used by some countries for the production of FMDV vaccines. Other viruses which have been inactivated with formaldehyde include polio, human rhinoviruses, Eastern and Western equine encephalitis, transmissible gastroenteritis, influenza A and B, mumps, measles, canine distemper and rinderpest viruses and various adenoviruses (2).

Other aldehydes have been used to inactivate viruses. Fontaine *et al.* (28) compared various inactivants including formaldehyde and ethyleneimine and concluded that glycidaldehyde was suitable for the inactivation of FMDV. Stott *et al.* (68) used glutaraldehyde to fix bovine nasal mucosa cells infected with respiratory syncytial virus and successfully inactivated the infectivity of the virus.

Operational Considerations. The tragic consequences of the initial use of inadequately inactivated formaldehyde-treated poliovirus vaccines focussed considerable attention on the mechanism of inactivation. Regrettably, the problem of residual infectivity following formaldehyde treatment had been appreciated by FMDV workers for many years. For example, Waldmann and Kobe (74) inactivated with 0.15% formalin for up to 72 hr at 25°C and reported that inactivations at pH 7.6 did not always go to completion, whereas pH 9.0 inactivations were apparently completely effective. It was also found that adsorption of FMDV to AL(OH)$_3$ gel prior to inactivation resulted in vaccines which were non-infectious in cattle, although a number of workers continued to dispute the innocuity of this type of vaccine. During their pioneering work with AEI, Brown and his colleagues (13) reported failure to completely inactivate FMDV with 0.05% formaldehyde for periods up to 144 hr at 26°C, pH 8.9.

Various workers have examined the possible influence of tissue culture proteins on the inactivation kinetics of formaldehyde. Lycke (49) reported that purified poliovirus types 1 and 3 were inactivated at the same rate as crude preparations. Pure type 2 virus was inactivated more rapidly than crude preparations of virus. He also demonstrated that glycine had a retarding effect on inactivation. Schaffer (64) reported that tris buffer and tryptophan conferred some protection to inactivation of poliovirus by formaldehyde, although the rates of inactivation of crude, partially purified and purified preparations were similar. High salt (1.2 M NaCl) and low concentrations of protein had little, if any, effect on inactivation of pure virus. It would appear, therefore, that low-molecular-weight sources of amino groups have a more significant retarding effect on inactivation than high-molecular-weight sources such as proteins. Nevertheless, thorough filtration of poliovirus preparations is considered desirable to remove unwanted protein and possible aggregates of virus which may protect some particles from the inactivant.

In the context of reactions with cell culture products, a disadvantage of the use of formalin has been reported by Capstick *et al.* (*14*). It was claimed that a formaldehyde–bovine serum complex was associated with a high incidence of anaphylactic reactions in cattle vaccinated against FMDV in some parts of Germany. Bovine serum has been widely used as a constituent of medium for the maintenance of BHK 21 suspension cells during virus production.

The exhaustive inactivation conditions practised by many laboratories for the production of poliovirus vaccine (e.g. 10 to 20 days at 37°C, 0.025% formaldehyde) do not appear to cause serious losses of immunogenic poliovirus particles. In recent work, we have not detected significant degradation of poliovirus types 1 and 3 with continued incubation at 37°C for 14 days following initial formaldehyde inactivation. Nor do inactivated preparations stored for 6 months show any evidence of degradation of the 155S particles. Poliovirus type 2 appears to be more labile to incubation at 37°C for 14 days (*22*). Although formaldehyde has been used extensively to prepare potent FMDV vaccines, glutaraldehyde treatment of the strain 0 BFS 1860 gave an essentially useless vaccine (T. R. Doel, unpublished results). The mechanism operating here would appear to be chemical modification of the major antigenic determinant rather than a disruptive effect on the virion (*21*). Certain strains of FMDV which are labile to AEI may be inactivated with formaldehyde to give potent vaccines (*62*). In studies from our laboratory, seven different strains of FMDV, including RNA-free particles, were more stable to thermal degradation following glutaraldehyde treatment (*21*).

3.1.3. *Other Chemical Methods of Inactivation*

β-Propiolactone (BPL).   The alkyl and acyl bonds at each end of the lactone structure make BPL highly reactive, and while the molecule is relatively stable in the pure and concentrated state, it degrades quickly in the presence of cellular debris and cell culture medium.

$$CH_2\text{—}CH_2\text{—}C\text{=}O$$

$$\text{L}\underline{}O\text{—}\text{⌐}$$

β-Propiolactone

Logrippo (*47*) reported that the half-life of BPL in plasma was between 24 and 32 min at 37°C and 16 and 20 hr at 4°C. At 37°C, hydrolysis was complete by 3 hr. BPL readily reacts with proteins and has a high affinity for nucleic acids (*29*). Although clearly very unstable compared with AEI, EI and formaldehyde, BPL has been reported to fully inactivate viruses such as Eastern equine encephalitis within 15 min at 37°C (*47*). The primary products of BPL hydrolysis are β-propionic and hydracrylic acid derivatives.

BPL has been used to inactivate a wide range of viruses including FMDV (*25*), Newcastle disease (*45*), polio, herpes simplex, influenza A and B, measles, mumps, rabies and Eastern and Western equine encephalitis (*47*),

African horse sickness (54) and bluetongue viruses (55). In the case of rabies and Newcastle disease virus, BPL was considered superior to formaldehyde with regard to immunogenicity of the finished vaccine (45, 70), although Appleton *et al.* (4) presented contradictory results with Newcastle disease virus. Nevertheless, BPL has been found wanting in a number of respects. Its ready reaction with proteins means that the inactivation kinetics will be influenced by the concentration of protein present in the culture. Furthermore, there are numerous reports of the tailing phenomenon with BPL inactivations, i.e. the persistence of residual infectivity. This problem may be offset by the combined use of UV light and BPL when considerably lower doses of each inactivant are required. Thus Logrippo (47) quoted one-sixth of the normal BPL dose and one-fourth of the normal UV dose in a combined inactivation procedure. The use of UV as an inactivant is also discussed in a later section.

A further problem associated with the use of BPL is its high toxicity, so much so that, until relatively recently, it was difficult to obtain. A point in favour of the reagent is its rapid breakdown to essentially innocuous products. Thus BPL has been used to sterilise serum for tissue culture work, and material sterilised with BPL produces no toxic or allergic reactions (47). The formation of acidic but otherwise innocuous products does present a problem with FMDV. The lability of the virus to slightly acidic conditions necessitates the continuous neutralization of the BPL products by buffers such as sodium glycollate (24).

Hydroxylamine. Hydroxylamine is thought to inactivate viruses by its action on the nucleic acid possibly by reacting with the C5 and C6 double bond of pyrimidines (29, 30). Fellowes (26) used a range of hydroxylamine concentrations to inactivate FMDV and found 0.25 M at 40° or 23°C to be optimum. However, Wittman and Bauer (77), while obtaining similar results to those of Fellowes, observed a significant deviation from first-order reaction kinetics within approximately 20 hr of the commencement of inactivation. They were also able to detect infectious virus after 42 hr of inactivation. Franklin and Wecker (30) reported that whereas swine influenza and fowl plague viruses were highly sensitive to hydroxylamine, herpes simplex, mumps and Newcastle disease viruses were relatively resistant. It was suggested that the mode of action of hydroxylamine was dependent on the configuration of the nucleic acid.

Miscellaneous Chemical Inactivants. A number of other chemicals have been used to inactivate viruses and deserve mention.

Crystal violet-inactivated hog cholera vaccines have proved successful over many years of use in the field, giving immunity up to 6 months (52).

Application of crystal violet to other viruses such as Newcastle disease virus has proved less satisfactory, although combined use with ethylene glycol was more successful (45).

An extremely novel method of inactivation was recently reported by Scodeller and his colleagues (65). It was found that FMDV incubated in the presence of ammonium ions was more rapidly inactivated than by classical aziridine methods. It would appear that the ammonium ions activate a virus-associated endogenous nuclease, which then degrades the viral RNA. The attractiveness of this procedure lies not just in its simplicity and low cost, but in the avoidance of large quantities of highly toxic chemicals.

Finally, a number of chemicals have been used in conjunction with light to produce photodynamic inactivation. For example, visible light has been used with methylene blue [canine distemper virus (57)], proflavine, rivanol or acriflavine [FMDV (31)], toluidine blue [influenza (75)], neutral red or methylene blue [FMDV (27)]. While many of these experiments produced acceptably immunogenic inactivated vaccines, the technical problems of assuring complete irradiation of large volumes of opaque cell culture fluids have almost certainly been a factor in inhibiting their use for commercial production of vaccines.

3.2. Physical

3.2.1. Ultraviolet Light

Ultraviolet light has been used to inactivate viruses with limited success. While its activity appears to be directed primarily against the nucleic acid of viruses, disulphide bonds of proteins are also readily excited (59). Hansen *et al.* (38) described the use of short (180 sec) exposures of FMDV to UV light to produce inactivated preparations which were antigenic in guinea pigs. Turner (70) cited data which indicated that UV inactivation of rabies virus produced a vaccine superior to others inactivated by some of the more conventional methods (Table II).

As mentioned earlier in this chapter, UV has been used in conjunction with β-propiolactone [e.g. FMDV (25); Newcastle disease virus (45)]. Under these circumstances the two inactivants complemented each other and eliminated, apparently, the tailing seen with each inactivant separately. Pollard (59) suggests that the tail of residual infectious virus seen with UV inactivation may be due to a mechanism in which the virus particles became less sensitive to further irradiation following initial irradiation.

There are, however, more significant problems associated with UV inactivation. The first is the phenomenon known as multiplicity reactivation, which is well documented with a wide range of micro-organisms [e.g. New-

TABLE II   Immunogenicity of Rabies Virus
Vaccines Produced by Different Inactivation
Methods[a]

Inactivant	Log_{10} PD_{50}[b]
Light + methylene blue	1.12
β-Propiolactone	1.01
UV light	1.24
Phenol	0.69
Formalin	0.69

[a] After Turner (70)
[b] Based on 50% protection of mice to a challenge
dose of 5–50 LD50.

castle disease virus (45)]. This will be discussed in depth in the next section. The second, and probably most relevant problem is that of the critical nature of the dose of UV required to inactivate. This is compounded by the technical difficulties involved in dosing large volumes of essentially UV-opaque solutions. Considering that the opacity of the virus preparation may vary significantly from batch to batch, the technical problems of UV inactivation are formidable. Overdosing a virus preparation is not the answer. Slightly too much irradiation may produce considerable losses in antigenicity (69).

3.2.2. X and γ Irradiation

Ionizing radiations inactivate nucleic acids by the formation of either cross-links or breaks in the structure (59). Just as with UV light, the disulphide bond of proteins is also thought to be sensitive.

Johnson (42) examined the influence of previously X-irradiated aqueous solutions on the infectivity of FMDV and vesicular stomatitis virus. He concluded that there were similarities between irradiated solutions and H_2O_2 in terms of inactivation of these viruses, although pure H_2O_2 was somewhat less effective. McCrea (53) cited many examples of viruses inactivated by ionizing radiations, in particular rabies virus vaccine, claimed to be more immunogenic than preparations made by phenol or UV radiation. Wild and Brown (76) demonstrated that while FMDV was rendered innocuous by X irradiation, the product was non-immunogenic. In the same year, Polatnick and Bachrach (58) came to similar conclusions with γ-irradiated FMDV preparations. They also observed that growth medium and cellular products protected the virus from inactivation, as did cysteine and gelatin when added to pure virus preparations. Polley (60) reported that histidine and sodium p-aminohippurate protected influenza A and mumps viruses from γ irradiation, although, unlike FMDV, efficacious inactivated vaccines were obtained.

While sample volume and opacity are very much less of a problem with ionizing radiation than with UV light, there is nevertheless the problem of multiplicity reactivation. Pavilanis *et al.* (56) have reported the phenomenon with γ-irradiated preparations of influenza A viruses. They observed the tailing phenomenon and attributed it to complementation between two (or more) differently inactivated influenza particles. Clearly this problem would be most likely with viruses such as influenza in which the genome is segmented, although it must be said that genome recombination has been demonstrated *in vitro* with single-stranded RNA viruses, i.e. polio and FMD viruses.

3.2.3. *Heat*

Although there are numerous reports on the use of heat to totally degrade virus, there has been little work aimed at preserving the essential immunogenicity while destroying the infectivity. Reesink *et al.* (61) have prepared apparently innocuous but immunogenic hepatitis B vaccine by heating HB_sAg isolated from plasma at 101°C for 1.5 min. However, this could be considered a special case in which the subunit of a virus is sufficiently immunogenic to protect the vaccinate against challenge by the whole particle. Nevertheless, it is worth considering that some viruses may have an extremely labile genome. Thus storage of FMDV preparations at 4° or 37°C results in a rapid decline in infectivity. Of course, temperature is often a critical factor in the inactivation of viruses by chemical methods.

4. TECHNICAL ASPECTS OF INACTIVATION

At this point, it must be said that I do not intend to devote very much attention to the hardware used for inactivation of viruses. For most of the inactivation procedures currently used, relatively unspecialised equipment is required. However, a number of aspects deserve mention.

For the chemical inactivation of viruses in suspension, relatively simple vessels are satisfactory. Considerable importance is attached to thorough stirring of the inactivant/virus mixture throughout the inactivation period to ensure that none of the virus can escape contact with inactivant. This is particularly important with inactivants which have relatively short half-lives, for example β-propiolactone. A common procedure to ensure thorough dispersion is to mix the inactivant with the virus suspension in one vessel and transfer the mixture to a second vessel. Thus droplets of non-inactivated virus will be left in the first vessel. Obviously, pH and temperature are also important conditions to monitor and control, not simply for consistency of inactivation conditions but also to preserve the immunogenicity of labile viruses such as FMDV as much as possible. The labile nature of some inactivants also demands that they are added with the minimum of delay to the virus suspension, which should have been equilibrated previously to the appropriate pH and temperature.

Because some inactivants react readily with non-viral proteins and some components of media, it is important to ensure that the levels of interfering materials do not vary greatly from one virus preparation to the next. On a similar note, it is common practice to remove cellular debris by filtration prior to inactivation and in the case of, for example, inactivated poliovirus, to filter again after inactivation to ensure the removal of aggregates of virus or virus and cellular debris which may contain protected particles of live virus.

Inactivation procedures based on UV light or visible light in conjuction with dyes such as methylene blue demand highly sophisticated equipment to allow the radiation to penetrate the virus suspension. For example, the source of radiation may be placed in the centre of a helix of plastic, radiation-transparent tubing or the virus suspension centrifuged as a thin film around the light source. Because of the usually critical nature of the exposure time, the rate of flow of the liquid through the system would require strict monitoring and control. The same would apply to ionising radiation inactivation.

5. INNOCUITY TESTING

5.1. Introductory Comments

The success of vaccines against diseases such as poliomyelitis and FMD has imposed an increased responsibility on the vaccine manufacturer to ensure that his vaccines are totally free of infectious virus. That is, in an endemic situation a low percentage of vaccine-associated disease might be tolerated provided there were obvious benefits from the use of the vaccine. With the considerable success of FMD vaccine in Western Europe, we now see a situation in which a number of outbreaks in recent years have been attributed either to escapes of virus from vaccine production plants or the use of inadequately inactivated vaccine (3). In fairness to the vaccine producers, it has to be said that innocuity testing can do no more than give a statistical probability of the level of infectivity in a given inactivated virus preparation. When the scale of vaccine application is also considered—approximately 500 million trivalent doses of FMD vaccine per annum in South America alone—it is hardly surprising that vaccine-associated outbreaks do occur from time to time. Thus considerable effort has been expended with both killed polio and FMD vaccines to assure as complete inactivation as possible.

5.2. Test Procedures

Henderson's pioneering work in the early 1950s (*39, 40*) provided the ground rules for innocuity testing of FMD vaccines in cattle. He demonstrated that the sensitivity of the lingual mucosa was up to 250,000 times greater than the subcutaneous route depending on the strain of virus used. With some viruses, there were no differences between the two routes. However, the lingual mucosa permitted the use of up to 100 different test sites per animal, whereas the subcutaneous route allowed only one observation per animal. Skinner's work with unweaned mice prompted their use for innocuity testing (*66*). However, Brown *et al.* (*13*) demonstrated that FMDV inactivated for 144 hr with formaldehyde still contained detectable levels of infectious virus in cattle, whereas 72-hr inactivated preparations appeared to be totally innocuous by the mouse test. They reported similar data with AEI-inactivated preparations, as did Dimpoullos *et al.* (*19*) with heat-treated preparations of the same virus. There appeared to be some evidence that killed virus interfered with the detection of live virus in the mouse test.

The early work with inactivated polio vaccine relied on innocuity tests in either tissue culture or monkeys. For a number of reasons, not the least of which was the statistical limitations, the monkey test has given way to innocuity testing in cultured cells. Van Steenis and Van Wezel (*72*) have evaluated subcultured monkey kidney cells and Vero cells and concluded that they were sufficiently sensitive to replace the commonly used primary monkey kidney cells. Furthermore, live virus could be detected for much longer periods in tissue culture than in monkeys. In the discussion of their paper, Melnick reflected on the Cutter incident in which live virus was detected in monkeys but not tissue culture, although developments in tissue culture have probably superseded this problem. Tissue culture tests are also used with inactivated FMDV preparations. Galuinas (*32*) reported that bovine kidney cell cultures in Povitsky bottles were similar to the lingual mucosa of cattle in terms of sensitivity to FMDV. Baby hamster kidney cell lines are also used extensively to detect residual infectivity in inactivated FMDV preparations (*1*).

5.3. Limitations and Remedies

An often quoted criticism of innocuity testing in either the host animal or a related species is that only relatively small samples of inactivated antigen may be tested in a few, highly expensive animals. Thus Henderson's cattle test for FMDV (*39, 40*) allows only approximately 10 ml of vaccine to be tested per animal, i.e. 0.001% of a 1000-litre production batch. Clearly, animal tests in which only one inoculation per animal is given are even less representative of a large batch of vaccine. At least with tissue culture methods, volumes of 1 litre are easily tested (in our laboratory 40 Roux flasks would accommodate this volume). Large volumes may also be tested in Povitsky bottles (32) or 1-litre suspension cultures (7). Suspension cultures have a number of advantages including simplicity of scale-up to 10 litres and testing of FMDV in the cell strain normally used for production. Unfortunately, there are a number of problems which are particularly relevant to tissue culture tests. In the first place, residual inactivants, e.g. AEI, may produce cytotoxic effects which either obscure cytopathic effects or render the test useless. This may be offset by the judicious use of reagents such as sodium thiosulphate to inactivate residual AEI and passage of extracellular fluids onto secondary and tertiary cell cultures. A further refinement is to freeze–thaw the cell sheets to release any infectious intracellular virus prior to further passage (48). A second problem is that the tissue culture test does not lend itself to innocuity testing of fully formulated vaccine. One approach used by some FMD vaccine producers is to elute the virus from the Al $(OH)_3$ gel, concentrate the eluate and use this to seed the cell monolayers (73). A complication of this procedure is that the use of certain concentrations of saponin in the vaccine prevents the complete elution of FMDV [10% or less with some strains of virus (23)]. Furthermore, it is not known whether live virus elutes to the same extent as inactivated virus.

The phenomenon of interference by high concentrations of inactivated virus is well documented. Böttiger et al. (10) reported that detection of live poliovirus could be inhibited by inactivated virus and suggested that this was due to an effect of formaldehyde on virus protein causing a delay of adsorption. Henderson's work with FMDV (39, 40) showed that the presence of inactivated virus did not appear to inhibit the detection of low levels of live virus, although Visser et al. (73) reported interference in cattle when relatively high concentrations of inactivated virus were included in the inoculum. Dinter (20) also demonstrated interference in cell cultures inoculated with a mixture of weak and strongly cytopathic strains of FMDV.

There is also the problem of variable susceptibility of different cell lines to different strains of virus. Thus Clarke and Spier (15) have demonstrated that BHK suspension cells from different laboratories vary in their susceptibility to a given strain of FMDV. Visser et al. (73) demonstrated that two strains of FMDV were more sensitive than a third strain to inhibition of infectivity in plaque assays by a range of concentrations of inactivated virus. Barteling (7) concluded that the total antigen (i.e. 146S particle) concentration should not exceed 1 $\mu g/10^6$ cells. Therefore, a 1-litre suspension culture would allow the testing of 1 to 2 mg of inactivated virus, i.e. approximately 300 doses of vaccine.

On a more optimistic note, it is probably true to say that serial passage of inoculum and regular testing of samples throughout the inactivation procedure gives an increased level of confidence to the interpretation of innocuity tests. This is particularly so with kinetic data which provide information on the rate of inactivation. Beale (8) described the development of innocuity testing of inactivated poliovirus in his laboratory. The final protocol described in his article involved the testing of duplicate 500-ml samples taken 3 days before the end and at the end of the inactivation. Each 500 ml was tested in monkey kidney cell cultures with subcultures being made at weekly intervals and all cultures being maintained for 28 days. During this work it was also noted that filtration through Seitz EKS2 filters was necessary to remove live virus presumably trapped in clumps of inactivated virus or cell debris.

6. CONCLUSIONS

While I have attempted to cover the field of virus inactivation, it is quite clear that vaccine production is dominated by the use of formaldehyde and the aziridines. Furthermore, there appears to be no immediate prospect of these reagents being usurped by different inactivants. To a large extent, this reflects the considerable experience gained with formaldehyde and aziridine inactivations, particularly with regard to confidence in the reliability of properly controlled inactivation procedures.

Both the future use of inactivants and the degree of effort devoted to improve on the current procedures will depend very much on the level of success achieved with vaccines based on virus-derived subunits, and genetically engineered or chemically synthesised materials.

ACKNOWLEDGMENTS

I would like to thank my colleagues, Dr. Noel Mowat and Dr. Ray Spier, for their advice and encouragement.

REFERENCES

1. Anderson, E. C., Capstick, P. B., Mowat, G. N., and Leech, F. B. (1970). In vitro method for safety testing of foot-and-mouth disease vaccines. *J. Hyg.* **68**, 159–172.
2. Andrewes, C., and Pereira, H. G. (1972). "Viruses of Vertebrates," 3 ed. Baillière, London.
3. Anonymous (1980). Report of the Executive Committee. Report of the Session of the Research Group of the Standing Technical Committee of the European Commission for the Control of Foot-and-Mouth Disease, Vienna, Austria (pp. 7 and 96–101).
4. Appleton, G. S., Hitchner, S. B., and Winterfield, R. W. (1963). A comparison of the

immune response of chickens vaccinated with formalin and BPL inactivated Newcastle disease vaccine. *Am. J. Vet. Res.* **24**, 827–831.

5. Bahnemann, H. G. (1973). The inactivation of foot-and-mouth disease virus by ethylenimine and propylenimine. *Zentralbl. Veterinaer Med.* **20**, 356–360.

6. Bahnemann, H. G. (1975). Binary ethyleneimine as an inactivant for foot-and-mouth disease virus and its application for vaccine production. *Arch. Virol.* **47**, 47–56.

7. Barteling, S. J. (1982). Safety control of foot-and-mouth disease (FMD) vaccines. II. Aziridine inactivated antigen produced in baby hamster kidney (BHK) cells. Report of the Session of the Research Group of the Standing Technical Committee of the European Commission for the Control of Foot-and-Mouth Disease. Pirbright, U. K.

8. Beale, A. J. (1982). Potency testing for killed poliomyelitis vaccine. *Dev. Biol. Stand.* **47**, 69–75.

9. Bittle, J. L., Houghten, R. A., Alexander, H., Shinnick, T. M., Sutcliffe, J. G., Lerner, R. A., Rowlands, D. J., and Brown, F. (1982). Protection against foot-and-mouth disease by immunization with a chemically synthesized peptide predicted from the viral nucleotide sequence. *Nature (London)* **298**, 30–33.

10. Böttiger, M., Lycke, E., Melén, B., and Wrange, G. (1958). Inactivation of poliomyelitis virus by formaldehyde. Incubation time in tissue culture of formalin treated virus. *Arch. Gesamte Virusforsch.* **8**, 259–266.

11. Brown, F., Cartwright, B., and Stewart, D. L. (1963). The effect of various inactivating agents on the viral and ribonucleic acid infectivities of foot-and-mouth disease virus and its attachment to susceptible cells. *J. Gen. Microbiol.* **31**, 179–186.

12. Brown, F., and Crick, J. (1959). Application of agar-gel diffusion analysis to a study of the antigenic structure of inactivated vaccines prepared from the virus of foot and mouth disease. *J. Immunol.* **82**, 444–447.

13. Brown, F., Hyslop, N. St. G., Crick, J., and Morrow, A. W. (1963). The use of acetylethyleneimine in the production of inactivated foot-and-mouth disease vaccines. *J. Hyg.* **61**, 337–344.

14. Capstick, P. B., Pay, T. W. F., and Beadle, G. G. (1969). Some studies on allergic reactions to foot-and-mouth disease vaccines in Lower Saxony, Germany. Report of the Session of the Research Group of the Standing Technical Committee of the European Commission for the Control of Foot-and-Mouth Disease. Brescia, Italy (pp. 213–218).

15. Clarke, J. B., and Spier, R. E. (1980). Variation in the susceptibility of BHK populations and cloned cell lines to three strains of FMDV. *Arch. Virol.* **63**, 1–9.

16. Crick, J. and Brown, F. (1971). An inactivated baby hamster kidney cell rabies vaccine for use in dogs and cattle. *Res. Vet. Sci.* **12**, 156–161.

17. Dempsey, T. F., and Mayer, V. (1934). Canine distemper vaccine. Two experiments with vaccines prepared by photodynamic inactivation of the virus with methylene blue. *J. Comp. Pathol.* **47**, 197–200.

18. Dermer, O. C., and Ham, G. E. (1969). "Ethylenimine and other Aziridines: Chemistry and Applications." Academic Press, New York.

19. Dimopoullos, G. T., Fellowes, O. N., Callis, J. J., Poppensiek, V. M. D., Edwards, A. G., and Graves, J. H. (1959). Thermal inactivation and antigenicity studies of heated tissue suspensions containing foot-and-mouth disease virus. *Am. J. Vet. Res.* **20**, 510–521.

20. Dinter, Z. (1958). Interferenz zirischen cytopathogenen und nichtcytopathogenen partikelch der virus der maul-and klauenseuche. *Arch. Gesamte Virus-forsch.* **8**, 42.

21. Doel, T. R., and Baccarini, P. J. (1981). Thermal stability of foot-and-mouth disease virus. *Arch. Virol.* **70**, 21–32.

22. Doel, T. R., Osterhaus, A., Van Steenis, G., and van Wezel, A. L. Unpublished results.

23. Doel, T. R., and Staple, R. F. (1982). The elution of foot-and-mouth disease virus from

vaccines adjuvanted with aluminium hydroxide and with saponin. *J. Biol. Stand.* **10,** 182–195.

24. Durand, M., Guilloteau, B., Giraud, M., Guerche, M., Pesson, M., and Prunet, P. (1968). A study of alkylating agents for the inactivation of foot-and-mouth disease virus and the preparation of a new inactivated vaccine. *Bull. Off. Int. Epizoot.* **69,** 429–465.

25. Fellowes, O. N. (1965). Comparison of the inactivation and antigenicity of foot-and-mouth disease virus by acetylethyleneimine and by combined effect of ultraviolet light and β-propiolactone. *J. Immunol.* **95,** 1100–1106.

26. Fellowes, O. N. (1966). Hydroxylamine as an inactivating agent for foot-and-mouth disease virus. *J. Immunol.* **96,** 772–776.

27. Fellowes, O. N. (1966). Inactivation of foot-and-mouth disease virus by the interaction of dye and visible light. *Appl. Microbiol.* **14,** 86–91.

28. Fontaine, J., Favre, H., Fargeaud, D., Roulet, C., and Dupasquier, M. (1974). Inactivation du virus aphteux influence des agents inactivants sur la conservation du virion. Report of the Session of the Research Group of the Standing Technical Committee of the European Commission for the Control of Foot-and-Mouth Disease, Lelystad, The Netherlands (pp. 14–41).

29. Fraenkel-Conrat, H. (1981). Chemical modification of viruses. *In* "Comprehensive Virology", (H. Fraenkel-Conrat and R. R. Wagner, eds.), Plenum, Vol. 17, pp. 245–283. New York.

30. Franklin, R. M., and Wecker, E. (1959). Inactivation of some animal viruses by hydroxylamine and the structure of ribonucleic acid. *Nature (London)* **184,** 343–345.

31. Galloway, I. A. (1937). The photodynamic action of dyes on the virus of foot-and-mouth disease. Progress Report of the Foot-and-Mouth Disease Research Committee (Appendix 3, Vol. 5, p. 342).

32. Galuinas, P. (1965). Detection of minimal quantities of foot-and-mouth disease virus with bovine kidney tissue cultures. *Appl. Microbiol.* **13,** 872–875.

33. Gard, S. (1960). Theoretical considerations in the inactivation of viruses by chemical means. *Ann. N. Y. Acad. Sci.* **83,** 513–760.

34. Gilman, A., and Philips, F. S. (1946). The biological actions and therapeutic applications of the β-chloroethylamines and sulfides. *Science* **103,** 409–415.

35. Graves, J. H. (1963). Formaldehyde inactivation of foot-and-mouth disease virus as applied to vaccine production. *Am. J. Vet. Res.* **24,** 1131–1136.

36. Graves, J. H., and Arlinghaus, R. B. (1967). Acetylethyleneimine in the preparation of inactivated foot-and-mouth disease vaccines. *Proc. 71st Annu. Meet. U.S. Livest. Sanit. Assoc. 1967,* pp. 396–403.

37. Gutekinst, D. E., and Pirtle, E. C. (1979). Humoral and cellular immune responses in swine after vaccination with inactivated pseudorabies virus. *Am. J. Vet. Res.* **40,** 1343–1346.

38. Hansen, A., Schmidt, S., and Holm, P. (1948). Irradiation par des ondes ultraviolettes (ultra-courtes) d'un suspension de virus aphteux adsorbe par l'hydroxyde d'aluminium. *C.R. Hebd. Seances Acad. Sci.* **227,** 1425–1427.

39. Henderson, W. M. (1952). A comparison of different routes of inoculation of cattle for detection of the virus of foot-and-mouth disease. *J. Hyg.* **50,** 182–194.

40. Henderson, W. M. (1952). Significance of tests for non-infectivity of foot-and-mouth disease vaccines. *J. Hyg.* **50,** 195–208.

41. Hiatt, C. W. (1964). Kinetics of the inactivation of viruses. *Bacteriol. Rev.* **28,** 150–163.

42. Johnson, C. D. (1965). The influence of previously X-irradiated aqueous solutions on the infectivity of the viruses of foot-and-mouth disease and vesicular stomatitis. *J. Gen. Microbiol.* **38,** 9–19.

43. Kaplan, C. (1969). Immunisation against smallpox. *Br. Med. Bull.* **25**, 131–135.
44. Kleid, D. G., Yansura, D., Small, B., Dowbenko, D., Moore, D. M., Grubman, M. J., McKercher, P. D., Morgan, D. O., Robertson, B. H., and Bachrach, H. L. (1981). Cloned viral protein vaccine for foot-and-mouth disease: Responses in cattle and swine. *Science* **214**, 1125–1129.
45. Lancaster, J. E. (1966). "Newcastle Disease, a Review 1926–1964", Can. Agric., Monogr. No. 3. Queen's Printer and Controller of Stationery, Ottawa.
46. Lei, J. C. (1981). Preparation of a 'Quil-A' adjuvanted inactivated porcine parvovirus vaccine. A summary. Paper presented at the Symposium on Vaccine Adjuvants, London, 1981, pp. 16–18. Superfos Export Company A/S, Denmark.
47. Logrippo, G. A. (1960). Investigations of the use of β-propiolactone in virus inactivations. *Ann. N. Y. Acad. Sci.* **83**, 578–594.
48. Lombard, M., Mougeot, H., and Favre, H. (1981). Safety tests of foot-and-mouth disease vaccine. Concentration of samples of inactivated antigens. Report of the Session of the Research Group of the Standing Technical Committee of the European Commission for the Control of Foot-and-Mouth Disease, Tubingen, FRG (pp. 46–48).
49. Lycke, E. (1958). Studies of the inactivation of poliomyelitis virus by formaldehyde. Inactivation of partially purified virus material and the effect upon the rate of inactivation by addition of glycine. *Arch. Gesamte Virusforsch.* **8**, 23–41.
50. Martinsen, J. S. (1962). Ph.D. Thesis, Syracuse University (*Diss. Abstr.* **23**, 12).
51. Martinsen, J. S. (1964). Inactivation of foot-and-mouth disease virus by glycidaldehyde. *Am. J. Vet. Res.* **25**, 1417–1432.
52. McBryde, C. N., and Cole, C. G. (1936). Crystal-violet vaccine for the prevention of hog cholera: Progress report. *J. Am. Vet. Med. Assoc.* **89**, 652–663.
53. McCrea, J. F. (1960). Ionizing radiation and its effects on animal viruses. *Ann. N. Y. Acad. Sci.* **83**, 692–705.
54. Parker, J. (1975). Inactivation of African horse sickness virus by betapropiolactone and by pH. *Arch. Virol.* **47**, 357–365.
55. Parker, J., Herniman, K. A. J., Gibbs, E. P. J., and Sellers, R. F. (1975). An experimental inactivated vaccine against bluetongue. *Vet. Rec.* **96**, 284–287.
56. Pavilanis, V., Gilker, J. C., Ghys, R., and Chagnon, A. (1969). Multiplicity reactivations in gamma irradiated myxoviruses. *Prog. Immunobiol. Stand.* **3**, 64–69.
57. Perdau, J. R., and Todd, C. (1933). Canine distemper. The high antigenic value of the virus after photodynamic inactivation with methylene blue. *J. Comp. Pathol.* **46**, 78–89.
58. Polatnick, J., and Bachrach, H. L. (1968). Irradiation of foot-and-mouth disease virus and its ribonucleic acid. *Arch. Gesamte Virusforsch.* **23**, 96–104.
59. Pollard, E. C. (1960). Theory of the physical means of the inactivation of viruses. *Ann. N. Y, Acad. Sci.* **83**, 654–660.
60. Polley, J. R. (1962). The use of gamma radiation for the preparation of virus vaccines. *Can. J. Microbiol.* **8**, 455–459.
61. Reesink, H. W., Reering-Brongers, E. E., Brummelhuis, H. G. J., Lafeberschat, L. J. T., Van Elven, E. H., Duimel, W. J., Balner, H., Stitz, L. W., Van Den Ende, M. C., Feltkamp-Vroom, T. M., and Cohen, H. H. (1981). Heat inactivated HBsAg as a vaccine against hepatitis B. *Antiviral Res.* **1**, 13–25.
62. Rowlands, D. J., Sangar, D. V., and Brown, F. (1972). Stabilizing the immunizing antigen of foot-and-mouth disease virus by fixation with formaldehyde. *Arch. Gesamte Virusforsch.* **39**, 274–283.
63. Salk, D. (1981). Herd effect and virus eradication with use of killed poliovirus vaccine. *Dev. Biol. Stand.* **47**, 247–255.

64. Schaffer, F. L. (1960). Interaction of highly purified poliovirus with formaldehyde. *Ann. N. Y. Acad. Sci.* **83**, 564–577.

65. Scodeller, E. A., Lebendiker, M. A., Dubra, M, S., La Torre, J. L., and Vasquez, C. (1982). An experimental vaccine for FMD. *Conf. Foot-and-Mouth Dis. Comm. 11th, 1982*, pp. 51–52.

66. Skinner, H. H. (1953). One week old white mice as test animals in foot-and-mouth disease research. *Proc. Int. Vet. Cong 15th, 1953*, Vol. 1, p. 195.

67. Spier, R. E. (1986). Vaccine production. *In* "Vaccine Preparation", Chapter I. Alan R. Liss, Inc., New York (to be published).

68. Stott, E. J., Taylor, G., and Thomas, L. H. (1981). Preliminary observations on 'Quil A' as an adjuvant for an inactivated respiratory syncytial virus vaccine. Paper presented at the Symposium on Vaccine Adjuvants, London, 1981, pp. 19–24. Superfos Export Company A/S, Denmark.

69. Traub, F. B., Friedemann, U., Brasch, A., and Huber, W. (1951). High intensity electrons as a tool for preparation of vaccines. *J. Immunol.* **67**, 379–384.

70. Turner, G. S. (1969). Rabies vaccine. *Br. Med. Bull.* **25**, 119–212.

71. Vallée, H., Carré, H., and Rinjard, P. (1925). Foot and mouth disease immunisation. *Bull. Soc. Cent. Med. Vet.* **78**, 297 (in French).

72. Van Steenis, G., and Van Wezel, A. L. (1981). Killed polio vaccine: An evaluation of safety testing. *Dev. Biol. Stand.* **47**, 143–150.

73. Visser, N., Woortmeijer, R., and Barteling, S. J. (1981). Safety tests of FMDV-vaccines. Report of the Session of the Research Group of the Standing Technical Committee of the European Commission for the Control of Foot-and-Mouth Disease, Tubingen, FRG (pp. 49–54).

74. Waldmann, O., and Kobe, K. (1938). Die aktive immunisierung des rindes gegen maul-und klauenseuch. *Berl. Tieraerztl. Wochenschr.* **46**, 317.

75. Wallis, C., Sakurada, N., and Melnick, J. L. (1963). Influenza vaccine prepared by photodynamic inactivation of virus. *J. Immunol.* **91**, 677–682.

76. Wild, T. F., and Brown, F. (1968). A study of the physical properties of the immunising antigen of foot-and-mouth disease virus and the effect of various inactivating agents on its structure. *Arch. Gesamte Virusforsch.* **24**, 86–103.

77. Wittman, G., and Bauer, K. (1968). Deviations from a first order reaction in the inactivation of foot-and-mouth disease virus by hydroxylamine. *Zentralbl. Bakteriorl. Parasitenkd., Infektionskr. Hyg., Abt. 1:* Orig. **207**, 259–261.

78. Wittman, G., Bauer, K., and Mussgay, M. (1972). Experiments on vaccination of pigs with ethylethyleneimine (EEI), diethylaminoethyl dextran (DEAE-D) foot-and-mouth disease vaccines. *Arch. Gesamte Virusforsch.* **6**, 251–264.

7

Purification of Products from Cultured Animal Cells

TERENCE CARTWRIGHT
M. DUCHESNE
Laboratoire Roger Bellon
Monts, France

1. INTRODUCTION

Animal cells offer an immensely versatile synthetic arsenal which is potentially of particular value when we wish to reproduce the natural products of

Animal Cell Biotechnology, Vol. 2

animal cells to permit the controlled use of these products for the manipulation or modification of the behaviour of other animal cells. The most readily
identifiable use of such products is in the clinical field, where the development of highly specific, preferably homologous therapeutic and prophylactic
agents is a major goal. It is the application of techniques of purification to this
goal which is the primary consideration in this chapter.

Currently, there are two major classes of products obtained from animal
cells. Of these, only one, viruses for the preparation of vaccine, is a commercial reality. The other class of product, biologically active proteins and peptides, is at present limited in application despite several early forays by
adventurous groups with human interferon and with plasminogen activator
(1–4). However, with our new capacity to manipulate the production of
proteins of interest by animal cells, this domain should not be neglected.
The production of usefully large amounts of antibody from hybrid cells is
already being achieved, and the genetic manipulation of animal cells may
yield advantages in the post-translational processing of proteins which may
mean that this is the eventual system of choice for the production of biologically active protein for human or veterinary use in the future.

In both cases the purification of the product is necessary to provide more
potent preparations, to eliminate the undesirable effects of contaminating
molecules and, importantly, to provide a defined entity with predictable
properties for therapeutic or prophylactic use. The synthesis of these objectives is to assure the safety and efficacy of the medicament for the user. This
chapter considers the methodology available for purification work for each
class of product and the considerations, ethical, scientific and ergonomic,
which will affect process selection.

Although the goals of purification of preparations of virus or of single
proteins may be identical as indicated above, the same methodology is not
appropriate in each case, nor are the same criteria of purity applied for each
type of product.

Purification of proteins is dealt with first here because the processes used
for the purification of single molecules are more readily visualised in physicochemical terms. When purification of virus is considered, we are looking
at the combined properties of an ensemble of protein, glycoproteins, nucleic
acid, and possibly lipid whose interaction with a separation system is perhaps less predictable, even when the same basic techniques are applied.

2. PRODUCT SAFETY AND PURIFICATION STRATEGY

As already mentioned, product safety is an important goal of purification
procedures for the products of animal cells which are intended to be used
clinically. It is legitimate and important to discuss safety at this point be-

cause the approach taken in the case of virus is now markedly different from that taken with defined proteins, and this separation in concept is likely to become more marked as more products from animal cells become commercialized.

The achievement of safety in virus preparations from animal cells may be divided into two parts. The question of biological safety, that is possible contamination by viruses other than the specific vaccination strain or by fragments of cell or viral DNA, is not traditionally addressed by purification primarily because of the extreme difficulty of differentially inactivating or separating such agents from the required virus.

Indeed systems have been evolved, discussed in Chapters 3 and 4, Vol. 1, to verify carefully that the cell stocks, virus seeds, media, sera and trypsin used in the preparation of virus lots do not themselves constitute a source of viral or other contamination nor of potential transfer of harmful genetic material. Thus biological "purity" is assured by careful biological control of the starting materials used in the preparation and by rigorous checks for in-process contamination.

Purification of virus as normally practised for vaccine production is, however, concerned with the removal of medium, serum and cellular components which may interfere with the desired effects of the vaccine, or, more probably, which may provoke immune responses unconnected with the vaccine's protective action, possibly resulting in allergy or sensitisation or even in the production of antibodies that could cross-react with host proteins.

In this context, then, the purification of virus may be considered as the removal of heterologous proteins or particles, particularly the removal of heterologous immunogenes. The concept of chemical purity is not normally evoked.

When we consider the purification of animal cell proteins, however, the situation is quite different. Although initial attempts were made to produce protein products in exhaustively tested "normal" cells using the same logic as is applied to virus production (5), it now appears clear that the animal cells capable of the production of useful quantities of proteins of interest will be abnormal, since normal cells under normal control do not overproduce such proteins. Cells may be genetically abnormal with tumourigenic potential, as is the case with the hybridomas used in the production of monoclonal antibody (6); they may be genetically abnormal, potentially tumourigenic and infected with pathogenic virus, as is the case for the Namalwa cells proposed for the production of human lymphoblastoid interferon (2); or they may be genetically altered by virus or other vectors, as will be the case in the genetically manipulated cell production lines now being proposed (7). In each case the potential biological hazard associated with the cells precludes them from the type of biological control followed by relatively non-rigorous purification which is acceptable for virus.

Fortunately, the techniques available for the purification and analysis of such protein products have advanced to the stage where the chemical purity of the product can, within reasonable limits of concern, be used to eliminate the biological hazard which may be associated with the cell culture material employed.

This approach is exemplified by the production of lymphoblastoid interferon from Namalwa cells by several groups. Here treatment of the fortunately robust interferon under strongly denaturing conditions and subsequent rigorous purification coupled with the careful demonstration that added viral contaminants and added DNA do not survive the extraction procedure have been shown to remove all reasonable doubt concerning the safety of lymphoblastoid interferon due to contaminating adventitious agents (8, 9).

On the basis of an assessment of the benefit/risk ratio in each individual application, it is now considered acceptable by the Bureau of Biologics to proceed with the use of "abnormal" cell lines for the development of biologics (10). Clearly, design and performance of the extraction and purification processes applied to such products are of paramount importance in the evaluation of potential risk.

3. ASSAY PROCEDURES

It is evident that the purification of proteins and viruses cannot be undertaken systematically unless a meaningful quantitative test procedure has been evolved by which the quantity of the required product may be estimated. It is desirable to have an assay which may be performed rapidly, this being more important than extreme accuracy during purification work.

Many methods are available for estimating the concentration of a given species and the assay may be based on physical determination of the product by methods such as electrophoresis or related techniques, HPLC or analytical ultracentrifugation or by serological determination by, for example, fixation of complement, ELISA or RIA (11–13). In the case of proteins with enzyme activity a measurement of specific activity in a specific enzyme assay is usually employed.

With viruses, infectivity assays may be used if there is insufficient knowledge of the physicochemical characteristics to permit the development of a test method based on these properties. Infectivity assays, however, combine the disadvantages of being very imprecise and of being long and tedious to perform. On this basis, time spent in the development of a physical assay is usually well spent.

Having mentioned a series of possible assay systems, it is important to emphasize that they do not all measure the same thing. Thus, the use of a serological method to follow the purification of a biologically active protein may be fast and convenient, but may not be able to distinguish between the active molecule and a denatured form. Equally, the enrichment of a virus band detected in the ultracentrifuge says nothing about its infectivity, which may have been completely destroyed during purification. Each proposed assay system must, therefore, be carefully validated to assure that the information it provides is appropriate to the end product sought.

4. SCALE

The applicability of different purification technologies to a given purification problem is governed by two related non-technical parameters: the scale of the operation and the value of the end product. In the part of the biotechnology market considered here, human and animal pharmaceuticals, we are concerned almost exclusively with low-volume, high-value products (in terms of grams of product required). As mentioned earlier, safety and efficacy considerations, especially in the human field, demand rigorous control and processing and hence relatively high recovery costs are implicit in the products' high value.

In consequence, we have assumed that, even on a commercial scale, production of the materials likely to be produced from animal cells will be at the grams to kilograms scale, and mainly at the lower end of this scale. The separation techniques discussed have been chosen accordingly. We have included methods which in other branches of biotechnology could only be considered as analytical procedures, because on this scale and with this emphasis on purity and definition of product, such high-resolution techniques can be integrated into a commercially viable purification strategy.

5. PURIFICATION OF PROTEINS FROM ANIMAL CELLS

5.1. Initial Steps—Extraction and Preliminary Concentration of Protein Molecules

Classically, the primary step in the purification of proteins derived from animal tissue has been the extraction of the protein from its intracellular sites by mechanical, enzymic or chemical treatments. Up to the present the useful products obtained from cultured animal cells have almost entirely been secreted proteins, which have been typically present in solution in the culture medium. Specific means of cell disruption are not, therefore, a

necessary part of their purification and direct treatment of the culture supernatant, perhaps after clarification if cell debris is a problem, may be
undertaken.

Ideally for an effective purification, the protein of interest should be present at high concentration in the starting fluid and should be purified to near
homogeneity in a single step. High concentrations of a desired protein are
achieved in genetically engineered bacteria and yeasts and may be obtainable in genetically modified animal cells. However, with unmodified cells the
concentration of secreted protein is very low, of the order of picograms per
cell, and the proportion represented by an individual protein of interest is
also low, often below 1%. In these circumstances, an initial enrichment and
concentration of the culture medium with respect to the desired product is
essential. This phase may well be simplified if the quantity of exogenous
protein in the culture can be reduced before the production phase of the cell
culture is entered. Thus it is usual to utilize protein-rich growth medium for
the "growth phase" of the tissue culture and to switch to a "production
medium" containing reduced protein, or better, no exogenous protein, for
the production phase. For each individual cell system it is necessary to
determine by experiment what protein is needed in the medium at each
phase to permit maximum yield of product. It should also be remembered
that a simple change of medium may be insufficient to eliminate the proteins
which were present in the growth medium. Cells bind for example serum
proteins and may liberate these later into a serum-free medium. Furthermore, this binding is selective, serum globulins being more firmly bound
and more slowly eluted than other proteins (14).

These observations serve to underline the fact that in reality there is no
absolute distinction between the synthetic process and subsequent purification; rather the overall process of production of a required product should be
viewed as a continuum in which all the parts interact. Thus, the production
system used will affect the purification methods chosen but equally the
purification requirements will feed back into the initial production process.

5.2. Solubility and Solution Methods for the Initial Extraction of Proteins

The choice of a preliminary method for the concentration and gross fractionation of crude culture fluids is often influenced greatly by the simplicity
of the methods available and their amenability for subsequent scale-up, the
use of complicated manipulation in sophisticated apparatus being avoided
when large volumes of crude fluids may be involved. This is particularly
important when rapid processing is required to recover a product which may
be unstable in dilute solution in a crude tissue supernatant. In this context

the use of relative differences in solubility under different physical conditions may be very useful. Proteins may be fractionated by stepwise variations of salt concentration (usually but not necessarily ammonium sulphate), by modification of the dielectric constant of the solution by addition of organic solvents, by change of pH or, less frequently, by differential precipitations by change of temperature.

Practical details for these traditional methods in protein chemistry are to be found in many standard text books; the essential elements are laid out perhaps most succinctly in Dixon and Webb (15). In such procedures the changes in physical conditions will certainly result in the denaturation of some proteins and possible alterations in the characteristics of the proteins of interest. Clearly, initial studies will have to be performed to verify that the required product is not damaged by the process, but fortunately with these techniques based on differential solubility it is a simple matter to perform the necessary pilot experiments.

In some circumstances where the required protein will withstand normally denaturing conditions, treatments of this sort may be useful for diminishing or eliminating contaminants which constitute a potential biohazard in the product, as discussed earlier. Thus in the case of the purification of interferon α from buffy coat leucocytes it is reassuring to be able to treat the crude interferon at pH 2 for several hours at 4°C, since under these conditions the paramyxovirus used for induction of the interferon may be shown to be rendered completely non-infective (16).

Even more extreme treatments were possible in the purification of human interferon α from lymphoblastoid cells where precipitation under strongly denaturing conditions was employed (17).

5.3. Initial Separation by Use of Adsorbants

In very dilute protein solutions the physical formation and sedimentation of protein precipitates may proceed inefficiently and yields of product may consequently be poor. In such circumstances, the other classic protein purification procedure, selective adsorption and de-sorption, may be employed. Like the precipitation methods, these techniques were introduced over 50 years ago but still find a place in protein purification technology.

The most widely used agents for the selective adsorption of proteins are certain metal oxides and salts such as calcium phosphate gel, silica gel, alumina, titanium dioxide and aluminosilicates. The principles governing the binding of proteins to these agents are still poorly understood and the development of any process based on them has to be derived largely empirically. With polar adsorbants such as those mentioned above, ion-exchange or partition is the dominant interaction and accordingly elution is

TABLE I   Purification and Concentration of Interferon β by Adsorption on Aluminosilicate [Alusil]

	IF titre (units/ml)	Volume (litres)	Total units	Protein (mg/ml)	Specific activity	Recovery (%)	Purification
Crude culture fluid[a]	4736	45	2.13×10^8	0.2	2.36×10^5	100	1
Alusil eluate[b]	1.5×10^5	1.35	2.03×10^8	5	4.0×10^7	95	169

[a] Crude culture supernatant derived from human fibroblasts and containing interferon β was cooled to 4°C, adjusted to pH 4.0 and contacted with Alusil (J. Crosfield and Sons, Warington, Lancs, England).

[b] The Alusil was allowed to sediment, recovered, washed and the interferon was eluted by 0.7 M phosphate, pH 8.0, at 4°C (18).

achieved either by increasing the concentration of other ionic species which bind competitively to the adsorbant, or by altering the strength of the interaction by modification of pH, dielectric constant or salt concentration. Polyvalent anions at high concentration will normally elute all bound proteins and classically phosphate at a concentration of 0.1 to 1.0 M; pH 7 to 9 is used for this purpose.

Table I illustrates the application of this type of adsorption process to the recovery of interferon from the culture supernatant of cultured human fibroblasts, using aluminum silicate. The method was applied here specifically to avoid the problems associated with poor precipitation of proteins from dilute solution (18).

Practical problems may be encountered with several of these adsorbant products since their composition is complex, variable and largely unknown. In particular, calcium phosphate gels are notoriously difficult to re-produce and in the past such obscure phenomena as aging gel preparations for several months were invoked as a means of improving reproducibility. Latterly, commercial preparations have become available that minimise variations in performance, but it is still prudent to examine carefully the properties of each batch before it is included in a purification process.

Another practical problem derives from the physical nature of the adsorbants which may be gelatinous or extremely finely divided (as is the case of Alusil). This means that sedimentation may be difficult and that column applications may be impossible. However, recent work by Kent et al. (19) has produced several adsorbants in the form of rigid microspheres which have excellent mechanical properties while retaining their resolving power. These improved mechanical properties mean that the supports become much more amenable to column use and in this case impressive protein separation may be achieved.

5.4. Chromatographic Methods

Once the volume and concentration of a protein or polypeptide solution have been rendered manageable, some form of liquid chromatographic separation is by far the most powerful and also the most frequently employed step for further purification. There are many chromatographic methods available which make use of different physicochemical properties of proteins to effect a separation; the method selected will, of course, depend on the particular product and starting material in each case.

5.4.1. Adsorption Chromatography

As mentioned in the previous section, differential adsorption of proteins may occur to a number of solid phases and these effects can be used in column applications. Figure 1 illustrates the separation of five distinct nucleases in a spleen extract by column chromatography on hydroxyapatite (20). The availability of supports with improved mechanical properties (19) may encourage the further use of this methodology.

5.4.2. Ion Exchange Chromatography

Separation on ion-exchange resins is perhaps the most widely used type of adsorption chromatography presently applied to protein recovery. In ion-

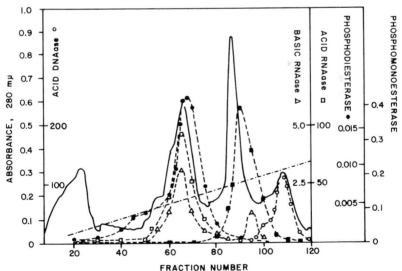

Fig. 1. Hydroxyapatite column chromatography of a mixture of spleen enzymes. Elution of five distinct nuclease activities by a linear phosphate gradient from 0.05 to 0.5 M. The solid line represents absorbance at 280 nm [from Bernardi et al. (20)].

exchange chromatography the proteins or polypeptides in solution are presented to an ion-exchange resin which bears charged groups. According to the pH of the solution, the proteins will themselves be more or less charged. Under conditions where they bear a charge opposite to that of the resin and at sufficiently low ionic strength, they may be bound to resin. Because attachment may occur via multiple sites this binding may be very strong.

Elution is achieved either by changing the pH to reduce the interaction between resin and protein or by increasing the salt concentration so that protein is displaced competitively. On columns it is usual to use gradient systems of pH, ionic strength or both to achieve this displacement, although stepwise changes of solvent may be used for both column and batch operation.

Ion-exchange resins are specifically designed polymer particles in which ion-exchange capacity, charge density, absence of non-specific adsorption and mechnical performance are all design properties which may be optimised. For protein work it is desirable to have lightly substituted resins, since this permits elution under mild conditions where there is less risk of denaturation. Another design parameter for protein and polypeptide work is the porosity of the support particle, which may determine the accessibility for ion-exchange of the charged groups to the macromolecules that it is wished to fractionate. The most widely used support polymers are currently substituted cellulose, dextrans and agaroses prepared by Whatman, Pharmacia and Biorad respectively, each of whom publishes useful application booklets for their products.

For a more general discussion of the different materials available and their practical application to protein and polypeptide separation, the reader is referred to Melling and Philips (*21*).

With ion-exchange chromatography in columns, very high resolution may be achieved particularly when closely controlled and reproducible programmed gradients are employed. Ion-exchange chromatography may also be employed as a batch process on a relatively large scale, using a basket-type centrifuge to recover and process loaded resin. Useful separation may still be achieved, and although some of the resolving power is of course lost, there exist examples where substantially purified products may be obtained by relatively simple batch operations of this type (*22*). This flexibility of scale of ion-exchange processes has meant that they are very frequently used in purification immediately after the initial concentration steps.

Other types of electronic interactions may also be used as the basis for chromatographic methods. Charge transfer chromatography, metal chelate chromatography and ion pairing chromatography all come into this category (*23*). Charge transfer reactions may occur between suitable electron receptors and donors. In proteins, electron-donating groups may be aromatic nuclei and the lone pair electrons from oxygen, sulphur or nitrogen atoms (24).

These interactions have recently been proposed as a basis for separation of proteins and peptides using transition state elements immobilized on gel supports such as Cu and Zn Sephadex (25). In certain circumstances the resolving power of this technology may be very good.

5.4.3. Hydrophobic Interaction Chromatography

Hydrophobic interactions result from the affinity between non-polar moieties in an aqueous environment and are thus in some senses perhaps the inverse of the ionic type of interaction discussed previously. When hydrophobic particles are introduced into aqueous solution, the structure of the water surrounding the particles becomes more ordered and the entropy of the mixture accordingly decreases. If several such particles coalesce, as might be the case with oil droplets, the quantity of orientated water molecules associated with the new particles is less than that associated with the several original smaller particles. This water is released from its ordered structure and the entropy of the system rises. It is this increase in entropy which provides a net negative free energy change when the hydrophobic interaction occurs. Hydrophobic sites are thus rather pushed together by the water that surrounds them than attracted directly to each other.

Although the "classic" structure of a protein involves a hydrophobic core surrounded by polar amino acids which are presented to the external aqueous environment, it is clear that there are both hydrophobic "islands" on the protein surface and hydrophobic "pockets" penetrating towards the interior of the molecule.

Several chromatographic supports exist with hydrophobic groups bound via flexible alkyl spacer arms which permit ready adaptation to the hydrophobic zones which may be presented by a protein.

Desorption in hydrophobic chromatography is favoured by lowered ionic strength or by decreasing the polarity of the solvent by such means as gradients of organic solvents. This approach to the chromatography of proteins has become particularly highly developed in HPLC applications.

5.4.4. Size Exclusion Chromatography (SEC)

In SEC (also called molecular exclusion chromatography, molecular sieving, gel permeation chromatography and gel filtration), molecules are passed through columns of a porous stationary phase into which small molecules may penetrate and thus be retained while large molecules are excluded and thus pass without hindrance. Ideally, the pore size of the support and the molecular weight of the molecules to be separated should be the only factors on which retention by the stationary phase depends. In practice, with many molecules adsorption and/or partition effects may occur which complicate this simple interpretation. However, these effects may be minimized so sucessfully that SEC is one of the most useful methods in general use for the determination of molecular weight. In addition to the separation of mixtures of macromolecules, SEC is widely used to eliminate low-molecular-weight contaminants of protein solutions, for example in desalting applications. The general field of SEC is discussed in detail by Yau *et al.* (26).

The development of stationary phases for SEC has been an area of intense activity. The first generation of supports were hydrated gels of dextran (Sephadex) and later of agarose (Sepharose) or polyacrylamide (Biogel). Problems arose with the fragile nature of such gels, especially when the pore size was increased to permit separation of such large moieties as viruses and subcellular particles. Under such conditions, the lack of mechanical strength of such inflated gels led to compacting of the gel bed and loss of flow rate. In particular, the tendency towards higher speeds of separation with the attendant higher operating pressures created a need for other support materials, of which the prototype was the controlled pore glass developed by Haller (27), and pore sizes up to 4000 Å corresponding to a molecular weight of 5×10^6 have been achieved.

In fact the silica-type supports contain surface hydroxyl groups able to interact with biopolymers, so that pure SEC on simple inorganic support phases is not practically achievable. In certain cases these seconday interactions have been used to achieve the primary separation (28).

The surface modification of glass and silica supports to optimise protein separations has become a development area in its own right and chemically modified silica supports are now very widely used in HPLC separations of biopolymers. This development will be discussed briefly in the section on HPLC.

5.4.5. Affinity Chromatography

In general usage, affinity chromatography is the system whereby a ligand with a biological specificity for the molecule of interest is fixed on an inert support in such a manner that it remains accessible for specific interaction with that molecule.

Thus a crude mixture containing the desired molecule may be run into an affinity column which will retain only that molecule, all impurities being removed by washing prior to specific elution of the product. Thus affinity chromatography comes closest to the ideal purification process where a single step can give a product which is substantially pure [reviewed by Lowe (29)]. It has been remarked by Porath and Dahlgren-Caldwell (30) that the term affinity chromatography may not be entirely appropriate for a system of chromatography based on biological recognition, such biospecific properties

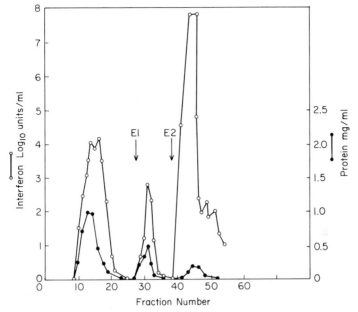

Fig. 2.   Purification of partially purified human fibroblast interferon by lectin affinity chromatography on concanavalin A–Sepharose. 2×10^9 units of interferon were loaded onto a 2-ml column in 0.05 M tris, 1.0 M sodium chloride, pH 7.3. Elution was in two stages, 0.1 M methyl mannoside (EI) and 0.1 M methyl mannoside in 25% propanediol (E2). Interferon elutes only at E2, suggesting a hydrophobic interaction with the lectin in addition to the sugar affinity. Pooled peak fractions had a specific activity of 10^8 units/mg, representing a purification of 200-fold (18).

resulting often from the combined effects of several of the types of molecular interactions or affinities which have been discussed so far. In some examples of affinity chromatography the affinity is type specific, for example the binding of glycoproteins to immobilized lectins (Fig. 2) or the binding of different enzymes to immobolized nucleotides (28). In these cases elution can be achieved by washing the column with buffer containing an excess of the bound sugar or nucleotide respectively.

The near-absolute specificity of some biological interactions may also be employed. Thus enzymes may be purified on columns whose stationary phase contains their specific substrate (31) or sometimes a specific inhibitor (Fig. 3) (32) and hormones have been purified on columns of their immobilized receptor and vice versa (33).

In the separation of biopolymers which do not have a known biospecific binding capacity or for which it is difficult or impossible to obtain the ligand in a pure state it is possible to use immunological recognition in an immunoaffinity column, where specific antibody to the molecule of interest is immobilized on the stationary phase (34).

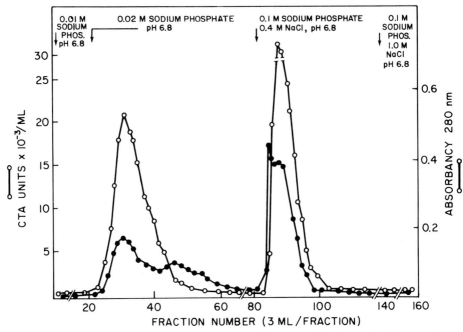

Fig. 3.   Affinity chromatography of urokinase using an immobilized specific inhibitor. 17.4 mg of urokinase in 0.01 M sodium phosphate, pH 6.8, was applied to a 0.9 × 20 cm column of Sepharose E–aminocaproyl–agmatine. Elution of different molecular species of urokinase was achieved by a stepwise increase in ionic strength as indicated [from Johnson *et al.* (32)].

In the situation where neither the ligand nor the required molecule may be obtained in a sufficiently pure state for the production of specific antibody, the new technology of monoclonal antibody production may provide a solution. Here hybrid antibodies producing cells are selected by cloning for their capacity to produce antibody of the required specificity (6). Accordingly, pure antigen is not essential for the production of monospecific antibody. This approach was applied by Secher and Burke (35) to the difficult problem of the purification of interferon α from lymphoblastoid cells. An impressive single-step purification of the order of 5000-fold was achieved (Fig. 4) and anti-interferon antibody produced by this technology is now available commercially and is widely used in interferon studies.

Generally speaking, one of the advantages of affinity chromatography is the mild conditions which can usually be employed for the desorption of bound molecules; as mentioned previously, under ideal conditions an excess of the ligand will produce specific elution without subjecting the column to extremes of pH or ionic strength. In the case of antibody affinity columns, that is, however, more difficult to accomplish since low pH, high salt or chaotropic agents are needed to disrupt the antigen/antibody complex.

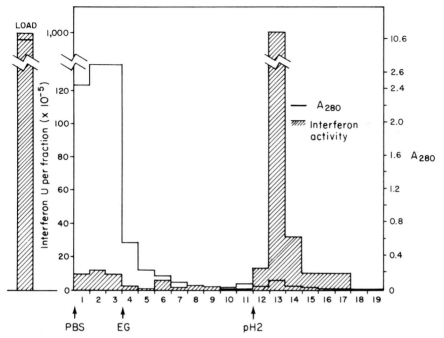

Fig. 4. Affinity chromatography of partially purified lymphoblastoid interferon using monoclonal anti-interferon antibody immobilized on CNBr-activated Sepharose. 10^8 units of Namalwa interferon was loaded onto a 0.5-ml column. The column was washed successively with PBS and with 9 M ethanediol, 0.34 M NaCl and 0.0075 M Na phosphate buffer, pH 7.4 (EG), and then the interferon was eluted with 9 M ethanediol, 0.3 M NaCl and 0.1 M citrate (pH 2). [From Secher and Burke (35) Reprinted by permission from *Nature*, **285**, 446–450. Copyright © 1980 by Macmillan Journals Limited.]

Recent developments have, however, made it possible for proteins to be eluted from antibody affinity columns in the same mild conditions as those used in, for instance, enzyme/substrate affinity columns (36). This is the result of the happy coincidence of the almost simultaneous realisation of several distinct technical approaches. Thus, using monoclonal antibodies which react with a single antigenic site on a protein molecule, it is possible to isolate the region of polypeptide chain recognized by the antibody. This will probably be 10–15 amino acids long and with modern sequencing techniques its primary structure may be rapidly determined on the picomolar scale (37). At the same time peptide synthesis technology is now sufficiently advanced to permit the production of useful quantitites of peptides of this length. The means therefore exist to achieve specific elution by soluble ligand as in the other cases mentioned. This approach has been achieved in practice in the purification of polyoma virus T antigen by Walter *et al.* (36)

and promises to be useful in the purification of rare proteins difficult to obtain by other means.

Having pointed out the advantages associated with affinity systems in terms of specificity, potential one-step operation and mild elution conditions, it is perhaps surprising that affinity systems are not much more widely used on an industrial scale. The reasons normally given for this are the relatively high cost of affinity supports, the relatively low capacity of many affinity supports and the observed fact that affinity columns do not always work well with crude starting materials. The first two of these disadvantages are, as already mentioned, less serious when we consider low-volume, high-value biologicals. It is significant that the first two commercialized products from animal cells, lymphoblastoid interferon and urokinase, both have affinity steps in their purification after initial clean-up and concentration processes (32, 38).

5.4.6. High-Performance Liquid Chromatography (HPLC)

All of the charomatographic modes discussed above may be operated using HPLC systems. The advantages of HPLC in the analytical field are well known and include great resolving power, reproducibility where conditions are appropriately controlled and rapidity of separation (minutes rather than hours).

Preparative adaptations of HPLC systems are becoming available, although their acceptance may be limited by price considerations since preparative columns of 1- to 4-cm diameter packed with 10-μm high-performance supports may easily cost up to \$10,000. For this reason it has been proposed that medium-performance liquid chromatography (mplc) supports will be developed for preparative use with particle sizes in the region of 37–74 μm in order to conserve many of the performance advantages of HPLC without incurring prohibitive costs (39). In any event it should be remembered that in terms of grams of proteins, a great mass is not required when we consider the purification, even on a commercial scale, of many of the very highly active biological mediators now being considered as candidate products from animal cells (for example interferons, vaccinating protein preparations etc.).

HPLC as applied to proteins is a relatively recent development. The principle and the desirability of this separation method have long been recognized but several different technical developments had to be awaited before the development of supports with appropriate adsorption properties became possible. Requirements for this application include physical stability to withstand high-pressure applications, and particle structures permitting adequate loading in adsorption chromatography and adequate penetration by large molecules in SEC. The application of HPLC techniques to protein separation has recently been reviewed by Clark and Kricka (11). An example

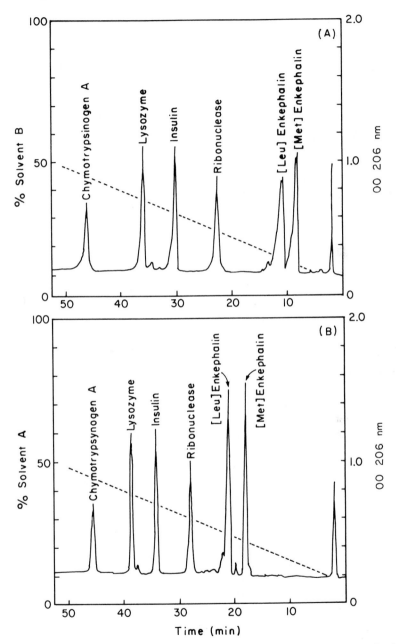

Fig. 5. Separation of a mixture of peptides and proteins containing 50 γ each of the components shown in the diagram on μ Bondapak phenyl alkyl support (A) and μ Bondapak C₁₈ (B). Solvent A, 0.05% TFA in water; solvent B, 0.05% TFA in acetonitrile; gradient, 10% solvent B to 60% solvent B over 1.0 hr at flow rate 2.0 ml/min [from Henderson *et al.* (*44*)].

of the high resolution and speed of separation achievable with HPLC of proteins is shown in Fig. 5.

As has already been mentioned in the section on SEC, direct use of inorganic supports for chromatography based on molecular size has met with limited success because of secondary interactions between the molecules to be separated and the column support material. Glass and silica supports, for example, have given problems due to the attractive or repulsive effects on biopolymers from negatively charged surface silanol functions. To avoid these difficulties, supports have been developed with chemically modified surfaces where the adsorption properties of the support are determined by groups, polar, apolar or ionisable, which are bonded to the particle surface. These "bonded phases" thus mask the underlying properties of the support particle and can be tailor-made for any separation application. Excess silanol groups, not needed for binding the new surface groups, are blocked by "end-capping" reactions.

It is usual to distinguish between two modes of operation: "normal phase," where the bonded phase is polar and the mobile phase less polar, and "reversed phase," where the opposite is true.

In normal phase work, the bonded phase may be used to provide a neutral hydrophilic surface that neither attracts nor repels biopolymers. Glyceryl-propyl bonded phases are widely used in this connection and are commercially available on silica (Syn Chropak GPC) or on glass (Glycophase CPG) supports. This type of neutral hydrophilic phase permits highly successful SEC of proteins. The TSK–SW gels from the Toya Soda Company, Japan, are also designed for this objective.

For ion-exchange work all of the usual ion-exchange groups may be bonded to inorganic or organic stationary phases. In reversed phase operation, hydrophobic moieties such as alkyl groups are bonded to the support surface (40). An example of a commercially available system is the Supelcosil range of stationary phases from Supelco, which is available with C8, C18 or diphenyl functional groups bonded to spherical 5-μm silica. Elution in the reverse phase mode is achieved by modifying the polarity of the mobile phase with, for example, propanol, isopropanol, acetonitrile or methanol. One of the problems with this approach is the denaturation of some proteins under these elution conditions and development work is proceeding in this area (41). However, many proteins do remain undenatured and impressive purifications of useful molecules have been reported including a highly successful purification of interferon from leukemic lymphocytes (42–44).

5.4.7. Preparative Electrophoresis

In general, electrophoretic systems, while very widely used in analytical work, have not found much application in preparative scale purification.

Preparative (hundreds of milligrams) scale isotachophoresis has been described but technical problems relating to sample recovery cannot be considered fully resolved (45). A truly preparative scale electrophoretic system has been developed by Mattock *et al.* (46), where separation is performed in free solution using an apparatus in which zone distortion due to convection in the liquid is eliminated by establishing laminar flow conditions in a narrow annulus. An interesting aspect of this system is its virtually complete independence of particle size, separation having been achieved under the same conditions of small peptides at one extreme and of mixed populations of viable lymphocytes at the other. Numerous protein separations have been achieved with this apparatus, including that of tissue culture disaggregating activity (47) and of factor VIII (48).

5.5. Sequence of Purification Steps

Obviously, the specific sequence of processes adopted will depend on the particular separation problems encountered, which will be determined by the scale of the operation, the concentration of the desired product, the nature of contaminants etc. Generally, however, the sequence of sections in this chapter will probably be followed, passing from clarification if needed to initial enrichment, an intermediate purification step on a batch scale probably by ion-exchange or adsorption, and finally purification by a high-resolution method to the standards of purity required for the particular product.

The ideal of a single-step purification will often be unobtainable because of low starting concentrations and large initial volumes. Clearly, any developments which improve concentration and purity in the starting material will simplify the extraction problem. It is in this direction that current studies on the genetic manipulation of animal cells promise to lead us.

6. PURIFICATION OF VIRUSES

6.1. Generalities

In the production of virus, as in the production of proteins, it is important to consider purification problems as part of an integrated production process rather than as an isolated technical exercise. Thus the most fundamental aspects of the process such as the choice of producing cells and of the production conditions are likely to have a major impact on the purification phases. Obvious factors to consider in the production of crude virus are the maximum obtainable virus concentration, the absence of impurities likely to give problems later on (such as high levels of bovine serum) and, above all, the absence of molecules of allergenic potential such as certain antibiotics, which may be effectively impossible to remove definitively by the purification methods available. Generally also, the cell culture used will have been chosen to avoid any overt biological hazard.

Purification of virus generally proceeds in three phases: (1) clarification of the cellular lysate, (2) concentration of the viral suspension, (3) purification proper on the concentrated suspension.

6.2. Clarification

Clarification to remove cell debris is essential, both to simplify subsequent purification and to permit the proper action of inactivating agents. Both filtration and centrifugation methods are used, although on the industrial scale filtration is often troubled by filter blinding, requiring either the use of filter aids or the pretreatment of the crude lysate. Often the two methods are used in series, continuous flow centrifugation being followed by filtration. The technical problems and possibilities of these methods are discussed fully in Chapter 5, of this volume.

6.3. Concentration and Purification

6.3.1. Precipitation Methods

Precipitation by neutral salts or by organic solvents was the earliest method used in virus purification, and these methods have been employed in relatively large-scale production of purified virus, although many types of virus are more or less unstable in the presence of organic solvents (49, 50).

Fractional precipitation by the milder polyethylene glycol (PEG) has also formed the basis of several separation processes (51) and has more recently been used on the industrial scale for the production of foot-and-mouth disease virus (FMDV) (52). In general, however, the precipitation methods do not find favour industrially, since they exhibit several important disadvantages:

1. They introduce extraneous components into the system which must then be removed at a later stage by dialysis or gel filtration.
2. The selectivity of the methods is not high and the purification achieved is often marginal.
3. Stability of the virus is often adversely affected.
4. On the industrial scale, the large quantities of precipitant required create practical problems.

6.3.2. Adsorption–Elution

In principle, the same range of non-specific adsorbants may be used with viruses as were discussed in the section on protein isolation, and agents such as calcium phosphate gel have been used since the earliest days of virus purification (53). Viruses do, however, exhibit rather specific binding to certain types of adsorbant and extensive use has been made of these.

Thus Adamowicz *et al.* (54) have developed a purification and concentration procedure based on the adsorption of virus particles to an insoluble complex formed from high-molecular-weight ethylene oxide polymer and calcium ions. This ternary complex is precipitated and recovered by centrifugation. The complex may be dissociated under very mild conditions, by the use of a calcium chelating agent such as EDTA or citrate. This method has been further developed for use on the several thousand litre scale for FMDV production. By the application of three adsorption/desorption cycles, it is possible to obtain virus 50% pure, 1000-fold concentrated in yields better than 90% (55).

The mechanism of this "calcium bridge" binding to polyoxyethylene has not been fully elucidated but it appears to be in some way analogous to the natural binding of virus to the surface of cells. This method appears not to be very specific for type of virus and has been used successfully in our laboratory for the purification of rabies virus, polio, and pseudorabies virus (Aujeszky's disease).

Ion-exchange procedures may also be applied to virus purification and some success has been claimed [for example with polio (56) and with a myxovirus (57)]. However, despite the relatively easy adaptation of ion-exchange separation to large-scale batch operation, these methods seem not to be used significantly. One of the reasons for this is certainly the irreversible binding to resin which has been observed with several virus types. However, use has been made of DEAE–Sephadex specifically for the removal of gamma globulin in viral preparations (57a).

6.3.3. Ultrafiltration

Ultrafiltration is the technique whereby a mixture of molecules of different molecular weights in a liquid phase are separated by passage under pressure across a membrane of defined pore size. Separation is thus achieved primarily on the basis of molecular size although shape and interaction with the membrane surface may be superimposed on this. Ultrafiltration may be used for three types of operation: (1) separations of macromolecules, (2) concentration of macromolecules, (3) changing the aqueous phase in which macromolecules are dispersed (diafiltration).

Generally the fractionation of similar macromolecules according to their molecular weight is seldom practised by using ultrafiltration because of the relatively large band width obtained with ultrafiltration membranes. The resolving power of the technique is relatively poor. However, ultrafiltration is very widely used for the concentration of viruses and for changing the aqueous phase of virus suspension prior to further purificaton. The process

TABLE II   Concentration of FMD Virus by Ultrafiltration[a]

Sample	Volume (litres)	Complement fixation titre	Infectivity titre
O_{VI} FMDV			
Clarified suspension	100.2	14	8.7
Concentration	2.1	572	10.2
C_8 FMDV			
Clarified suspension	100	14	7.5
Concentration	2.5	480	9.1

[a] Virus—clarified supernatant from BHK suspension culture; Millipore Pellicon cassette system—two cassettes, filter surface 0.9 m[2]; flow rate, 40 litres/hr; concentration time, 2–3 hr (59).

of dialysis may be operated either discontinuously (with successive concentrations of the virus and subsequent dilutions in fresh buffer) or continuously with solvent feed to maintain constant volume. The latter approach, known as diafiltration, is to be preferred since, although it results in greater consumption of solvent, it avoids the problems of membrane blinding and viscosity which may arise with highly concentrated viral suspensions.

Ultrafiltration is the method of choice for the concentration and dialysis of virus on a large scale. Among the important advantages are:

1. There is no addition to extraneous reagents which require subsequent removal.

2. Defined conditions of pH and ionic strength may be maintained throughout the process.

3. Suitable apparatus exists for the large-scale, continuous operation of the technique under contained conditions [Amicon hollow fibre system or Millipore "Pellicon" cassette system (58)].

4. Membranes may be re-used several times without major loss of performance (it is normal practice to prefilter clarified viral suspensions through a 1-μm filter to avoid possible premature blockage of the ultrafiltration membrane).

Tables II and III show results obtained in our laboratory for the concentration by Pellicon ultrafiltration of FMDV and Aujeszky virus respectively (59). Similar results may be obtained with the Amicon hollow fibre system (60–62).

6.4. Purification Proper

6.4.1. Purification by Density Gradient Centrifugation

Once the virus suspension has been clarified and concentrated to a suitable volume for further purification, the technique of density gradient cen-

TABLE III   Concentration of Aujeszky Disease Virus by Ultrafiltration[a]

Sample	Volume (litres)	Complement fixation titre	Infectivity titre
Preparation I			
Clarified suspension	83	14	8.6
Concentrate	2.26	514	9.9
Preparation II			
Clarified suspension	40	8	8.7
Concentration	1.56	205	10.1

[a] Virus—clarified suspension from BHK suspension culture; Millipore Pellicon cassette system—one cassette, filter surface 0.45 m^2; flow rate, 21 litres/hr (59).

trifugation is arguably the most useful separation tool available. For practical purposes, the preparative use of centrifugational methods is at its most developed for the separation of viruses and subcellular particles.

The rate of sedimentation of particules in a centrifugal field is influenced by particle size and shape. Better resolution of particles of similar sedimentation rate may be achieved if use is made of a density gradient. This has the dual advantage of stabilizing the zones against distortion due to convection and adding, as a further separation parameter, the buoyant density of the particle.

In the density gradient, particles continue to sediment until they reach their isopycnic density. Before they arrive at this level in the gradient, separation will occur according to their sedimentation rate, which depends on the square of the particle diameter, the buoyant density and, to a lesser extent, the shape of the particle. The theory of density gradient separations has been discussed in detail, initially by Svedberg and Pedersen (63).

It is usual to distinguish two modes of operation in density gradient centrifugation. In the *rate zonal technique* the sample is layered into a pre-formed gradient, the density of which nowhere exceeds that of the particles to be separated. Particles thus sediment through the gradient according to their sedimentation rate. Centrifugation must be stopped before they reach the bottom of the tube or the wall of the rotor. In the *isopycnic technique* the gradient is sufficiently dense that the particles of interest will arrive at their isopycnic density and form stable zones within the gradient. Separation depends only on density difference and is independent of time.

In fact, in most practical separations these two effects are combined, some particles banding at their isopycnic densities while others continue to move through the field. The successful application of density gradient methods to the purification of virus is possible largely because of the differences in density between the various virus groups and the likely contaminating material derived from tissue culture cells (Table IV).

TABLE IV Approximate Densities of Macromolecules
in Sucrose Solutions[a]

Macromolecules	Density (g/cm^3)
Golgi apparatus	1.06–1.10
Plasma membranes	1.16
Smooth endoplasmic reticulum	1.16
Intact oncogenic viruses	1.16–1.18
Mitochondria	1.19
Lysosomes	1.21
Peroxisomes	1.23
Plant viruses	1.30–1.45
Soluble proteins	1.30
Rhino and enteroviruses	1.30–1.45
Nucleic acids, ribosomes	1.60–1.75
Glycogen	1.70

[a] After Griffiths (64).

Several media are used for the generation of density gradients, the most usual being sucrose and caesium chloride. Since the density of most viruses is less than 1.3, sucrose solutions (density 1.32 at 66% at 20°C) are very widely used for forming gradients for viral separations. Sucrose has the advantage also of being cheap and of maintaining the stability of most viruses. In certain circumstances the relatively high viscosity of sucrose gradients may be a problem. High viscosity tends to minimize zone broadening due to diffusion, but also prolongs centrifugation time. Caesium chloride is used when more dense solutions or lower viscosities are required and finds particular use in isopycnic centrifugations.

An important practical point in the preparation of gradients is the use of sufficiently pure reagents to avoid spurious adsorption at the wavelengths used for the detection of viral bands. Also, in the case of sucrose, grades must be used which assure the absence of contaminating nucleases.

There are two types of rotor available for preparative scale centrifugation of virus preparations: zonal rotors and continuous flow rotors. Each of the principal manufacturers produces a range of rotors of different capacities, maximum speeds of flow characteristics (Table V). In zonal rotors the capacity of the original bucket-type ultracentrifuges has been increased by producing a large cylindrical cavity to replace the original discrete tubes. The cylindrical cavity is divided into sectors by vanes attached to the rotor core, the design of which varies according to the specific hydrodynamic properties required. The cylinder is closed by a lid, which in turn carries a rotating seal assembly, either fixed or removable, through which fluid may be passed

TABLE V   Some Examples of Zonal and Continuous Flow Rotors

| | A. Zonal rotors | | | | |
Rotor	Maximum speed (rpm)	Maximum (g)	Sample volume (ml)	Rotor capacity (ml)	Path length (mm)
Beckman					
JCF-Z	20,000	40,000	50–300	1,900	69
Z 60	60,000	256,000	10–15	330	48.2
Ti 14	48,000	172,000	20–50	665	53
Al 14	35,000	91,300	20–50	665	53
Ti 15	32,000	102,000	50–200	1,675	75
Al 15	22,000	48,100	50–200	1,675	75
M.S.E.					
B 14 Ti	47,000	165,000	20–50	650	54
B 15 Ti	35,000	121,800	50–200	1,670	76
B 29	35,000	62,000	50–150	1,430	76

| | B. Continuous flow rotors | | | | |
Rotor	Maximum speed (rpm)	Maximum (g)	Flow rate (litres/hr)	Rotor capacity (ml)	Path length (mm)
Beckman					
JCF-Z	20,000	40,000	~2	660	17
CF 32 Ti	32,000	102,000	<9	430	9
M.S.E.					
B 20	35,000	122,000	—	305	9.5
Electronucleonics					
K II	35,000	90,000	10–15	3,200	11.4

while the rotor is spinning. Normally, gradient forming, loading of sample and sample unloading are done at a low speed sufficient to stabilize the gradient, say 2000–3000 rpm, and the rotor is accelerated to its operating speed of 20,000–60,000 rpm for the separation proper. Specific instructions for the use of individual rotors are given in the manufacturers' handbooks.

Several different types of rotor core are available which permit the use of different operating techniques (Table VI). With the standard rotor core, the gradient is loaded at the periphery of the rotor and sample and overlay solutions are applied at the centre. Samples are then recovered from the centre by displacement by a heavy fluid pumped in at the periphery. With the B29 rotor the sample may be recovered at the centre or at the periphery since the rotor design permits non-turbulent flow of liquid through the edge parts of the rotor (64). An advantage of this system is that the separated sample may be displaced from the centre by a light solution, which results in important economies of the density gradient material. The third type of

TABLE VI   Different Types of Zonal Rotor Core

Type of core	Gradient loading	Sample loading	Recovery of gradient	Loading speed (rpm)
Standard core	Peripheral	Central	Central	2000–3000
B29 core	Peripheral	Central	Central or peripheral	2000–3000
Re-orienting gradient core	Central	Central	Central	Static

core, the re-orienting gradient core, permits loading and unloading of the rotor at rest, the gradient being automatically re-oriented in the radial direction when the rotor is spinning. This approach is used to avoid destruction of particles which would be damaged in rotating seal assemblies and may also limit the risks associated with the centrifugation of hazardous materials. However, to obtain satisfactory resolution, long and critically controlled periods of acceleration and deceleration are required so that dynamic loading is the method of choice where possible (64).

Either rate zonal or isopycnic separations may be performed in zonal rotors. Sample size is limited by the rotor volume (Table V) and resolution is limited by the thickness of the band of sample that is applied. This latter imposes further limitations on sample size. For larger-scale separations, continuous flow rotors must be used. Continuous flow rotors also comprise a spinning cylinder, but in this case core design is quite different to permit the continuous delivery of sample to the light end of the gradient while the centrifuge is at full operating speed. The particles move from this fluid stream into the gradient and migrate to their isopycnic density while the depleted fluid stream is continuously eliminated from the rotor. Clearly a balance has to be achieved between sample flow rate and the rate of removal of particles from the input stream (the actual banding, being isopycnic, is not time-dependent). If necessary the sample may be recycled through the centrifuge to achieve this result. Because of the long run time that may be involved, adequate cooling must be provided for the sample and for the centrifuge.

Separation path length is very short in continuous flow rotors (Table V) and consequently resolution of multicomponent mixtures cannot normally be achieved, products being recovered from the gradient as a single band or, exceptionally, as a pellet from the rotor wall. Pelleting is rarely performed due to the inability of some viruses to withstand this treatment, but also because of the difficulty of resuspension of pelleted virus and its recovery from the rotor.

Purification of virus by density gradient techniques is widely practiced in many laboratories and examples of the use abound in the literature. Thus, for example, commercial scale purification of rabies virus has been performed using a zonal method (65) and also using isopycnic banding in a continuous flow system (66). An interesting and complex application of centrifugational technique has been used in the preparation of "first-generation" hepatitis B vaccine from blood (67). Here multiple centrifugation steps are employed interspersed by ultrafiltration steps to reconcentrate the harvested viral zones. Separations of pathogenic viruses of this type give rise to a serious biological hazard. By its nature the centrifuge is prone to mechanical mishaps such as seal failure or occasionally rotor failure and, because of the high kinetic energy involved, containment of the resultant debris and aerosols constitutes a major engineering problem. In addition, the complex nature of the equipment requires regular overhaul and the subsequent possible exposure of maintenance staff to hazardous conditions. Considerable effort has been devoted to the development of automatic and fail-safe systems. A useful discussion of this subject has been published by Hellman et al. (68).

6.4.2. Affinity Chromatography of Viruses

When the volume and protein load of a virus preparation are sufficiently reduced, column techniques may be considered for the later purification steps.

As with simple proteins, affinity chromatography offers the possibility of a substantial single-step purification. Affinity separations have been attempted for several viruses on the basis of lectin binding of their glycoproteins, Thus concanavalin A has been successfully used for the concentration and purification of enveloped RNA viruses (69) and oncornaviruses (70).

More recently a purification has been achieved with a togavirus (hog cholera) which proved otherwise difficult to purify. In the case there was no affinity for concanavalin A (which binds mannoside residues) but a substantial affinity for the galactoside binding lectin obtained from the castor bean (71). Chromatography of the PEG-precipitated virus preparation was performed on Sepharose 4B–lectin columns, and specific elution was achieved using 0.1 M galactose. Good purification and retention of infectivity are reported.

At present totally specific affinity chromatography of viruses is limited to the use of antibody affinity columns, although the probable identification of a specific cellular receptor in the case of rabies virus (the acetylcholine receptor) and the existence of a competing ligand (acetylcholine) (72) suggest the possibility of eventual separation based on immobilized receptor proteins.

Antibody affinity columns have not so far found great use in virus purification work largely because many viruses are not stable in the conditions

TABLE VII   Concentration and Purification of Polivirus Type I by Immune Adsorption and Elution[a]

	Volume (ml)	Du/ml	Serum content (μ/ml)		Recovery (%)
			Albumin	Immune globulin	
Virus suspension (17× concentrated)	600	1675	±6600	±1400	100
Immune adsorbant eluate	190	3311	<0.06	3.75	63
DEAE–Sephadex eluate	194	2926	<0.06	<0.5	60

[a] From van der Marel et al. (57a).

needed to dissociate the antigen/antibody complex to achieve elution (high salt, low pH, chaotropic agents). As discussed in Section 5.4.5, the advent of readily available small peptide specific eluting agents may change this, and use has already been made of this approach in the purification of at least one viral product. Thus Walter et al. (36) have used a synthetic hexapeptide corresponding to the C terminal of the polyoma virus medium tumor (T) antigen to obtain specific elution of the antigen from immunoaffinity columns. In this case, however, detergent was also needed to release the bound antigen.

The problem of instability has been ingeniously solved by some workers without the use of such sophisticated and, at present, costly eluants by making use of the two columns in tandem. The eluted virus passes immediately into a gel filtration column which permits rapid removal of the potentially inactivating buffer system. Van der Marel et al. (57a) have successfully applied this method to the purification of poliovirus. As shown in Table VII, a very substantial purification with respect to serum protein was achieved by this technique and recoveries of up to 80% were reported. An additional ion-exchange step on DEAE–Sephadex was also shown to further reduce the level of contaminating gamma globulin.

A problem reported in this study is the great difficulty involved in sterile working with the column system when the support medium cannot be effectively subjected to sterilizing agents, as is the case with bound antibody and bound lectin. Even the supports themselves in the case of Sephadex, Sepharose and Agarose pose problems because of their instability to effective sterilization. With such chromatographic systems there is no easy solution to this problem, although the use of sterile technique and of sterile depyrogenated equipment and buffers, and prolonged washing of the columns with these buffers, will reduce the bacterial loading to levels which, depending on the system, will probably be acceptable. Obviously each separation problem will have to be specifically evaluated in this regard.

6.4.3. SEC of Viruses

Column separations based on particle size may be performed with viruses as with soluble proteins (Section 5.4.4). Separation of viruses from relatively small molecules may be easily achieved. Thus gel columns may be used in place of dialysis or diafiltration for changing the salt concentration and composition of virus suspensions or for removing excess sucrose. Equally, gel filtration columns have been used to remove serum proteins from poliovirus (61) and from rabies virus (60) using Sepharose 4B.

Separation of populations of virus has been difficult hitherto because of the lack of gels with a sufficiently large pore size and with sufficient mechanical rigidity. Also, as mentioned previously, the difficulty of sterilization of polysaccharide-based gels is a serious drawback to their use in virus purification. Recently, however, large-pore silica-based particles have become available (e.g. Spherosil, Rhone-Poulenc) which permit chromatography of viruses, are sufficiently rigid to permit operation at a high flow rate and may be steam-sterilized. Further interest in the use of this type of material is to be expected. An industrial process has been described using large columns of Spherosil for the automated purification of influenza virus from allantoic fluid (73).

7. EVALUATION OF PURITY

The determination of the purity of complex entities such as protein molecules or virus particles presents a number of technical difficulties. The very complexity of the system means that the assignment of discrete physicochemical properties may not be unequivocally achieved, and the discriminatory power of the analytical methods available may not permit the detection of low levels of contaminants. Thus when we consider biological products, purity is not always definable in absolute terms.

Ultimately purity is a concept which has to be related to the eventual use of the product, and when we consider that the achievement of safety in clinical use is a primary goal in the purification of biologicals, the situation may appear clearer. Thus we normally define purity in these agents as the absence of components in the final preparation which may contribute to the development of undesirable secondary responses in the recipient.

Thus in the case of virus preparations, where the very complex nature of the particle may make many physicochemical criteria of purity difficult to evaluate, it is primarily the removal of unwanted immunogenic or infectious material which is the aim of purification. In this case, serological methods to illustrate the absence of potential contaminants such as animal sera may be very useful, and may, coupled with other validated physical, chemical or biological tests, form the basis for a specification of the product (5).

At present the majority of cell substrates used in virus vaccine production are the products of cell lines which have been carefully screened to establish the absence of any biological hazard associated with the cells themselves (5). However, several examples now exist where virus is produced in cells other than the "normal" primary and diploid cell strains. Thus Horodniceau *et al.* (74) have proposed the use of HeLa cells for poliovirus production while Merten *et al.* have proposed the production of hepatitis B surface antigen in the human hepatoma cell PLC/PRF/5 (75).

In both of these cases, extensive analysis of the purified product has shown the absence of exogenous nucleic acids in the vaccine preparations within the present limits of detection. However, such approaches always leave the possibility of doubt regarding absolute safety because they are inherently limited by the sensitivity of imperfect assay procedures. In this connection it is of interest to note that it has been suggested that Q_β replicase is capable of autocatalytic synthesis of RNA because synthesis may proceed in the absence of added template. It is only very recently that the enzyme has been sufficiently purified under denaturing conditions to indicate that the so-called autocatalytic synthesis was indeed template directed by a previously undetected contaminating functional RNA (76).

With single proteins or peptides it is more feasible with the available purification and characterization methods to move towards obtaining a defined chemical entity as end product. Although this may not yet be practical in many cases, instances exist where very high standards of purity are demanded for clinically used proteins.

In the case of insulin, for example, it has been recommended that, in addition to a biological test of potency, an HPLC method should be used to verify identity and determine purity, and that "at least two radio-immunoassays should be used to identify specific extraneous materials to define purity" (77). Impurity limits in insulin have been set as low as 1 ppm pancreatic polypeptides and 10 ppm proinsulin.

In the case of interferon from lymphoblastoid cells, a specification at the limits of detection (around 1 ng per 10^6 units) has been set for the presence of DNA in the protein preparation (78).

In general, with the battery of extremely high-resolution techniques now available, and the existence of exquisitely sensitive methods to examine the purity of protein products, the tendency in the future must be towards chemically pure entities to guarantee the safety and efficacy of molecules derived from genetically modified animal cells.

REFERENCES

1. Finter, N. B., Fantes, K. H., and Johnson, M. (1977). Human lymphoblastoid cells as a source of interferon. *Dev. Biol. Stand.* **38**, 343–348.
2. Biliau, A., Joniau, M., and De Somer, P. (1973). Mass production of human interferon in diploid cells stimulated by poly IC. *J. Gen. Vitrol.* **19**, 1–8.
3. Cartwright, T. (1980). The clinical potential of human fibroblast interferon. *Dev. Antiviral Ther.* 249–264.
4. Barlow, G. H., Reuter, A., and Tribby, I. (1977). Biosynthesis of plasminogen activator by tissue culture techniques. *Vasc. Surg.* **11** (6), 406–412.
5. Cartwright, T. (1976). Characterisation of the Searle 17/1 fibroblast cell line. *Proc. NIH Workshop Cell Substrates Vaccine Prod., 1976*, pp. 184–197.
6. Kohler, G., and Milstein, C. (1975). Continuous culture of fused cells secreting antibody of pre-defined specificity. *Nature (London)* **256**, 495–497.
7. Wigler, M., Sweet, R., Sim, G. K., Wold, B., Pellicer, A., Lacy, F., Maniatis, T., Silverstein, S., and Axel, R. (1979). Transformation of mammalian cells by genes from prokaryotes and eukaryotes. *Cell* **16**, 777–785.
8. Finter, N. (1979). *In* "Human Interferon in the Clinic: Guidelines for Testing," NIH Workshop, 1979, pp. 3–21.
9. Zoon, K. C., Buckler, C. E., Bridgen, P. J., and Gurari-Rotman, D. (1978). Production of human lymphoblastoid interferon by Namalwa cells. *J. Clin. Microbiol.* **7**, 44–51.
10. Petricianni, J. C., Salk, P. L., Salk, J., and Noguchi, P. D. (1982). Theoretical considerations and practical concerns regarding the use of continuous cell lines in the production of biologics. *Dev. Biol. Stand.* **50**, 15–25.
11. Clark, P. M. S., and Kricka, L. J. (1981). High resolution analytical techniques for proteins and peptides and their applications in clinical chemistry. *Adv. Clin. Chem.* **22**, 247–296.
12. Dellamonica, C., Baltassat, P., and Collombel, C. (1977). Les dosages immunologiques: Principes et applications. *Lyon Pharm.* **28** (4), 289–303.
13. Moynihan, M., and Petersen, I. (1982). The monitoring of antigen levels during inactivated poliovirus vaccine production: Evaluation of filtration techniques. *Dev. Biol. Stand.* **50**, 243–249.
14. Bonin, O., Schmidt, K., Schmidt, R., Mauler, R., and Grushkaa, H. (1974). Calf serum content of culture grown viral vaccine and possibilities for its reduction. *J. Biol. Stand.* **2**, 139–141.
15. Dixon, M., Webb, E. C., Thorne, C. J. R., and Tipton, K. F. (1979). "The Enzymes," 3rd ed. Academic Press, New York.
16. Strander, H., and Cantell, K. (1966). Production of interferon by human leukocytes in vitro. *Ann. Med. Exp. Biol. Fenn.* **44**, 265–273.
17. Bridgen, P. J., Anfinsen, C. B., Corley, L., Bose, S., Zoon, K. C., and Ruegg, U. T. (1977). Human lymphoblastoid interferon: Large scale production and partial purification. *J. Biol. Chem.* **252**, 6585–6587.
18. Cartwright, T., and Thompson, P. (1979). Unpublished results.
19. Kent, C. A., Miles, B. J., Laws, J. F., and Thomson, A. R. (1982). Biological process separations using organic adsorbants. *Inst. Chem. Eng. Symp., 1982*, "Bioprocessing in the Eighties," Chap. 1.

20. Bernardi, G., Bernardi, A., and Chersi, A. (1966). Studies on acid hydrolases. 1. A procedure for the preparation of acid deoxyribonuclease and other acid hydrolases. *Biochim. Biophys. Acta* **129**, 1–11.

21. Melling, J., and Phillips, B. W. (1975). "Handbook of Enzyme Biotechnology," pp. 58–88, 181–202. Ed Awilson, Ellis Herwood, Chichester.

22. Stanworth, D. R. (1960). A rapid method for preparing pure serum gamma globulin. *Nature (London)* **188**, 156–158.

23. Lawrence, J. F., Frei, R. W. (1977). "Chemical Derivatisation in Liquid Chromatography." (Journal of Chromatography Library, Vol. 7) Elsevier, Amsterdam.

24. Poráth, J. (1979). *In* "Les Colloques de l'INSERM sur chromatographie d'affinté et intéractions moléculaires" (J. M. Egly, ed.), pp. 79–89. INSERM, Paris.

25. Poráth, J., Carlsson, J., Olsson, I., and Belfrage, G. (1975). Metal chelate affinity chromatography, a new approach to protein fractionation. *Nature (London)* **258**, 598–599.

26. Yau, M. W., Kirkland, J. J., and Bly, D. D. (1979). "Modern Size-exclusion Chromatography." Wiley (Interscience), New York.

27. Haller, W. (1965). Chromatography on glass of controlled pore size. *Nature (London)* **206**, 693–696.

28. Lowe, C. R., and Dean, P. D. G. (1971). Affinity chromatography of enzymes on insolubilized cofactors. *FEBS Lett.* **14**, 313–316.

29. Lowe, C. R. (1979). "An Introduction to Affinity Chromatography." Elsevier, Amsterdam.

30. Poráth, J., and Dahlgren-Caldwell, K. (1977). Protein immobilization and chromatography. *In* "Biotechnological Application of Protein and Enzymes" (Z. Bohak and N. Sharon, eds.), pp. 83–102. Academic Press, New York.

31. Berg, R., and Prockop, D. J. (1973). Affinity column purification of procollagen proline hydroxylase from chick embryos and further characterisation of this enzyme. *J. Biol. Chem.* **248**, 1175–1182.

32. Johnson, A. J., Soberano, M., Org, E. B., Hevy, M., and Schoellman, G. (1977). Urinary urokinase, two molecules or one. *In* "Thrombosis and Urokinase" (R. Paoletti and S. Sherry, eds.), pp. 59–67. Academic Press, London.

33. Vauquelin, N. H., Geynet, P., Hanoune, J., and Strosberg, A. D. (1977). Isolation of adenylate cyclase-free, β-adrenergic receptor from turkey erythrocyte membrane by affinity chromatography. *Proc. Natl. Acad. Sci. U.S.A.* **74**, 3710–3714.

34. Folkersen, J., Teisner, B., Ahrons, S., and Svehag, S. E. (1978). Affinity chromatographic purification of the pregnancy zone protein. *J. Immunol. Methods* **23**, 17–25.

35. Secher, D. S., and Burke, D. C. (1980). A monoclonal antibody for large scale purification of human leucocyte interferon. *Nature (London)* **285**, 446–450.

36. Walter, G., Hutchinson, M. A., Hunter, T., and Eckhart, W. (1982). Purification of polyoma virus medium size tumor antigen by immunoaffinity chromatography. *Proc. Natl. Acad. Sci. U.S.A.* **79**, 4025–4029.

37. Hunkapiller, M. W., and Hood, L. E. (1983). Protein sequence analysis: Automated microsequencing. *Science* **219**, 650–659.

38. Johnston, M. D., Christofinis, G., Ball, G. D., Fantes, K. H., and Finter, N. B. (1978). A culture system for producing large amounts of human lymphoblastoid interferon. *Dev. Biol. Stand.* **42**, 189–192.

39. Freeman, D. H. (1982). Liquid chromatography in 1982. *Science* **218**, 235–241.

40. Regmier, F. E., and Gooding, K. M. (1980). High performance liquid chromatography of proteins. *Anal. Biochem.* **103**, 1–25.

41. Barford, R. A., Sliwinski, B. J., Breyer, A. C., and Rothbart, H. L. (1982). Mechanism of protein retention in reversed-phase high performance liquid chromatography. *J. Chromatogr.* **235**, 281–288.

42. Van Der Rest, M., Stolle, C. A., Prockop, D. J., and Fietzek, P. P. (1982). Separation of human pro α1 (I) and pro β2 (I) procollagen chains by reverse phase HPLC. *Collagen Relat. Res.: Clin. Exp.* **2**, 281–285.

43. Rubenstein, M., Rubenstein, S., Familetti, I. C., Gross, M. S., Miller, R. S., Waldman, A. A., and Peska, S. (1979). Human leucocyte interferon: Production, purification to homogeneity and initial characterization. *Proc. Natl. Acad. Sci. U.S.A.* **76**, 640–644.

44. Henderson, L. E., Sowder, R., and Oroszlan, S. (1981). Protein and peptide purification by reversed-phase high pressure chromatography using volatile solvents. *In* "Chemical Synthesis and Sequencing of Peptides and Proteins" (Liu, Schechter, Herivikson, and Contiffe, eds.), pp. 251–260. Elsevier/North-Holland Amsterdam.

45. Hampson, F., and Martin, A. J. P. (1979). Displacement electrophoresis in gel as a technique for separating proteins on a preparative scale. *J. Chromatogr.* **174**, 61–74.

46. Mattock, P., Aitchison, G. F., and Thomson, A. R. (1980). Velocity gradient stabilised, continuous, free flow electrophoresis. *Sep. and Purif. Methods* **9**(1), 1–68.

47. Dickerson, C. H., Birch, J. R., and Cartwright, T. (1980). A novel, rapid and continuous method for the resolution of cell dispersal activities in crude trypsin preparations. *Dev. Biol. Stand.* **46**, 67–74.

48. Cartwright, T., Dickerson, C. H., and Austen, D. E. G. (1979). Free zone electrophoresis of clotting factors. *Thromb. Haemostasis* **42**, 70.

49. Bachrach, H. L., and Schwerdt, C. E. (1952). Purification studies on Lansing poliovirus: pH stability, CNS extraction and butanol purification experiments. *J. Immunol.* **69**, 551–561.

50. Markham, R. (1959). The biochemistry of plant viruses. *In* "The Viruses" (F. M. Burnet and W. M. Stanley, eds.), Vol. 2, pp. 35–125. Academic Press, New York.

51. Polson, A., and Deeks, D. (1962). Electron microscopy of neurotropic African horsesickness virus. *J. Hyg.* **61**, 149–153.

52. Lei, J. C. (1974). Application of PEG in the preparation of concentrated purified FMD vaccine. Present status of research. *Bull. Off. Int. Epizoot.* **81**, 1169–1199.

53. Stanley, W. M. (1936). Chemical studies on the virus of tobacco mosaic. *Phytopathology* **26**, 305–320.

54. Adamowicz, P., Legrand, B., Guerche, J., and Prunet, P. (1974). Un nouveau procédé de concentration et de purification de virus. Application au virus de la Fièvre Aphteuse produit sur cellules BHK 21 pour l'obtention de vaccins hautement purifiés. *Bull. Off. Int. Epizoot.* **81** (11–12), 1125–1150.

55. Duchesne, M., Guerche, J., Legrand, B., Proteau, M., and Colson, X. (1982). The use of highly concentrated, purified (by a large scale method) and long term liquid nitrogen stored FMD viruses for the preparation of vaccines: Physico-chemical quality controls and potency tests after storage. *Dev. Biol. Stand.* **50**, 249–259.

56. Levintow, L., and Darnall, J. E. (1960). A simplified procedure for the purification of large amounts of poliovirus: characterization and amino acid analysis of type I poliovirus. *J. Biol. Chem.* **235**, 70.

57. Laver, W. G. (1962). The structure of influenza viruses. 1. N-Terminal amino acid analysis. *Virology* **18**, 19–32.

57a. Van der Marel, P., van Wezel, A. L., Hazendonk, A. G., and Kooistra, K. (1980). Concentration and purification of poliovirus by immune adsorption on immobilized antibodies. *Dev. Biol. Stand.* **46**, 267–273.

58. Van der Marel, P., and van Wezel, A. L. (1978). Isolation of biologically active components from rabies and other enveloped viruses. *Dev. Biol. Stand.* **42**, 93–98.

59. Duchesne, M., and Legrand, B. (1982). Unpublished results.

60. Panon, G., and Atanasiu, P. (1980). Concentration and purification of human rabies vaccines: Results. *Dev. Biol. Stand.* **46**, 257–264.

61. van Wezel, A. L., Van Herwaarden, J. A. M., and Van de Marrel de Rijk, F. W. (1979). Large scale concentration and purification of virus suspension produced on microcarrier cells for the preparation of inactivated virus vaccine. *Dev. Biol. Stand.* **42**, 65–69.
62. Trudel, M., Trepanier, P., and Payment, P. (1983). Concentration and analysis of labile viruses by hollow fibre ultrafiltration and ultracentrifugation. *Process Biochem.* **18** (1), 2–9.
63. Svedberg, T., and Pedersen, K. O. (1940). "The Ultracentrifuge." Oxford Univ. Press (Clarendon), London and New York.
64. Griffiths, O. (1979). "Techniques of Preparative, Zonal and Continuous Flow Centrifugation." Beckman Instruments Inc.
65. Atanasiu, P., Tsiang, H., Lavergne, M., and Chermann, J. C. (1977). Purification par centrifugation zonale d'un vaccin antirabique humain obtenu sur cellules rénales de foetus bovin. *Ann. Microbiol. (Paris)* **128B**, 297–302.
66. Benmansow, A. (1982). Purification du virus rabique par centrifugation isopynique à flux continu sur rotor JCF-Z. *Ann. Virol.* **133E**, 273–279.
67. Adamowicz, P., Gerfaux, G., Platen, A., Muller, L., Vacher, B., Mazer, M. C., and Prunet, P. (1980). Large scale purification of hepatitis B vaccine. *In* "Proceedings of the International Symposium of Hepatitis B Vaccine" (P. B. Maupas and P. Guery, eds.), pp. 37–49. INSERM, Paris.
68. Hellman, A., Oxman, M. N., and Pollack, R., eds. (1973). "Biohazards in Biological Research." Cold Spring Harbor Lab., Cold Spring Harbor, New York.
69. Becht, H., Rott, R., and Klenk, H. D. (1972). Effect of concanavalin A on cells infected with enveloped viruses. *J. Gen. Virol.* **14**, 1–8.
70. Stewart, M. L., Summer, D. F., Screiro, R., Fields, B. M., and Maizel, J. V. (1973). Purification of oncornavirus by agglutination with concanavalin A. *Proc. Natl. Acad. Sci. U.S.A.* **70**, 1308–1315.
71. Neukirch, M., Moenning, V., and Leiss, B. (1981). A simple procedure for the concentration and purification of hog cholera virus (HCV) using the lectin of *Ricinus communis*. *Arch. Virol.* **69**, 287–290.
72. Lentz, T. L., Burrage, T. G. Smith, A. L., Crick, J., and Tignor, G. H. (1982). Is the acetylcholine receptor a rabies receptor? *Science* **215**, 182–184.
73. Elf Aquitaine-Institut Pasteur (1977). Séparation et purification des protéines par chromatographie. French Patent FR 771,278.
74. Horodniceau, F., Crainic, R., and Barme, M. (1981). Cell substrate and risk in killed poliomyelitis vaccine. *Dev. Biol. Stand.* **47**, 35–39.
75. Merton, O. W., Reiter, S., Scheirer, W., and Katinger, H. (1982). Purification of HBsAg produced by the human hepatoma line PLC/PRF/5 by affinity chromatography using monoclonal antibodies. *Dev. Biol. Stand.* **55**, 121–129.
76. Hill, D., and Blumenthal, T. (1983). Does Qβ replicase synthesise RNA in the absence of template. *Nature (London)* **301**, 350–352.
77. Skyler, J. (1982). "FDA/USP Workshop on Drug and Reference Standards for Insulins, Somatotropins and Thyroid-axis Drugs."
78. Lazar, A., Marcus, D., Reuveny, S., Grosfeld, H., Traub, A., Feinstein, S., and Mizarahi, A. (1982). Human lymphoblastoid interferon for clinical trials: Large scale purification and safety tests. *Dev. Biol. Stand.* **55**, 231–241.

8

Concentration

PIETER VAN DER MAREL*

Rijksinstituut voor Volksgezondheid en Milieuhygiëne
Bilthoven, The Netherlands

1. INTRODUCTION

The concentration of products (generally proteins) in animal cell cultures is generally low, certainly when compared to product concentrations in cultures of bacteria and yeasts. For instance, under optimal conditions the amount of poliovirus produced in animal cell cultures or the production of monoclonal antibodies by hybridoma cell lines is in the order of micrograms per milliliter of culture volume. Although production of a number of substances may be increased to a considerable extent by selection and even more by application of recombinant DNA techniques, the latter approach especially is still in its infancy and certainly not possible for all purposes. As a

*Present address: Intervet International, W. de Köverstraat 35 5830 AN Boxmeer, The Netherlands.

Animal Cell Biotechnology, Vol. 2

consequence large cell cultures yielding large amounts of culture fluids have to be generated and handled to obtain the components of interest on a production scale. Examples are the large-scale production of interferon in suspension culture in 1000- or 2000- liter fermenters and the cultivation of poliovirus on a 300- to 1000-liter scale for the production of inactivated polio vaccine. In terms of animal cell technology such culture volumes are really big. Besides interferon and poliovirus, other viruses, plasminogen activator(s), growth factors, monoclonal antibodies, lymphokines and many other cellular products are good candidates for large-scale production. Depending on the production system and the application of the products, a more or less extensive purification will be needed. Especially when the products are to be administered to humans purification should be extensive to meet the requirements set by the World Health Organization (WHO) and national control authorities. One of the first steps in downstream processing of dilute solutions or suspensions is reduction of their volumes to an extent which allows further purification within a reasonable time. Therefore concentration procedures are basically intended to reduce the volume of a solution containing the component of interest. As a consequence, contaminants are also concentrated at this stage. However, by carefully choosing both the concentration method and the experimental conditions it is possible to also obtain some purification in this stage. Concentration can be carried out in many different ways, but all are based on the following basic methods:

1. Precipitation
2. Adsorption/elution
3. Centrifugation
4. Ultrafiltration

Precipitation and elution/adsorption techniques have been in use for decades, whereas centrifugation and ultrafiltration represent more recent developments. In this chapter the methods are evaluated with emphasis on their applicability in large-scale procedures.

2. PRECIPITATION

2.1. Salt

2.1.1. General

Upon addition of salt to a protein-containing solution, first, at low salt concentration, there is an increase in protein solubility. At higher salt concentration the solubility decreases, finally resulting in precipitation of the protein. The protein is salted out. This technique has been important in protein concentration for many years and is still often applied for this purpose. Several comprehensive reviews on the theory of salting out have been published (*16, 21*). The solubility of a protein in a salt solution depends on several factors, the most important ones being pH, temperature and type of salt. Generally proteins are least soluble at their isoelectric point (IEP). Solubility may increase rapidly at pH values either at the acid or the alkaline side of the IEP. The IEP may be determined by isoelectric focussing. Often it is hardly possible to carry out pH measurements in concentrated salt solutions. Dilution to a concentration below 0.2 M is necessary to obtain reliable values.

The effect of temperature on the solubility of proteins in salt solutions may be of two kinds. Normally, in dilute salt solutions solubility increases with increasing temperature. However, in concentrated salt solutions solubility commonly decreases at increasing temperature, e.g. carboxyhemoglobin is ten times as soluble in ammonium sulphate at 0°C as it is at 25°C, which means that heating such a solution from 0° to 25°C results in precipitation of about 90% of the protein. Especially when a solution contains few different proteins this negative temperature effect may be applied advantageously.

The third important factor which determines the solubility of a protein in solution is the type of salt. In concentrated electrolyte solutions the solubility of a protein depends linearly on the ionic strength of the solution. Since monovalent salts do not easily give solutions of high ionic strength these are relatively inefficient precipitants. Salts of higher valence are much better precipitants. Widely used are ammonium sulphate, $(NH_4)_2SO_4$, sodium sulphate, Na_2SO_4, and phosphates. The latter ones have the advantage of being buffers, so pH can be controlled easily. Although sodium sulphate is still more effective than ammonium sulphate in precipitation, its solubility and consequently range of application is less.

2.1.2. Ammonium Sulphate

Ammonium sulphate is by far the most commonly used salt for protein precipitation. It is very soluble in aqueous solution (3.9 M at 0°C and 4.1 M at 25°C). Further, it may have some protective influence on the biological activity of proteins. The substance itself is slightly acidic, even when pure. Control of pH in buffered solutions is hampered by loss of ammonia. For this salt a nomogram has been designed from which it is easy to read how much salt has to be added to a solution to get the desired concentration (Fig. 1) (*15*). In practice, ammonium sulphate precipitation is applied both for concentration of (dilute) protein solutions and for an initial fractionation of the protein mixture. Salt is added to a concentration at which the proteins of interest are still not precipitated. Contaminating proteins may be precipitated at this stage and removed. Addition of salt has to be carried out slowly

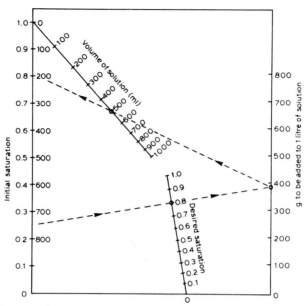

Fig. 1. Nomogram for ammonium sulphate precipitation (15). Reprinted by permission from *Biochem. J.* **54,** 457–458, copyright © 1953 The Biochemical Society, London. A straight line through the initial saturation and the desired saturation gives the amount of $(NH_4)_2SO_4$ to be added to 1 liter of the solution. A line from this point passing through the volume of the solution gives the amount required for this volume. An example is indicated by the dashed line.

and under constant stirring to prevent local high salt concentrations and uncontrollable coprecipitation. Then more salt is added until all proteins of interest are precipitated. The precipitate is collected; the supernatant containing residual contaminants is discarded. During all steps pH and temperature are carefully controlled. A span of 5–10 in ammonium sulphate saturation percentage is considered a good compromise between yield and precipitation. Once the precipitation range of the protein under investigation has been determined empirically, it should be possible to duplicate the experiment. In determining the best conditions for precipitation the protein concentration should be taken into account. The fractionation limits of a protein vary with its concentration. Lowering the protein concentration results in a shift of the precipitation range to higher salt concentrations. Variation of the protein concentration may lead to an optimal precipitation of the desired protein, i.e. less precipitation of unwanted components and maximal recovery. Excessive dilution should be avoided, especially in large-scale work, because then very large amounts of salt are needed to precipitate the proteins. From the foregoing it is obvious that in salt precipitation of proteins not only pH and temperature, but also the concentration of the desired protein, should be standardized continuously in order to obtain reproducible precipitation and fractionation results.

2.1.3. Applications

Ammonium sulphate precipitation has been mainly used in small-scale (laboratory) processes. Examples are the concentration of proteins such as interferons (6, 47, 48), interleukin (41), plasminogen activator (11) and viruses. The stability of foot-and-mouth disease virus decreased after ammonium sulphate precipitation (5). Morrow et al. (46) preferred sodium sulphate to the ammonium salt for precipitation of this virus. For rabies virus a good recovery of protective activity after ammonium sulphate precipitation was reported (1).

Because high salt concentrations, sometimes up to 80–90% saturation, are needed for protein precipitation, the method is less suitable for the processing of large volumes. In such cases the bulk volume is reduced first by a more suitable method, such as ultrafiltration (cf. Section 5). The resulting crude suspension may subsequently be concentrated further by differential salt precipitation with a simultaneous initial purification of the proteins of interest. Such a process has been in use for a long time in the production of bacterial vaccines against diphtheria and tetanus (65).

2.2. Polyethylene Glycol (PEG)

As early as the beginning of the 1950s PEG was used for precipitating organic macromolecules from a chloroplast suspension (44). In 1963 Hebert used this polymer for the precipitation of plant viruses (24). Since that time PEGs have been widely applied for separation and concentration of proteins, bacteriophages and viruses [for a review, see Vajda (62)].

Polyethylene glycols are linear condensation polymers of ethylene glycol with the general structural formula:

$$HOCH_2 —(CH_2 — CH_2 — O)n—CH_2OH$$

They are nonionic and water-soluble. The molecular weight of the polymer may vary from a few hundred to ten thousands depending on the polymerization conditions. Because of this wide range of possible sizes the molecular weight of the product is usually defined, e.g. PEG 300, PEG 6000, PEG 40,000. PEGs do not interact with proteinaceous materials and produce minimal denaturation (50). Therefore they can be applied at room temperature. This is in contrast to organic solvents such as ethanol and acetone, which can only be used at low temperatures to avoid denaturation. For the precipitation of proteins and viruses PEGs with molecular weights of

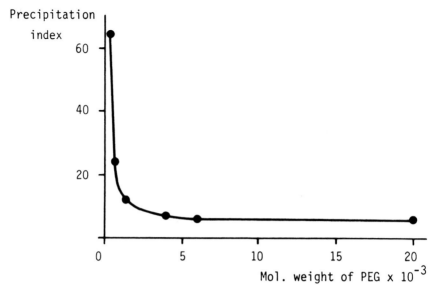

Fig. 2. Precipitation index of PEG with different molecular weights. The precipitation index is defined as the polymer concentration required to precipitate 50% of the protein from a 1% solution (50).

2–6 kilodaltons (PEG 2000–6000) are mostly used. PEGs with a higher molecular weight give viscous solutions which are almost impossible to work with. For PEGs with a lower molecular weight the precipitation index varies considerably with the molecular weight (Fig. 2). PEG probably acts by excluding particles sterically from part of the solution (30). Protein concentration has some effect on the precipitation behaviour, but only to a limited extent (33). For very large particles, e.g. viruses, the excluded volume is almost independent of the particle concentration, which means that in this case precipitation behaviour is independent of virus concentration (71). As also found in salt precipitation, the most critical factor in PEG precipitation is pH. Shifting pH to the IEP of a protein or virus particle results in a decrease of the PEG (and salt) concentration needed for precipitation. There also exists an inverse relationship between PEG and salt concentration, which means that at the same pH and temperature under high salt concentration, less PEG is needed for precipitation than under conditions of low salt (24, 39). Large particles, such as viruses, are completely precipitated over a very narrow concentration range once precipitation has started (33). This aspect may be used to one's advantage in fractionation procedures, e.g. the separation of a virus from contaminating smaller proteins.

After resuspension of the precipitate a small amount of PEG will be present in the concentrate. The easiest way to get rid of this residual PEG is to include a gel filtration step in the process. Under suitable conditions concentration factors of 50–100 times may be obtained (*19, 45, 71*).

2.2.1. Applications

In animal cell biotechnology PEG precipitation has been used mainly for precipitation of viruses, i.e. the separation and concentration of relatively large particles from a solution containing numerous small components. By choosing proper conditions for precipitation a substantial purification may be achieved at the same time. Precipitation with PEG is a rapid alternative for ultracentrifugation of aqueous virus suspensions. With regard to the concentration of proteins and/or fractionation of protein mixtures it should be mentioned that rather sophisticated systems have been developed for the fractionation of plasma proteins and/or immunoglobulins (*30, 50, 69*) as an alternative to the cold ethanol procedure developed by Cohn (*14*). Small-scale concentration of viruses has been reported for a number of animal viruses, e.g. FMD (*2, 62*), measles (*62*), rubella (*19, 62*), rabies (*45*). Also, influenza virus has been concentrated from allantoic fluids (*13, 62*). Barteling (*7*) described a closed system for concentration of FMD virus on a 300-liter scale which permits operation under sterile conditions. The last example underlines the feasibility of PEG for the concentration of both small and large volumes of virus harvests.

2.3 Other Precipitants

2.3.1. Organic Solvents

Because of their tendency to denature proteins, especially at temperatures above 0°C, these substances are generally less suitable for preparative processes. In this respect the blood proteins have a special position. Cohn *et al.* (*14*) developed a method for fractionating plasma proteins in several classes by differential precipitation with increasing concentrations of ethanol at low temperatures.

2.3.2. Acids

Although organic acids, together with organic solvents, are frequently used to precipitate proteins for analytical purposes, e.g. gel electrophoresis, they also display strongly denaturing effects. Again, one exception has to be mentioned. Interferon (obtained from leucocytes, tissue culture or by rDNA technology) is acid stable and may be precipitated by trichloroacetic acid (*18, 32, 34*).

2.3.3. Metal Ions

The former application of zinc as a precipitant for proteins—interferon (48)—and viruses—rabies (56)—has now been replaced by the introduction of metal chelate affinity chromatography (see Section 3).

2.4. Affinity Precipitation

In affinity precipitation a ligand is covalently coupled to a soluble polymer carrier giving a polyfunctional derivative, which can be precipitated by simple changes in environmental conditions, e.g. a change of pH. After complete formation, the polyfunctional derivative is forced to precipitate, and the product of interest can be recovered from the precipitate by a suitable extraction/elution procedure. Although the technique seems very suitable for industrial scale concentration and purification of proteins, its application is not wide-spread thus far (30a, 40a) but the increasing number of animal cell products certainly will stimulate further development of this technique.

3. ADSORPTION/ELUTION

Adsorption

The interaction of soluble macromolecules (proteins) with a solid matrix takes place mainly by electrostatic and hydrophobic forces. Binding of proteins to anion and cation exchange matrices is dependent on the negative and positive charges of the macromolecules; hydrophobic matrices interact with hydrophobic domains on the molecules. Interactions between proteins are seldom purely electrostatic or purely hydrophobic but rather a combination of the two. However, for matrices it is desirable that they display only one type of interaction, e.g. an ion exchanger should not bind proteins by (aspecific) hydrophobic forces.

A relatively recent development is metal chelate affinity chromatography (49, 51). Many proteins have a high affinity for heavy metal ions, such as Cu^{2+} and Zn^{2+}, which can form complexes with imidazole and thiol groups. So, immobilized metal ions may bind proteins by the formation of coordination compounds with histidine and cysteine.

Because the interactions mentioned above are based on physical forces, all macromolecules which share certain properties will display affinity for a particular matrix. By careful selection of matrix and binding conditions (pH, ionic strength) it is possible to obtain a more or less preferential binding of components of interest. On the other hand, the biological or immunological properties of a macromolecule can be used to tailor-make matrices. For example coupling antibodies to an inert carrier will give a matrix which is highly specific in binding only those proteins/antigens against which the antibodies are directed, and immobilization of substrates or inhibitors will yield matrices which are specific for the enzymes which use these substrates or are blocked by the inhibitors. This method of adsorption, which is generally referred to as affinity chromatography, is very suitable for the adsorption of macromolecules from dilute suspensions.

Thus, in theory, adsorption and affinity chromatography are convenient methods for the concentration of proteins. In practice, several complications have to be taken into account:

1. Often harvests containing animal cell products also contain varying amounts of cell debris. These suspensions must be clarified by centrifugation or filtration before the adsorption step is initiated, certainly when column adsorption is used.

2. The fraction of solute to be concentrated in relation to the total amount of protein is important. Often impurities are also adsorbed to some extent to a matrix, which may eventually be "poisoned" by the contaminants. For instance the presence of serum proteins in cell culture media containing very low concentrations of the component of interest may be problematic. By careful selection matrices and/or binding conditions can be chosen which give a preferential binding of this component. In this way concentration goes together with partial purification. As a practical problem it must be emphasized that the matrices which are currently available in high quantities at moderate prices (e.g. ion exchangers, silicates) generally display little specificity in binding proteins. More specific matrices are expensive and therefore less attractive for large-scale processes. The general-purpose matrices can be best used in suspensions with few contaminating proteins, e.g. serum-free cell culture media containing cellular products.

3. Depending on the affinity of a macromolecule for a matrix, adsorption proceeds quickly or more slowly. The existing range of matrices, however, is so wide that for almost every component a matrix can be found with a high affinity.

4. A choice must be made between batchwise and column adsorption. In the batchwise process matrix particles are suspended by stirring the solution to be processed. After a sufficient contact time the matrix is allowed to settle down by gravity or separated from the liquid by filtration and processed further. In the column adsorption process the matrix is packed in a column which is perfused with the harvest solution. Generally adsorption occurs more quickly in columns than in batchwise processes. Because of the high capacity of many matrices large volumes of harvest have to be pumped through small columns. This requires high flow rates. Most of the current affinity matrices consist of relatively soft beads (derivatives of agarose, cellulose, polyacrylamide) which are compressed by high flow rates, and subse-

quently the columns will block. To some extent this can be avoided by using very wide, short columns. Currently manufacturers are actively developing new matrices which combine a high binding capacity and a high mechanical stability. For the processing of relatively small volumes both methods can be applied. The handling of larger volumes (>50–100 liters) can still be more quickly and easily done with the batchwise method (*30a*).

5. Capacity: The amount of product which can be adsorbed per unit of matrix depends to a large extent on the area of the matrix which is accessible for the product. For this reason matrices with an open-pore structure, allowing penetration of particles with molecular weights > 10–20×10^6 daltons, are often used for ligand attachment. However, the coupling procedure may reduce the effective pore size and consequently the accessible surface area. Very large particles, such as viruses, are adsorbed at the surface and hence the capacity of affinity matrices for such particles is limited.

Elution

Once adsorbed, the macromolecules must be eluted. To obtain high concentration factors elution should take place in a volume as small as possible. Further, the conditions of elution (composition of eluents and contact time) must be harmless for the products of interest. When harsh elution conditions are required it is important to reduce contact time between eluents and product as much as possible. In small-scale experiments this does not give problems, but on a large scale contact time almost inevitably increases. Elution in a column may be advantageous because a gradient with increasing concentration of desorbing solution can be used. In this way additional purification is feasible.

Sterility

The column process for adsorption/elution can easily be designed as a closed circuit, which can be operated under sterile conditions. In the ideal situation the whole system is sterilized *in situ* by a suitable disinfectant such as formalin. However, many disinfectants are reactive with matrices used in adsorption and affinity chromatography. The choice of disinfectant depends on the physicochemical properties of the matrix and is often a compromise between effectiveness and harmlessness for the matrix. As an alternative the column can be sterilized separately, permitting sterilization of the rest of the system with the most effective substance. General-purpose adsorption matrices, such as ion exchangers based on e.g. cellulose, cross-linked dextran and agarose, are autoclavable. When these kinds of materials are used for the column process the matrix must be autoclaved in a separate container, whereafter the column must be filled under sterile conditions. In principle, sterile operation in a batchwise process is also possible, but is more difficult to achieve. Matrix must be added as a slurry, which increases the volume to be processed. The process vessel must be equipped with a stirrer to keep the matrix particles in suspension. Further, it must be possible to separate matrix from liquid and to perform elution in a small volume. For small-scale operations these requirements are difficult to meet. For large-scale processes the equipment can be specially designed for each particular process and operation can be performed under sterile conditions.

Applications

It is impossible to give even a rough outline of adsorption/elution or affinity chromatography procedures in use for the concentration and purification of products from animal cells. Therefore, in this section only examples on the more or less large-scale concentration of proteins or viruses with the adsorption methods mentioned above will be outlined.

Non-specific Adsorption

INORGANIC SALTS AND POLYELECTROLYTES   Adamowicz *et al.* (2) used an insoluble complex of calcium and ethylene oxide (MW > 100,000) for the 1000–1500-fold concentration of foot-and-mouth disease virus on a 1000-liter scale in a batch process. At the same time >99% of contaminating proteins were removed. Insoluble aluminum and calcium salts, e.g. aluminum phosphate and hydroxide and calcium phosphate, have a high capacity for adsorbing viruses even from highly dilute solutions (68). Schneider *et al.* (55) developed a procedure for the batchwise adsorption of rabies virus on aluminum phosphate which on a 3-liter scale gave a 300-fold concentration and 2000-fold purification. Elution occurs under mild conditions, e.g. addition of EDTA to the calcium–ethylene oxide complex or by changing pH and salt concentration in the case of the insoluble salts.

HYDROPHOBIC MATRICES.   With hydrophobic matrices usually more harsh elution conditions have to be applied, e.g. low pH or high concentrations of chaotropic salts. For some proteins these may be harmful, but for others they may even exert a stabilizing effect and protect them from inactivation. This is illustrated by Bock *et al.* (12) for controlled pore glass (CPG) beads. CPG, first introduced by Huller (29), has been used for concentration and purification of viruses [e.g. rabies (36)], proteins [e.g. interferons (10, 17, 37)] and interleukin 2 (25). Elution of bound proteins from CPG requires strongly chaotropic conditions. Mees *et al.* (44a) circumvented this problem by using carboxymethyl derivatized CPG beads for concentration/purification of the IFN-α. Especially for interferons numerous concentration/purification processes with hydrophobic matrices have been described [for reviews, see Pestka (48) and Baron *et al.* (6)]. This illustrates the fact that hydrophobic interaction chromatography is still highly empirical. For each application suitable matrices and optimal conditions for adsorption and elution have to be determined.

Specific Adsorption

IMMUNE ADSORBENTS.   As already mentioned, matrices with very well defined specificities are obtained by coupling antibodies to a carrier. The antibodies can be obtained from hyperimmune animal sera or may be monoclonal. Van der Marel *et al.* (64) demonstrated the high binding capacity of such a matrix for poliovirus. One milliliter of immune adsorbent bound the amount of virus from 1 liter of infected cell culture medium. For immediate

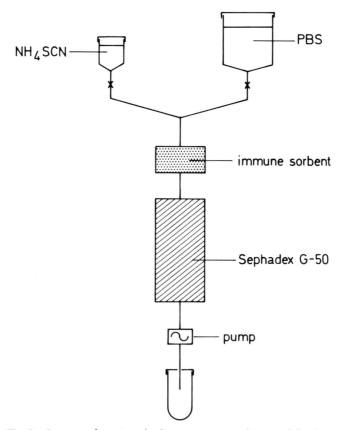

Fig. 3.   Immune adsorption of poliovirus: one-step elution and desalting.

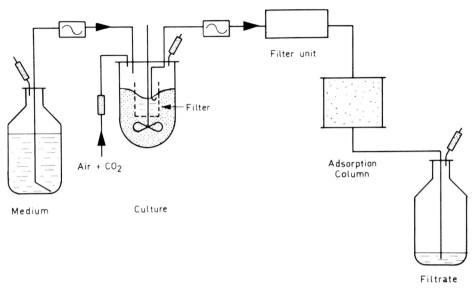

Fig. 4.   In-line production, clarification and concentration of excreted products from animal cell cultures (35).

removal of the chaotropic salt used for elution an in-line gel filtration process was used (Fig. 3). Such a procedure is generally applicable in column adsorption/elution procedures.

AFFINITY ADSORBENTS.   Particularly interesting are proteins which are produced continuously during a prolonged time by cell cultures, such as tissue-type plasminogen activator (t-PA) by melanoma cells (23, 53). t-PA, even when present in very low concentrations, can be adsorbed effectively by metal chelate affinity chromatography to immobilized zinc. Batches of several hundred liters of serum-free cell culture harvests have been processed in this way (35). However, the adsorption process is lengthy because of limitations in flow rate. This drawback can be overcome by the installation of the zinc column in line with the cell cultivation equipment (Fig. 4). Perfusion rate of the cell culture with fresh medium is dependent on the production rate of t-PA, but is sufficiently low to permit direct introduction of clarified medium into the adsorption column. When the column is saturated it can be replaced by a fresh one, the first one already being processed when production is still going on. Such a system is generally applicable, not only for natural cell metabolites such as t-PA, growth factors and lymphokines, but also for components which are excreted by genetically engineered animal cells.

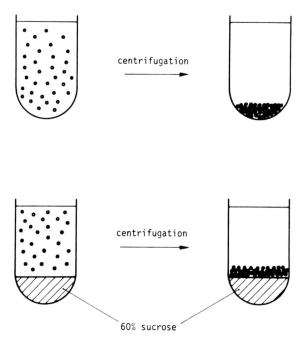

Fig. 5.   Concentration of macromolecules by centrifugation methods.

4. CENTRIFUGATION

General

A widely used technique for the concentration and/or purification of particles such as viruses, subcellular organelles or large molecules is sedimentation in the ultracentrifuge. In its simplest form concentration is performed by pelleting molecules on the bottom of a centrifuge tube (Fig. 5) and resuspending them in a small volume of buffer. However, in this way particles are heavily compressed, which may result in a considerable loss of (biological) activity after resuspension. Therefore pelleting on a cushion of a high-density solution (usually 60% sucrose) is often used as an alternative (Fig. 5). The particles are concentrated at the interphase between the upper and lower liquids and stay in solution.

Whether or not ultracentrifugation is suitable for concentration of particles or macromolecules largely depends on their hydrodynamic properties such as size and density. The sedimentation rate, and hence the time needed for pelleting, depends not only on the properties of the particles, but also on those of the solution in which they are suspended (22). Generally not the sedimentation rate, but rather the sedimentation coefficient, which is de-

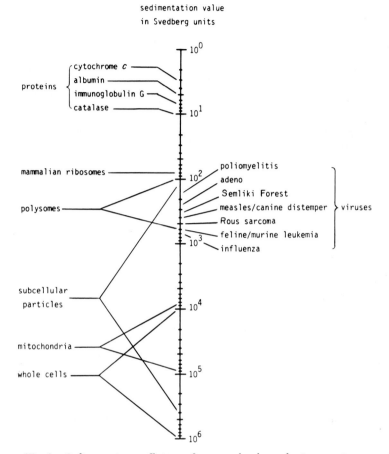

Fig. 6. Sedimentation coefficients of macromolecules and microorganisms.

fined as the sedimentation rate per unit of centrifugal force, is used to characterize the hydrodynamic properties of a particle. The sedimentation coefficient is expressed in svedberg (S) units. In Fig. 6 the S coefficients for a variety of molecules and particles are indicated. Whether or not particles can be pelleted within a reasonable time can be estimated by using so-called k factors. Each rotor has its specific k value. Equation (1) shows the relationship between k value and t (time in hours) required to pellet a particle of known S coefficient in an aqueous solution from top to bottom of a centrifuge tube at maximum rotor speed.

$$t = \frac{k}{S_{20,w}} \qquad (1)$$

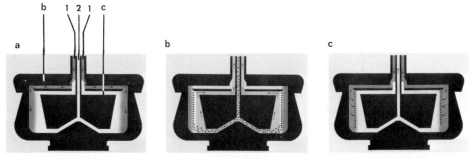

Fig. 7. Continuous flow isopycnic centrifugation. The rotor is loaded at low speed (about 2000 rpm) with a gradient of high-density liquid, which serves as a cushion for isopycnic centrifugation (a). Then the rotor is filled further with buffer and simultaneously accelerated to operation speed. At this point sample is introduced (b). When all the sample has been pumped through the rotor centrifugation is continued for some time to allow the particles to reach their equilibrium density, feeding the rotor with buffer. Finally, it is decelerated to low speed and the contents are displaced by introducing a high-density solution (c).

The lower the k value, the faster pelleting occurs. Calculation of the time for a particle to sediment in a sucrose gradient or in some other viscous solution is also possible, but requires more complicated formulas (22). The k factors are generally in the order of 10–50 for conventional fixed-angle high-speed rotors (>50,000 rpm), 100–1000 for conventional fixed-angle large-volume and zonal rotors and 40–100 for high-speed continuous flow rotors. These values demonstrate that the use of centrifugation for concentration purposes is practically limited to particles with S coefficients exceeding 100–200, i.e. larger viruses, various subcellular components and whole cells. Fixed-volume rotors are only suitable for concentration of relatively small volumes, e.g. a few liters. When volumes exceeding 2–5 liters have to be processed continuous flow centrifugation is to be preferred. The technique is very suitable for the concentration of bacteria, tissue homogenates, subcellular organelles and large viruses.

A continuous flow rotor basically exists of two parts: bowl and core (Fig. 7). Buffer or sample can be introduced in the rotor either directly into the bowl (b) or through the core(c). During operation the sample solution is pumped into the rotor through the "center line" (2) continuously while the rotor is spinning at operation speed. Particles in the sample solution will sediment to the rotor wall whereas the "supernatant" flows along the core to the periphery line (1). The path length between core and rotor wall is relatively small. The sedimentation area is located rather far away from the center of the rotor in order to generate a high centrifugal force and hence a low k factor. The sedimentation of the particles may take place either directly to the wall of the rotor (pelleting) or on a cushion of high-density liquid (equilibrium or isopycnic centrifugation). The latter procedure is depicted in Fig. 7. It should be taken into account that the flowing sample liquid constantly washes out small amounts of the sucrose solution due to diffusion and turbulence. In practice this means that the volume of suspension which can be processed is limited. Because the path length of sedimentation is small the resolving power of most continuous flow rotors is not very high and will greatly depend on the total particle load. Almost every high-speed or ultracentrifuge can be equipped with a continuous flow rotor. For high-speed machines (maximum speed about 20,000 rpm) the application of this type of rotor is largely confined to clarification of suspensions or concentration of very large particles such as bacteria or animal cells. In clarification the centrifugal step removes large-sized contaminants whereas the particles of interest, of much smaller size, remain in the flow-through fraction. In this way large volumes can be clarified in a short time. Although the continuous flow rotor in high-speed centrifuges can be used for the concentration of large viruses, in these cases the low maximum flow rate—about 1 liter hr^{-1}—limits its use to small-scale operations.

Higher flow rates can be obtained with continuous flow rotors designed for ultracentrifuges. The flow rate for isopycnic centrifugation of viruses in sucrose gradients varies from 1.5 liters hr^{-1} for small viruses such as adenovirus to almost 4 liters hr^{-1} for larger viruses such as leukemia viruses and influenza virus.

As already mentioned, the short path length of most continuous flow rotors, although allowing concentration of particles, precludes the simultaneous efficient removal of contaminants. McAleer et al. (43) developed a core that accommodated the Beckman CF-32Ti rotor, but had a greatly increased path length of 4.5 cm in comparison with 1.4 cm for the standard core. This combination gave excellent results for simultaneous concentration and purification of influenza B and herpes simplex viruses. With conventional ultracentrifuges volumes of 10–20 liters may be processed within a reasonable time, e.g. a working day. For larger volumes specially designed equipment is needed. Such equipment became available more than a decade ago. At the end of the 1960s scientists at the Oak Ridge National Laboratory in Tennessee, U.S.A., developed a new group of centrifuges, designated K series, suitable for the purification and concentration of large quantities of viruses and subcellular particles (3). The machines were air-driven and could accommodate a 3.6-liter continuous flow rotor with a length of 76 cm and a sedimentation path length of 1.15 cm. In contrast to conventional continuous flow rotors the inlet and outlet were at opposite ends of the rotor (Fig. 8). The K centrifuges were commercialized by Electronucleonics in New Jersey, U.S.A. Nowadays, several rotor types are available, including a rotor with built-in clarification system which may serve as an alternative to a

IN

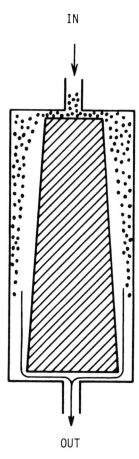

OUT

Fig. 8.   Continuous flow rotor from the K-series centrifuges.

separate clarification step. Extensive investigations showed that loading and unloading the rotor for isopycnic centrifugation at rest instead of spinning at 2000–3000 rpm resulted in sharper bands (3, 38, 52).

Containment

In zonal and continuous flow operations liquid is pumped into and out of the rotor while it is spinning at low speed (zonal centrifugation) or high speed (continuous flow centrifugation). The flow of liquid is allowed by a rotating seal assembly which connects the spinning rotor with the static pumping equipment. The rotating seal is a very critical piece, because it must operate almost without friction and yet may not leak. If it is not functioning properly liquid will escape from the seal as an (almost) invisible aerosol or droplets. It is virtually impossible to control the actual condition of a rotating seal visually, and even when this piece of equipment is polished or replaced regularly the risk of aerosol generation is great. Also leakage may occur when the seal is not correctly put on or removed from the rotor for loading or unloading. Therefore, processing of pathogenic materials, e.g. infectious viruses, requires extreme precautions to prevent any release of pathogens to the environment. In fact the only reliable solution is the installation of the complete centrifugation unit in a separate containment room. Remote control of the process is essential in this design. Whereas the price of continuous flow equipment as such is already considerable, these precautions require another high investment. For these reasons continuous flow equipment, certainly at production scale, is not widely applied.

Sterility

Although it is virtually impossible to prevent escape of material from the rotor to the environment, the opposite—preventing material from outside to enter the rotor—is possible and the rotor content may be processed under sterile conditions. Most rotors can be sterilized by steam and sterilization of the delivery system also will not give many problems. Connecting tubes and placing the rotating seal in its position are the most risky maneuvers. As an alternative, the whole system, fully assembled, may be sterilized *in situ* by chemical means, e.g. 1–5% (w/v) formalin. In fact this is the only way to sterilize the large-scale K-series centrifuges. However, the sterilizing agent should be removed completely by washing with sterile buffer before any of the process solutions can be introduced. In the literature a contamination rate on production scale of less than 20% has been reported (26), but probably for most procedures the value is lower. This indicates that sterility may not be considered as a problem in continuous flow centrifugation.

Applications of Continuous Flow Centrifugation

Johnson *et al.* (31) clarified 60- to 120-liter lots of Rauscher murine leukemia virus at a flow rate of 30–35 liter hr^{-1} at 20,000 g in a high-speed centrifuge. On the other hand, the flow rate for isopycnic centrifugation of rabies virus in a sucrose gradient using a similar centrifuge was limited to at most 1 liter hr^{-1} (9). These figures underline the limited application range for continuous flow centrifugation in high-speed machines.

Continuous flow rotors designed for ultracentrifuges have been applied for further concentration and purification of influenza virus from allantoic fluid for vaccine preparation (63). In this case bulk concentration of the allantoic fluid was carried out by ultrafiltration. They have also been used for the concentration of volumes in the 10-liter range of a.o. Sindbis virus (67) and hepatitis B virus (58) at flow rates of 1.5–2.5 liter hr^{-1}, which actually limits the use of this equipment to the volume range mentioned before.

The application range for the K-series ultracentrifuges is much wider.

Reimer *et al.* (52) produced experimental influenza vaccine from allantoic fluid on a 150-liter scale using prototypes of this equipment. Lavender and Frank (38), Hilfenhaus *et al.* (26) and Atanasiu *et al.* (4) processed rabies virus obtained from cell cultures for vaccine purposes. Large amounts of several RNA tumour viruses were produced by Toplin and Sottong (60) and of influenza virus by Gerin and Anderson (20) and Hilfenhaus *et al.* (26, 27). Mostly the viruses were concentrated isopycnically in a sucrose gradient. Depending on the sedimentation coefficient of the virus flow rates ranging from 5 liters hr^{-1} (Semliki forest virus) to 12–18 liters hr^{-1} (mumps and vaccinia virus respectively) were obtained (28). So, in one working day (supposing 6 hr of effective centrifugation time) 30–>100 liters of virus suspension can be processed. Because of the washing out of sucrose solution by the flowing stream the maximum volume which can be processed in one run is restricted to 150–200 liters. Several hundredfold concentration factors have been achieved, depending on the feeding volume (26–28, 52). At the same time purification factors up to 100-fold, based on protein determination, have been found in particular applications (27). Data on recoveries in the continuous flow system show large variations. Especially for the smaller viruses, such as Semliki forest virus, the recovery in the concentrate may be as low as about 30% even when low flow rates are used, but for larger viruses recoveries from 70 to 100% are easily obtainable, which makes the system attractive for large-scale concentration and purification of these viruses in one step.

5. ULTRAFILTRATION

General

Filtration is usually applied for removal of particulate material from solutions or suspensions. Pore sizes in normal filtration range from tens of micrometers down to 0.22 or 0.15 μm used for sterile filtration. Although bacteria, moulds and cellular debris are effectively retained by these filters, viruses and other macromolecules will pass practically unhindered. By lowering the pore size of a filter to a sufficient extent it becomes "impermeable" also for these molecules and even much smaller ones. Ultrafiltration is the process in which molecules above a certain size are retained on a filter whereas smaller ones may pass. The dimensions of the filter pores are no longer indicated in micrometers, but rather the terms "exclusion limit" and "cut-off" are used. The actual filter is a very thin membrane (about 1 μm thick). Because of its fragility it is strengthened by a support layer which is much thicker and highly permeable. Today a wide range of ultrafilters with exclusion limits ranging from 500 to 10^6 daltons is available. In theory, all molecules with sizes below the exclusion limit pass the filter freely, whereas those with sizes above this value are fully retained. In practice, ultrafilters have no such sharp cut-off, but depending on the quality of the filter, the solution to be filtered and the experimental conditions, a more or less wide range of molecules with sizes well below the exclusion limit is retained. The lower the exclusion size the sharper the rejection characteristics, but for example a membrane with a 10,000-dalton cut-off may retain more than 60% of molecules in the 5000-dalton range. Whereas the above holds true for very dilute solute suspensions, more concentrated suspensions cause another problem: concentration polarization.

Concentration Polarization

This is a major and still not fully resolved problem in all ultrafiltration systems. When left undisturbed initially, a transmembrane flow of solvent and small solute molecules (<cut-off) occurs when pressure is applied to the filtration system (Fig. 9a). Larger particles are trapped at the surface of the membrane. This results in a reduction of transmembrane flow and rejection of solute molecules which, under optimal conditions, would have passed the membrane. As more molecules are trapped on the membrane a layer of protein finally occurs and the membrane is blocked almost totally (Fig. 9b). Several systems have been developed to reduce or prevent this unwanted phenomenon. They all are based on the generation of turbulence and high shear forces in the liquid. When the solution is circulated over the membrane with a high velocity the transmembrane flow of solute molecules is counteracted by a tangential flow (Fig. 9c). At low protein concentration and a sufficient tangential flow rate no solute will be trapped at the membrane surface. When the protein concentration increases, tangential flow cannot be increased sufficiently to keep the membrane clean and subsequently protein is deposited on the membrane surface. Due to high tangential flow and corresponding shear forces, deposited protein molecules may have more or less the same orientation. A gel-like protein layer results, which phenomenon has been characterized as concentration polarization. At high solute concentration (5–40% protein) effective pore size and transmembrane flow are so low that further processing is useless and hence this is the concentration limit.

Equipment

Depending on the volume to be concentrated different equipment is chosen. The concentration of very small volumes, generally used for analytical purposes, will not be considered in this context. Volumes in the range 10–2000 ml can readily be concentrated in stirred cells (Fig. 10a). The cell can be used separately or in combination with a reservoir. Concentration

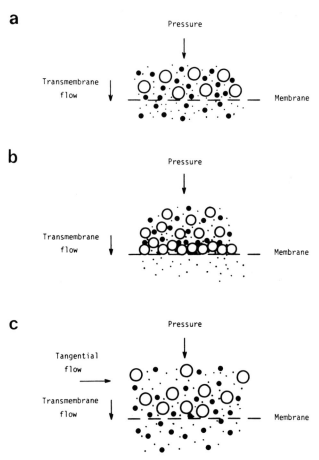

Fig. 9.   Ultrafiltration under conditions without (a and b) and with (c) circulation of process solution.

polarization is reduced in this system by the stirring of a magnetic bar which is positioned at a small distance from the filter. Concentration factors of 100× can be obtained, but the maximum attainable solute concentration is about 10%. When a higher concentration is desired a more efficient reduction of concentration polarization is required. This is obtained in the so-called thin-channel system, as produced by Amicon (Fig. 10b). The solution is circulated with a high velocity over the membrane area through a system of narrow channels by means of a pump. This method dates from the mid-1950s. Strohmaier (57) wrote a review on the application of such systems which is still interesting. In this way protein concentrations up to 40% can be obtained. The maximum process volume for this system is limited to

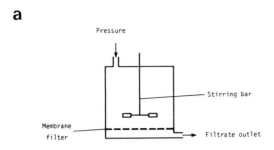

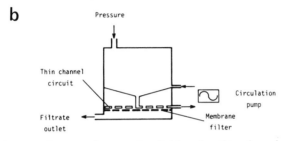

Fig. 10. Schematic drawing of a stirred cell (a) and a thin channel system (b) for ultrafiltration.

the 10-liter range. All the above-mentioned systems use membrane filters. A wide series of membranes is available including special ones with a low capacity for protein adsorption. The cell may be equipped with the membrane of choice and may be used for concentration of a variety of molecules or particles. In contrast to this versatile application range, the cells have a fixed membrane surface which limits the amount of solution to be concentrated. For the concentration of much larger volumes, in the range of tens to hundreds of liters, two systems are available: (1) the hollow fiber system and (2) the flat membrane (cassette) system.

Hollow fibers are fine cylindrical tubes with an internal diameter of 0.2–1.1 mm or, for special purposes, higher. The 0.2-mm fibers are most suitable for processing aqueous solutions. More viscous solutions or cases where the final protein concentration should be high (>20%) require fibers with a larger diameter. The fiber consists of the same material as a membrane filter. The actual filter membrane is on the inside of the fiber, the support layer on the outside. Bundles of hundreds of hollow fibers are sealed together in a cartridge, which results in a very high surface-to-volume ratio (Fig. 11a). When a suspension is circulated through the fibers under pressure, microsolute and solvent pass the membrane and may be collected through the outlet port, whereas macrosolute (>cut-off) is concentrated in the lumen of

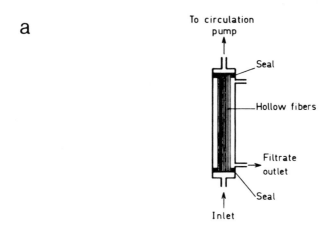

a

To circulation
pump

Seal

Hollow fibers

Filtrate
outlet

Seal

Inlet

b

To circulation pump → Inlet
Membrane
Filtrate → Filtrate outlet
outlet

Fig. 11. Schematic drawing of a hollow fiber cartridge (a) and a unit of the membrane cassette system (b).

the fibers. A big advantage of hollow fibers over flat membranes in stirred cells is that, due to their cylindrical form, liquid flow may be either from inside to outside or outside to inside, so-called backflushing. When the flux in the system decreases during operation due to concentration polarization the membrane surface may be cleaned by backflushing with suitable buffers.

Hollow fiber cartridges are available with exclusion sizes from 1,000 to over 1,000,000 daltons. Maximum allowable pressure at the inlet is usually less than 2 bars. The membrane surface area varies from several hundred centimeters squared for the smallest to almost 5 m^2 for the largest cartridge. Typical flow rates for deionized water vary from 0.15 to 1.0 ml cm^{-2} min^{-1}. With solute-containing liquids the flow rate decreases with increasing solute concentration.

The basic unit of the flat membrane (cassette) system consists of a membrane supported at both sides by a fine mesh screen. The unit is covered with an impermeable seal and has inlet and outlet ports (Fig. 11b). In a cassette a number of units are stacked together in such a way that all inlet and outlet ports are interconnected. One cassette accommodates a number of units corresponding to several square meters of surface area. Membranes are available in the same exclusion size range as hollow fibers. In contrast to hollow fiber cartridges, membrane cassettes can be operated at fairly high pressures, up to 10 bars. Also, the cassette system can be backflushed for removal of adsorbed solute molecules. Normalized flow rates are similar to those for hollow fibers.

At the end of a concentration process much solute is often adsorbed to the membrane despite circulation of the solution. Besides concentration polarization most membranes will adsorb proteins to some extent. In order to prevent loss of concentrated solute, stirred cells or thin channel systems must be operated unpressurized for some time at the end of the process. When there is no transmembrane flow the adsorbed proteins will be sheared from the membranes. For hollow fiber and cassette systems this procedure is not possible because the flow resistance of the cartridge or cassette causes a pressure build-up which increases with increasing circulation rate. Extensive backflushing with sterile buffer at the end of the process will clean the membrane surface. The backflushed volume finally can be removed again by careful ultrafiltration at low pressure and low recirculation velocity.

Containment and Sterility

The hollow fiber, cassette and thin channel systems can be operated in a fully closed system. Sterilization may be carried out by treatment with 1–5% (w/v) formalin, which allows subsequent sterile processing of suspensions without risk of aerosols. After processing, the entire system can be disinfected by circulation of proper disinfecting agents. Finally, the system can be cleaned with enzyme, detergent, acid or alkali solutions according to the manufacturer's instructions. Upon storage the equipment should be filled with disinfectant, e.g. formalin. The systems can be reused over a very prolonged time without risk of improper functioning. Lifetimes of over 5 years with frequent usage are not exceptional.

The Pump

In principle, any pump which is compatible with the suspension to be concentrated can be used. Desired capacity (liters per hour) and operating pressure (bars) depend on the ultrafiltration system. Several points should be kept in mind in choosing a pump:

Shear. In order to avoid shear, which may be harmful for the proteins to be concentrated, the desired pumping capacity should be reached at a low rotation speed.

Speed Adjustment. The speed of the pump should be easily adjustable down to very low circulation rates.

Dead Volume. The internal volume of the pump (and also the rest of the concentration system) should be small enough to permit high concentration factors.

Design. The propulsion should be absolutely leakproof; no unwanted particles (from the pump itself or from outside) should enter the pumping
circuit. Tubing and magnetically driven pump types are convenient for
this purpose.

It is recommended, certainly when dealing with possibly infectious
agents, that the pump be secured against overpressure or that such a provision be built into the circulation system.

Applications

Ultrafiltration is widely used for the concentration of small proteins and
large viruses on both small and large scales. Rommerts *et al.* (*54*) used a
single hollow fiber (cut-off 10,000 daltons) for the concentration of products
in cell culture media by 100- to 1000-fold. Other applications are the concentration of different viruses, e.g. influenza in allantoic fluid (*63*), or viruses
in cell culture media: polio, rabies (*8, 66*), rubella (*61*) and C-type viruses
(*59, 70*).

Low-molecular-weight products can also be processed in a special way.
Marquardt *et al.* (*42*) used the property of a small growth factor (molecular
weight about 14 kilodaltons) to associate with a large protein at neutral pH to
concentrate the combination with a filter with a high cut-off, after which the
complex was dissociated.

The opposite is also possible; e.g. when the desired component is relatively small, large contaminants can be removed by rejection on a membrane
which allows passage of the smaller particles to the ultrafiltrate, which is
concentrated in a second run. Leuthard and Schuerch (*40*) used this procedure for the concentration of human leucocyte and lymphoblastoid interferon and Maizel *et al.* (*41*) for the concentration of interleukin 1.

With both methods some purification may be obtained at the concentration stage. However, together with the desired component(s), contaminating
molecules will also generally be concentrated.

In large-scale operations (100–1000 liters) concentration factors up to 500
times in one step (*66*) or up to 1000 times in two steps (*61*) could be obtained
in a couple of hours. Recoveries, as measured by protein content or biological activity, are usually better than 60% and often as high as 100%. Detailed
information including literature references on the application range of ultrafiltration equipment is available from the manufacturers. The prices of modern ultrafiltration equipment are relatively low. Except for cleaning there is
almost no cost for maintenance and the lifetime is very long. In our institute
the same ultrafiltration equipment has been used for more than five con-
secutive years in the production of inactivated polio vaccine on a 100- to 500-
liter scale. Scaling up an ultrafiltration process consists of simply extending
the amount of units in the circulation loop and perhaps exchanging the

circulation pump for a larger one. Doubling the filtration area results in processing twice as much solution in the same time.

Recent developments have resulted in the construction of ultrafilters with a more or less uniform exclusion range of 10^6 daltons. With such filters virus suspensions can be freed from residual serum proteins and other contaminants originating from the tissue culture medium, making subsequent purification superfluous. Together with research on membranes with low adsorptive capacity, this development will certainly increase general interest in the application of ultrafiltration procedures in animal cell biotechnology.

A technique has been described which combines the advantages of ultrafiltration as a concentration step with the powerful purifying potential of affinity chromatography: "ultrafiltration affinity purification" (51a). Using a modified soluble polymer as affinity matrix and a hollow fiber unit with a molecular weight cut-off of 10^6 daltons the product of interest can be concentrated, whereas contaminants are washed-off through the large pores. Although thus far, only the principle has been demonstrated and several problems must be overcome (40a), this system deserves further development.

6. SUMMARY AND DISCUSSION

For the concentration of suspensions containing proteinaceous material different methods are available: precipitation, adsorption/elution, centrifugation (continuous flow) and ultrafiltration. In Table I advantages and disadvantages of the four systems have been put together. It can be seen that precipitation, adsorption/elution and centrifugation have a limited potential for scaling up. An exception is precipitation with PEG, because this method requires relatively low concentrations of the polymer for complete precipita-

TABLE I   Evaluation of Different Concentration Methods

	Ultrafiltration	Continuous flow centrifugation	Adsorption/ elution	Precipitation
Equipment	Simple	Complex	Simple	Simple
Cost				
Investment	Low	High	Low[a]	Low[b]
Maintenance	Low	High	Low[a]	Low[b]
Scaling up	Excellent	Limited	Limited	Limited
Purifying potential	Poor	Good	Excellent	Intermediate
Sterility/containment	Good	Special precautions necessary	Good	Good

[a] Price of matrix not included.
[b] Price of precipitant not included.

tion of proteinaceous particles. On the other hand, when relatively small volumes are to be concentrated these three methods can readily be applied. They offer the additional advantage of simultaneous moderate or considerable purification of the component of interest. This is most notable for adsorption/elution (affinity chromatography) and centrifugation. With regard to centrifugation the range of application is restricted to fairly large particles with sedimentation coefficients exceeding about 200S, such as viruses. The cost of investment and maintenance of centrifugation equipment is prohibitive for its widespread use.

For the concentration of very large volumes ultrafiltration is the method of choice. With basically the same equipment its capacity can be increased from small to very large by simply extending the number of units in the system. The equipment is not expensive and can be operated for many years without significant costs. With ultrafiltration contaminants are also concentrated to a greater or lesser extent, although the development of new membranes may reduce this problem. For the processing of bulk amounts of cell culture media ultrafiltration and possibly PEG precipitation are most useful for reducing the volume considerably in a first step. Subsequently, a further concentration together with a purification may be obtained by application of one of the other methods.

ACKNOWLEDGMENTS

I am indebted to Ir. A. L. van Wezel for helpful comments and critical reading of the manuscript. Mrs. B. v. d. Laan and S. Vermeij for design of the drawings and Karin de Froe for invaluable secretarial help.

REFERENCES

1. Aaslestad, H. G., and Wiktor, T. J. (1972). Recovery of protective activity in rabies virus vaccines concentrated and purified by four different methods. *Appl. Microbiol.* **24**, 37–43.
2. Adamowicz, P., Legrand, B., Guesche, J., and Prunet, P. (1974). Un nouveau procède de concentration et de purification du virus. Application au virus de la Fievre Aphteuse produit sur cellules BHK21, pour l'obtention de vaccins hautement purifiés. *Bull. Off. Int. Epizoot.* **81**, 1125–1150.
3. Anderson, N. G., Waters, D. A., Nunley, E. C., Gibson, R. F., Schilling, R. M., Denny, E. C., Cline, G. B., Babelay, E. F., and Perardi, T. E. (1969). K-series centrifuges. I. Development of the K-II continuous-sample-flow-with-banding centrifuge system for vaccine purification. *Anal. Biochem.* **32**, 460–494.
4. Atanasiu, P., Tsiang, H., Lavergne, M., and Chermann, J. C. (1977). Purification par centrifugation zonale d'un vaccin antirabique humain obtenu sur cellules renales de foetus bovins. *Ann. Microbiol. (Paris)* **128B**, 297–302.
5. Barlow, D. F. (1972). The aerosol stability of a strain of foot-and-mouth disease virus and

the effects on stability of precipitation with ammonium sulphate, methanol or polyethylene glycol. *J. Gen. Virol.* **15,** 17–24.

6. Baron, S., Dianzani, F., and Stanton, G. J., eds. (1981–1982). "The Interferon system: A review to 1982," Part I, Sect. V (Tex. Rep. Biol. Med., Vol. 41).

7. Barteling, S. (1979). Some aspects of FMDV-production in growing cells and a closed system for concentration of FMDV by polyethylene glycol. *Dev. Biol. Stand.* **42,** 71–74.

8. Barth, R., Gruschkau, H., Jaeger, O., and Milcke, L. (1982). Purification, concentration and evaluation of rabies virus antigen. *Comp. Immunol. Microbiol. Infect. Dis.* **5,** 211–216.

9. Benmansoeur, A. (1982). Purification du virus rabique par centrifugation isopycnique a flux continu sur rotor JCF-Z. *Ann. Virol.* **133E,** 273–279.

10. Billiau, A., Van Damme, J. Van Leuven, F., Edy, V. G., De Ley, M. Cassiman, J.-J., Van den Berghe, H., and De Somer, P. (1979). Human fibroblast interferon for clinical trials: Production, partial purification and characterization. *Antimicrob. Agents Chemother.* **16,** 49–55.

11. Bobbitt, J. L., Clavin, S. A., Hutchins, J. F., Arnett, G. C., and Hull, R. N. (1980). Studies on plasminogen activator from pig kidney cell cultures. III. Purification and characterization. *Thromb. Res.* **18,** 315–332.

12. Bock, H. G., Skene, P., Fleischer, S., Cassidy, P., and Harshman, S. (1976). Protein purification: Adsorption chromatography on controlled pore glass with the use of chaotropic buffers. *Science* **191,** 380–382.

13. Cockerell, R. M., and Mostratos, A. (1982). Concentration of influenza virus using polyethylene glycol. *Acta Virol. (Engl. Ed.)* **26,** 292–294.

14. Cohn, E. J., Gurd, F. R. N., Surgenor, D. R. M., Barnes, B. A., Brown, R. K., Derouaux, G., Gillespie, J. M., Kahnt, F. W., Lever, W. F., Liu, C. H., Mittelman, D., Mouton, R. F., Schmid, K., and Uroma, E. (1950). A system for the separation of the components of human blood: Quantitative procedures for the separation of the protein components of human plasma. *J. Am. Chem. Soc.* **72,** 465.

15. Dixon, M. (1953). A nomogram for ammonium sulphate solutions. *Biochem. J.* **54,** 457–458.

16. Dixon, M., and Webb, E. C. (1961). Enzyme fractionation by salting-out: A theoretical note. *In* "Advances in Protein Chemistry" (C. B. Anfinsen, Jr., M. L. Anson, K. Bailey, and J. T. Edsall, eds.), Vol. 16, pp. 197–219. Academic Press, New York.

17. Edy, V. G., Braude, J. A., de Clercq, E., Billiau, A., and De Somer, P. (1976). Purification of interferon by adsorption chromatography on controlled pore glass. *J. Gen. Virol.* **33,** 517–521.

18. Fantes, K. H. (1966). Purification, concentration and physico-chemical properties of interferons. *In* "Interferon" (N. B. Finter, ed.), Chapter 5, pp. 119–179. North-Holland Publ., Amsterdam.

19. Filczak, K., and Korbecki, M. (1978). Preparation of rubella virus haemagglutinating antigen by polyethylene glycol precipitation. *Acta Virol.* **22,** 70–73.

20. Gerin, J. L., and Anderson, G. (1969). Purification of influenza virus in the K-II zonal centrifuge. *Nature (London)* **221,** 1255–1256.

21. Green, A. A., and Hughes, W. L. (1955). Protein Fractionation on the basis of solubility in aqueous solutions of salts and organic solvents. *In* "Methods in Enzymology" (S. P. Colowick and N. D. Kaplan, eds.), Vol. 1, pp. 67–138. Academic Press, New York.

22. Griffith, O. M. (1979). "Techniques of Preparative, Zonal and Continuous Flow Ultracentrifugation." Beckman Instruments Inc., Palo Alto, California.

23. Gronow, M., and Bliem, R. (1983). Production of human plasminogen activators by cell culture. *Trends Biotechnol.* **1,** 26–29.

24. Hebert, T. T. (1963). Precipitation of plant viruses by polyethylene glycol. *Phytopathology* **53,** 362.

25. Henderson, L. E., Heweton, J. F., Hopkins, R. F., Sowder, R. C., Neubauer, R. H., and Rabin, H. (1983). A rapid, large scale purification procedure for gibbon interleukin 2. *J. Immunol.* **131**, 810–815.

26. Hilfenhaus, J., Kohler, R., Barth, R., Majer, M., and Mauler, R. (1976a). Large-scale purification of animal viruses in the RK-model zonal ultracentrifuge. I. Rabies virus. *J. Biol. Stand.* **4**, 263–271.

27. Hilfenhaus, J., Kohler, R., and Behrens, F. (1976b). Large-scale purification of animal viruses in the RK-model zonal ultracentrifuge. II. Influenza, mumps and Newcastle disease viruses. *J. Biol. Stand.* **4**, 273–283.

28. Hilfenhaus, J., Kohler, R., and Gruschkau, H. (1976c). Large scale purification of animal viruses in the RK-model zonal ultracentrifuge. III. Semliki forest virus and vaccinia virus. *J. Biol. Stand.* **4**, 285–293.

29. Huller, W. (1967). Virus isolation with glass of controlled pore size: MS2 bacteriophage and Kilham virus. *Virology* **33**, 740–743.

30. Iverius, P. H., and Laurent, T. C. (1967). Precipitation of some plasma proteins by the addition of dextran or polyethylene glycol. *Biochim. Biophys. Acta* **133**, 371–373.

30a. Janson, J. C. (1984). Large-scale affinity purification—state of the art and future prospects. *Trends Biotechnol.* **2**, 31–38.

31. Johnson, R. W., Perry, A., Orson, R. R., and Shibley, G. P. (1976). Method for reproducible large-volume production and purification of Rauscher murine leukemia virus. *Appl. Environ. Microbiol.* **31**, 182–188.

32. Johnston, M. D., Christofinis, G., Ball, G. D., Fantes, K. H., and Finter, N. B. (1978). A culture system for producing large amounts of human lymphoblastoid interferon. *Dev. Biol. Stand.* **42**, 180–192.

33. Juckes, I. R. M. (1971). Fractionation of proteins and viruses with polyethylene glycol. *Biochim. Biophys. Acta* **229**, 535–546.

34. Klein, F., Ricketts, R. T., Jones, W. I., De Armon, I. A., Temple, M., Zoon, K. C., and Bridgen, P. J. (1979). Large-scale production and concentration of human lymphoid interferon. *Antimicrob. Agents Chemother.* **15**, 420–427.

35. Kluft, C., van Wezel, A. L., Van der Velden, C. A. M., Emeis, J. J., Verheijen, J. H., and Wijngaards, G. (1983). Large scale production of extrinsic (tissue-type) plasminogen activator from human melanoma cells. *Adv. Biotechnol. Processes* **2**, 98–109.

36. Krasilnikow, I. V., Aksenova, T. A., and Mchedlishvili, B. V. (1981). Purification and concentration of rabies virus by chromatography on chemically modified porous silicates. *Acta Virol.* **25**, 205–212.

37. Langford, M. P., Georgiades, J. A., Stanton, G. J., Dianzani, F., and Johnson, H. M. (1979). Large-scale production and physico-chemical characterization of human immune interferon. *Infect. Immun.* **26**, 36–41.

38. Lavender, J. F., and van Frank, R. M. (1971). Zonal-centrifuged purified duck embryo cell culture rabies vaccine for human vaccination. *Appl. Microbiol.* **22**, 358–365.

39. Lebermann, R. (1966). The isolation of plant viruses by means of "simple" coacervates. *Virology* **30**, 341–347.

40. Leuthard, P., and Schuerch, A. R. (1980). A simple and rapid method for concentration of interferon and removal of concentrated inducing virus. *Experientia* **36**, 1447.

40a. Lowe, C. R. (1984). New developments in down stream processing. *J. Biotechnol.* **1**, 3–12.

41. Maizel, A. L., Mehta, S. R., Hauft, S., Franzini, D., Lackman, L. B., and Ford, R. J. (1981). Human T-lymphocyte/monocyte interaction in response to lectin: Kinetics of entry in the S-phase. *J. Immunol.* **127**, 1058–1064.

42. Marquardt, H., Wilson, G. L., and Todaro, G. J. (1980). Isolation and characterization of a

multiplication-stimulating activity (MSA)-like polypeptide produced by a human fibrosarcoma cell line. *J. Biol. Chem.* **255**, 9177–9181.

43. McAleer, W. J., Hurni, W., Wasmuth, E., and Hilleman, M. R. (1979). High resolution flow–zonal centrifuge system. *Biotechnol. Bioeng.* **21**, 317–321.

44. McClendon, J. H. (1954). The physical environment of chloroplasts as related to their morphology and activity in vitro. *Plant Physiol.* **29**, 448–458.

44a. Mees, I., Chin, D., Fox, F., and Krim, K. (1984). Purification of human leukocyte interferon alpha by carboxymethyl controlled pore glass bead chromatography. *Arch. Virol.* **81**, 303–311.

45. Mikhailovsky, E. M., Tsiang, H., and Atanasiu, P. (1971). Concentration du virus rabique par le polyethylene glycol. *Ann. Inst. Pasteur, Paris* **121**, 563–568.

46. Morrow, A. W., Whittle, C. J., and Eales, W. A. (1974). A comparison of methods for the concentration of foot-and-mouth-disease virus for vaccine preparation. *Bull. Off. Int. Epizoot.* **81**, 1155–1167.

47. Osborne, L. C., Georgiades, J. A., and Johnson, H. M. (1979). Large-scale production and purification of mouse immune interferon. *Infect. Immun.* **23**, 80–86.

48. Pestka, S., ed. (1981). "Methods in Enzymology," Vol. 78, Part A, Sects. II and V. Academic Press, New York.

49. Pharmacia (1982). "Chelating Sepharose 6B. For Metal Affinity Chromatography" (brochure). Pharmacia Fine Chemicals, Uppsala, Sweden.

50. Polson, A., Potgieter, G. M., Largier, F. J., Mears, G. E. F., and Joubert, F. J. (1964). The fractionation of protein mixtures by linear polymers of high molecular weight. *Biochim. Biophys. Acta* **82**, 463–475.

51. Porath, J., Carlsson, J., Olsson, I., and Belfrage, G. (1975). Metal chelate affinity chromatography, a new approach to protein fractionation. *Nature (London)* **258**, 598–599.

51a. Ramstorp, M. (1983). Novel affinity techniques. Ph.D. thesis. University of Lund, Sweden.

52. Reimer, C. B., Baker, R. S., van Frank, R. M., Newlin, T. E., Cline, G. B., and Anderson, N. G. (1967). Purification of large quantities of influenza virus by density gradient centrifugation. *J. Virol.* **1**, 1207–1216.

53. Rijken, D. C., and Collen, D. (1981). Purification and characterization of the plasminogen activator secreted by human melanoma cells in culture. *J. Biol. Chem.* **257**, 7035–7041.

54. Rommerts, F. F. G., Clotscher, W. F., and Van der Molen, H. J. (1977). Rapid concentration and dialysis of proteins with single hollow fibers: Possible applications in analysis of protein secretion by isolated cells and steroid radioimmunoassays. *Anal. Biochem.* **82**, 503–511.

55. Schneider, L. G., Horzinek, M., and Matheka, H. D. (1971). Purification of rabiesvirus from tissue culture. *Arch. Gesamte Virusforsch.* **34**, 351–359.

56. Sokol, F. Kuwert, E., Wiktor, T. J., Hummeler, K., and Koprowski, H. (1968). Purification of rabiesvirus grown in tissue culture. *J. Virol.* **2**, 836–849.

57. Strohmaier, K. (1967). Diffusion. *In* "Methods in Virology" (K. Maramorosch and H. Koprowski, eds.), Vol. 2, pp. 245–274. Academic Press, New York.

58. Takahashi, T., Nakagawa, S., Hashimoto, T., Takahashi, K., Imai, M., Miyakawa, Y., and Mayumi, M. (1976). Large scale isolation of Dane particles from plasma containing hepatitis-B antigen and demonstration of a circular double-stranded DNA molecule extruding directly from their cores. *J. Immunol.* **117**, 1392–1397.

59. Tamma, T.-A., and Takano, T. (1978). A new, rapid procedure for the concentration of C-type viruses from large quantities of culture media: Ultrafiltration by Diaflo membrane and purification by Ficoll gradient centrifugation. *J. Gen. Virol.* **41**, 135–141.

60. Toplin, I., and Sottong, P. (1972). Large-volume purification of tumour viruses by use of zonal centrifuges. *App. Microbiol.* **23**, 1010–1014.

61. Trudel, M., and Payment, P. (1980). Concentration and purification of rubella virus hemagglutinin by hollow fiber ultrafiltration and sucrose gradient density centrifugation. *Can. J. Microbiol.* **26**, 1334–1339.

62. Vajda, B. P. (1978). Concentration and purification of viruses and bacteriophages with polyethylene glycol. *Folia Microbiol. (Prague)* **23**, 88–96.

63. Valeri, A., Gazzei, G., Botti, R., Pellegrini, V., Corradeschi, A., and Soldateschi, D. (1981). One-day purification of influenza A and B vaccines using molecular filtration and other physical methods. *Microbiologica (Bologna)* **4**, 403–412.

64. Van der Marel, P., van Wezel, A. L., Hazendonk, A. G., and Kooistra, A. (1980). Concentration and purification of poliovirus by immune adsorption on immobilized antibodies. *Dev. Biol. Stand.* **46**, 267–273..

65. Van Ramshorst, J. D. (1963). The fractionation of diphtheria toxoid by means of ammonium sulphate. *Antonie van Leeuwenhoek* **29**, 101–111.

66. van Wezel, A. L., Van Herwaarden, J. A. M., and Van de Heuvel-de Rijk, E. W. (1979). Large-scale concentration and purification of virus suspension from microcarrier culture for the preparation of inactivated virus vaccines. *Dev. Biol. Stand.* **42**, 65–69.

67. Von Bonsdorff, C.-H., and Harrison, S. C. (1978). Hexagonal glycoprotein arrays from Sindbis virus membranes. *J. Virol.* **28**, 578–583.

68. Wallis, C., and Melnick, J. L. (1967). Concentration of viruses on aluminum and calcium salts. *Am. J. Epidemiol.* **85**, 459–468.

69. Warrington, R. E., and Morgan, D. O. (1971). Foot-and-mouth disease virus in cattle and pigs: Use of polyethylene glycol or dextran for purifying 19S M immunoglobulin from sera. *Arch. Gesamte Virusforsch.* **33**, 134–144.

70. Weiss, S. A. (1980). Concentration of baboon endogenous virus in large-scale production by use of hollow fiber ultrafiltration technology. *Biotechnol. Bioeng.* **22**, 19–31.

71. Yamamoto, K. R., Alberto, B. M., Lawhorne, L., and Treiber, G. (1970). Rapid bacteriophage sedimentation in the presence of polyethylene glycol and its application to large scale virus purification. *Virology* **40**, 734–744.

9

Virus Fractionation and the Preparation of Sub-unit Vaccines

B. ROBERTS

Glaxo Animal Health Ltd.
Uxbridge, Middlesex, United Kingdom

1. INTRODUCTION

A conventional virus vaccine consists basically of a suspension of virus particles, either attenuated or inactivated, formulated at a concentration

Animal Cell Biotechnology, Vol. 2

which will produce the required immune response in the recipient. The vaccine may be preserved by freeze-drying or it may contain an adjuvant to enhance its potency. It will have been exhaustively checked for freedom from contaminating biological agents and shown to be safe and effective when correctly administered. Despite the care that goes into its production it will remain, from the biochemical point of view, a very impure preparation of which only a small part can be regarded as the active ingredient.

An important difference between live attenuated and inactivated virus vaccine lies in the amount of viral material that constitutes an immunizing dose. Generally speaking, live attenuated vaccines require only sufficient virus in a dose to establish a sub-clinical infection in the recipient. They rely on the subsequent multiplication of the virus contained in the dose to provide the necessary immunogenic stimulus. In contrast, a dose of inactivated vaccine needs to contain enough viral material from the start in order to stimulate the host's immune system and so confer protection against a future infection. Unless there is sufficient antigenic mass contained in the dose, the vaccine will not be effective.

This need to ensure that sufficient viral antigen is incorporated into a dose of inactivated vaccine presents a major problem in preparing these vaccines. Whereas a virus titre of about 10^3 TCD_{50} may be sufficient for a dose of live attenuated vaccine, it may be necessary to use the equivalent of $10^7–10^8$ TCD_{50} for an inactivated vaccine to be effective. Hence, the concentration of virus in a dose of inactivated vaccine may be 10 to a 100,000 times that in a dose of live attenuated vaccine. It is not always possible to find acceptable cell systems in which the vaccine virus can be propagated to yield titres high enough for the direct preparation of an inactivated vaccine. In such cases it may be necessary to resort to using concentrating techniques in order to obtain virus of sufficient titre. As a consequence, the final dosage form of the vaccine will contain not only large amounts of various viral proteins but also protein concentrated from the substrate on which it was propagated.

This means that inactivated vaccines will inevitably contain a considerable amount of irrelevant protein which does not contribute to the potency of the vaccine but may lead to undesirable side effects in the patient. Various purification procedures used during manufacture may minimise the amount of unwanted protein retained from the substrate but if whole virus is used as the active constituent there will always be excess viral protein present in every dose of vaccine.

The concept of a biochemically pure vaccine can be realised if the components of the virus particle which stimulate the formation of neutralizing antibody in the recipient can be identified. A preparation containing these components alone could be regarded as a purified antigen preparation. It would be expected to induce highly specific immunity, free from any of the side effects attributable to viral nucleoprotein or residual substrate protein, which might result from a conventional whole virus preparation.

This chapter describes the progress that has been made towards the objective of producing purified preparations of biologically active viral sub-units and the application of these techniques to improving inactivated viral vaccines.

2. METHODS OF FRACTIONATION

2.1 Solubilization

The sensitivity of numerous viruses to detergents and lipid solvents has long been known. Many viruses lose their infectivity when treated with one of these agents or lose their ability to stimulate the formation of neutralizing antibody (9, 19). Electron microscopy shows that susceptible viruses are disrupted by this treatment. In fact, these techniques have been used as a tool to investigate virus morphology and structure (19, 38). Sensitivity to lipid solvents is a characteristic by which viruses are classified, as it identifies those viruses with an outer lipoprotein envelope. Indeed, it is from among the enveloped viruses that most progress with sub-unit extraction has been made. It was the demonstration that a mixture of diethyl ether and Tween would disrupt the lipid membrane of the influenza virion (19) that led the way to numerous studies with a variety of agents.

Organic Solvents. Diethyl ether, with and without Tween, and tri(n-butyl) phosphate have been described for the disruption and sub-unit preparation of influenza (19, 31, 42) and rabies viruses (32). Chloroform and acidified methanol have been used, not very successfully, to prepare sub-unit infectious bovine rhinotracheitis vaccine (15) (a herpesvirus vaccine) but generally organic solvents have not proved to be a tool of choice for sub-unit preparation.

Detergents. The observation that treatment of influenza vaccine with sodium deoxycholate or with sodium dodecyl sulphate (42) reduced its pyrogenicity drew attention to these agents. However, their rather severe action tends to destroy sub-unit antigenicity and their value in the preparation of vaccines is not convincing. Selective solubilization of glycoprotein from para-influenza virus has been claimed for the cationic detergent cetyltrimethylammonium bromide (37). Far more widely used for sub-unit solubilization are thenon-ionic detergents such as Triton X-100 and Nonidet P 40. These agents are sufficiently mild in their action not to harm the sub-units, which consequently retain their antigenicity and are suitable for vaccine preparation. A somewhat unusual detergent, the zwitterionic detergent Empigen BB (16), has been used successfully to solubilize glycoprotein subunits from para-influenza virus.

Enzymes. Proteolytic enzymes have been used for sub-unit preparation. Bromelain (5) has been described for the preparation of influenza virus haemagglutinin and yields a product free of the neuraminidase which would normally be present after extraction with detergent. Trypsin (9) will remove the surface projections of vesicular stomatitis virus and a protease obtained from *Streptomyces fradiae* (17) will remove haemagglutinin and neuraminidase from influenza virus.

Other Agents. Sub-unit antigens have been extracted directly from infected cells still attached to the growth surface, by treatment with 4 M urea (26). The 12 S sub-unit antigen of foot-and-mouth disease virus can be prepared by dialysis against a citrate buffer at pH 4.5 (13).

TABLE I Agents Used in the Solubilization of Viral Sub-units

Solubilizing agent	Viruses used	Reference
Organic solvents		
Diethyl ether (with	Vesicular stomatitis	9
Tween)	Influenza	19,42
	Semliki Forest	30
	Herpes simplex	45
Chloroform and methanol	Bovine rhinotracheitis	15
Tri(n-butyl) phosphate	Influenza	31
	Rabies	32
Detergents—ionic		
Sodium deoxycholate	Vesicular stomatitis	9
	Influenza	20,42
	Semliki Forest	30
Sodium dodecyl sulphate	Influenza	42
	Herpes simplex	21
Myristyl trimethylammonium bromide	Parainfluenza	37
Cetyltrimethylammonium bromide	Parainfluenza	37
Detergents—non-ionic		
Nonidet P40	Vesicular stomatitis	9
	Measles	10
	Rubella	6
	Semliki Forest	1,30

(continued)

A selection of solubilizing agents and the viruses with which they have been used to prepare sub-units is given in Table I.

2.2. Separation

The treatment of a virus suspension with a suitable solubilizing agent will satisfactorily disrupt the virus particles to release the required sub-units. The result is a mixture of viral cores and surface sub-units from which pure preparations of sub-units can be obtained only by applying an appropriate separation procedure. Although some workers (30) have studied the properties of sub-units without undertaking this step, it is preferable to do so, especially if the objective is to produce a purified non-reactogenic vaccine.

One of the simplest methods of separating sub-units from unwanted viral debris is to adsorb the mixture with aluminium hydroxide (20). As, for vaccine purposes, it may ultimately be necessary to formulate the final dosage form with an adjuvant, the use of aluminium hydroxide as a means of separating the sub-units has obvious attractions. The hydroxide gel is added

TABLE I (*Continued*)

Solubilizing agent	Viruses used	Reference
	Herpes simplex	7,22,34,40,41
	Bovine rhinotracheitis	25
	Marek's disease	24
Triton X-100	Parainfluenza	39
	Semliki Forest	3,29,39
	Aujeszky's disease	35
	Bovine rhinotracheitis	25
	Rabies	2
Triton N101	Influenza	4,11
Saponin	Rabies	38
	Semliki Forest	30
Detergents—zwitterionic		
Empigen BB	Influenza	16
	Parainfluenza	16
	Newcastle disease	16
Enzymes		
Trypsin	Vesicular stomatitis	9
Proteolytic ex *Streptomyces fradiae*	Influenza	17
Bromelain	Influenza	5
Others		
Urea	Herpes simplex	26
pH	Foot-and-mouth disease	13

to the sub-unit mixture at a suitable pH and then centrifuged. The supernatant fluid is discarded and the pellet of aluminium hydroxide with adsorbed sub-units is resuspended in an appropriate diluent. A disadvantage of this method is that the sub-units are not available in a free form. This makes accurate assay very difficult and presents severe problems when attempts are made to standardise the product.

Column chromatographic methods have been used on a preparative scale and DEAE–Sephadex A50 columns have proved satisfactory for the separation of hexons and fibre sub-units of adenovirus type 5 (12). Influenza virus, when adsorbed onto a calcium phosphate column and then eluted with tri(n-butyl) phosphate, yields a pure preparation of sub-units and the viral cores remain on the column (31).

By far the most widely favoured method of separating sub-units is density gradient centrifugation, usually using a sucrose gradient. The method has the advantage of being applicable both to preparative and to large-scale operations. Sub-units are usually less dense than the virus cores from which they have been detached. Therefore when a solubilized mixture is centrifuged through a sucrose density gradient, sub-units and cores will form distinct and well-separated bands which can be collected as separate fractions.

A further advance on the density gradient technique has found an application in the commercial manufacture of a sub-unit influenza vaccine (4). The non-ionic detergent Triton N101, when incorporated into a sucrose gradient loaded into a zonal centrifuge rotor, forms a band of detergent micelles in the gradient under the influence of centrifugation. When influenza virus particles are centrifuged through this gradient they are disrupted when they reach the detergent band. The surface glycoproteins, haemagglutinin and neuraminidase, become detached from the virus cores and form a band in the gradient close to the detergent. The viral cores move to a denser region of the gradient so that a separation is effected. This technique solubilizes and separates the sub-units in one operation and collection of the appropriate fractions from the gradient provides a sub-unit suspension almost entirely free of viral core material.

2.3 Purification

Of the methods described above for separating viral sub-units from viral cores, those involving adsorption and elution have the advantage of providing a sub-unit preparation free from the solubilizing agent. Unless, for the purposes for which the sub-unit preparation is required, the presence of the solubilizing agent is not harmful, some method for its removal must be employed.

One variation on the use of an adsorption and elution method is to use Bio-Beads SM2 (2), a styrene divinylbenzene polymer which has an affinity for non-ionic detergents such as Triton. Sub-unit preparations are simply mixed with an appropriate quantity of Bio-Beads. The detergent is preferentially absorbed by the polymer and the supernatant fluid, free of detergent, is then decanted and collected.

Triton can also be removed by taking advantage of the cloud point phenomenon (4). Fractions collected from a density gradient separation containing sub-units and Triton are pooled together and 2 M phosphate buffer at pH 7.0 is added to raise the molarity of the mixture to 0.5 M. At room temperature under these conditions the cloud point of the detergent is exceeded. On standing for 18 hr a phase separation occurs in which the detergent floats to the top of the mixture and the sub-units remain in the lower layer, from which they can be collected free of detergent.

3. APPLICATIONS

3.1. Influenza Vaccine

Numerous methods for disrupting influenza virus have been described and the most important ones have been referred to above. However disruption, even if followed by an efficient separation and purification technique, is not necessarily going to provide an effective vaccine. It is essential to ensure that the sub-units are released from the virus and collected in such a way that their antigenicity is retained (4, 10, 39).

Treatment with organic solvents such as diethyl ether will break up the lipid membrane of the virion and release the internal nucleoprotein. The pieces of lipid membrane, still retaining their surface glycoprotein sub-units, form small rosettes which are antigenic and can form the basis of a vaccine (19). Enzymes such as bromelain will cleave the rod-like sub-units at the point where they are inserted into the lipid membrane, leaving the virus core intact. Non-ionic detergents such as Triton N101 will free the glyoprotein sub-units intact from the base of the structure without any cleavage (4, 11). Again, the virus core remains intact. The antigenicity of the sub-units is particularly well retained when the last of these three methods is used.

Examination by electron microscopy of sub-unit preparations prepared with Triton N101 reveals a probable explanation for this observation (4). The glycoproteins projecting from the surface of the influenza virion are of two kinds—haemagglutinin and neuraminidase. Treatment with Triton N101 releases these sub-units intact and in aqueous suspension they reaggregate into small clusters of their own kind. Hence, groups of haemagglutinin rods

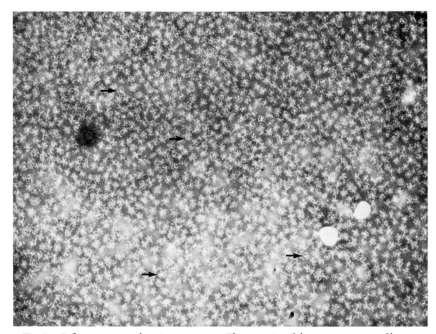

Fig. 1. Influenza virus sub-unit preparation. The majority of the aggregates are of haemag-glutinin. Aggregates of neuraminidase are arrowed.

will adhere to each other by the 'sticky' end originally inserted into the lipid membrane of the virion. Similarly, the neuraminidase sub-units will form aggregates which, because of the bulbous outer ends of the rod structure of this glycoprotein, are easily recognised as distinct from haemagglutinin aggregates (Fig. 1). This ability to form aggregates, which is not shared by sub-units released by bromelain, appears to be of critical importance in retaining sufficient antigenicity for vaccine purposes.

The objective when developing a sub-unit influenza vaccine is to produce a vaccine which is as immunogenic as conventional whole virus vaccine but less reactogenic (27). Comparative studies with modern vaccines of the two kinds have tended to show that this objective is obtained, but the differences have not been dramatic (27, 28). It was originally considered necessary to enhance the antigenicity of sub-unit influenza vaccine by adsorption onto aluminium hydroxide (4) but, by doing this, some of the benefit of decreased reactogenicity was lost. It is now accepted that the aluminium hydroxide adjuvant is not necessary and comparison of adsorbed with aqueous sub-unit vaccines has shown equal immunogenicity and an even lower incidence of reactions for the aqueous sub-unit preparation (28).

Purified sub-unit influenza vaccines were the first vaccines prepared to a high level of biochemical purity to be made commercially available.

3.2. Para-influenza and Measles

Like influenza virus, the paramyxoviruses such as measles, Sendai virus and the para-influenza viruses are all enveloped viruses. They mature from the cell membranes of infected cells and express two glycoproteins identified as fusion protein (F) and either HN glycoprotein (para-influenza virus) or HA glycoprotein (measles). HN glycoprotein shows both haemagglutination and neuraminidase properties, whereas HA glycoprotein consists only of haemagglutinin (16).

Isolation and purification of these glycoprotein sub-units have been achieved with the zwitterionic detergent Empigen BB and the crude preparation so obtained was purified on a DEAE–Bio-GelA column. The glycoproteins are identified by sodium dodecyl sulphate polyacrylamide gel electrophoresis (SDS–PAGE), using carbohydrate staining techniques. The concentration of Empigen BB may be critical; 2% is commonly used, but for Newcastle disease virus 0.1% gave maximum glycoprotein yields (16). The method is very efficient and almost complete haemagglutinin and neuraminidase recovery has been reported. Preparations of sub-units made by this method have been shown to be immunogenic but the suitability of the method for producing sub-unit vaccines has still to be demonstrated.

The preparation of measles virus glycoproteins followed by the additional step of reconstituting them into artificial lipid membranes to give 'virosomes' with projecting glycoprotein spikes has also been described (10). Initial disruption of purified virus was achieved with Nonidet P40 and the solubilized glycoproteins were immobilised on a lentil lectin column. This allowed the Nonidet P40 to be substituted by octyl β-D-glucopyranoside, which is fully dialysable. The glycoprotein sub-units were eluted with α-methyl mannoside, in a form suitable for making virosomes, by treatment with phospholipid. This technique has been proposed as a useful tool for the analysis of the interaction of the virus with complement, lymphocytes and cells (10). As such it should enable further advances in understanding the human disease but its application in the field of vaccination has not been investigated.

A novel form of sub-unit preparation, described as an immuno-stimulating complex (iscom), can be formed from the envelope glycoproteins of para-influenza 3 or measles viruses by the action of the glycoside Quil-A, a purified form of saponin (29a). These complexes have been shown to be up to 10 times more immunogenic than conventionally aggregated glycoprotein sub-unit preparations and also to stimulate the formation of antibody to fusion (F) protein. Electron microscopy shows iscoms to have a characteristic cage-like structure about 12 nm in diameter.

3.3. Rabies and Vesicular Stomatitis

The rhabdoviruses, of which both rabies and vesicular stomatitis are examples, are characterised morphologically as bullet-shaped particles with one end rounded and the other end flattened. One of the most notable features shown by negative staining electron microscopy is a covering of surface projections located on a double membrane over all of the particle except the flat end. These projections are about 7 nm long and have been identified as glycoprotein (38).

Highly purified preparations of this surface glycoprotein have been obtained by treating rabies virus with Nonidet P40, tri(n-butyl) phosphate or bromelain (38). These preparations contain small amounts of membrane protein and final purification of the glycoprotein sub-unit can be obtained by isoelectric focusing (38). These preparations have been shown to induce the formation of neutralizing antibodies in rabbits and to protect mice against an intracerebral lethal challenge with virulent virus (14, 38). The success of this work has led to serious suggestions that such preparations should be considered as vaccines for human use (14, 38). The choice of cell substrate for producing such a vaccine need not be limited to the few normally accepted for vaccine production, because the final product would be highly purified and free from cellular and other viral components (38).

Unlike whole rabies virus, purified glycoprotein does not exhibit haemagglutination with goose red blood cells. Treatment of whole virus with saponin will yield a soluble rabies virus haemagglutinin which still protects mice against virulent challenge (38). Treatment of the saponin extract with sodium deoxycholate causes a loss of haemagglutinating activity, which suggests that the haemagglutination by rabies virus is dependent on the integrity of the viral membrane (38).

Similar studies with vesicular stomatitis virus have demonstrated that a stepwise degradation of the virus particle is possible (9). The surface projections can be removed by treatment with trypsin, which destroys the ability of the virus to induce the formation of neutralizing antibodies. The immunizing antigen can be released by treatment with Tween and ether or Nonidet P40 or sodium deoxycholate.

Sucrose gradient centrifugation can be used to separate the surface subunits from the virus cores after detergent or trypsin treatment. The resulting components are found in different positions in the gradient according to the treatment used. Hence, Nonidet P40 produces immunizing antigens which remain near the top of the gradient and viral 'skeletons' which sediment to the bottom. The ribonucleoprotein released by sodium deoxycholate sediments to the 140S position, whereas the projection-free component obtained from trypsin treatment sediments to approximately the same position as untreated virus (9).

Purified glycoprotein from vesicular stomatitis virus has also been incorporated into phospholipid vesicles to form liposomes. The method involves treatment with dimyristoylphosphatidylcholine in cholic acid. These preparations have been used to elicit anti-vesicular stomatitis cytotoxic T lymphocytes for use in fundamental studies on the immune response to viral antigens (36).

3.4 Herpes Simplex Virus

Of all the virus infections against which the production of sub-unit vaccines has been considered, herpes simplex virus provides the strongest argument for this approach. The use of a live attenuated vaccine against herpes simplex raises concern over the ability of the vaccine virus to establish a latent focus with possible later complications of reactivation. Furthermore, particularly with herpes simplex type 2, it has been postulated that the virus may be oncogenic in humans (18, 43). Therefore, sufficient doubt exists over the safety of live attenuated herpes simplex vaccines to discourage their development. Furthermore, as it is not necessary for virus to be infectious in order to transform cells, the concern over the oncogenicity of live vaccine can be extended to any viral preparation containing viral DNA (18, 43). It is argued, therefore, that the only safe way of vaccinating against herpes simplex is to use a purified, DNA-free, sub-unit vaccine.

Simple detergent-treated preparations of inactivated herpes simplex type 1 virus, purified by density gradient centrifugation to remove viral DNA, were first shown to confer protection in mice against challenge with virulent virus (21). This protection was demonstrated despite neutralizing antibodies being either undetectable or at a very low level. A similar type 1 vaccine prepared with non-ionic detergent has been shown to protect mice against herpes simplex type 2 (40). Sub-unit preparations obtained from Nonidet P40 treatment of inactivated virus suspensions have also been found satisfactory for raising neutralizing antibodies in guinea pigs (22). Restriction of a latent herpes simplex infection in rabbits has been demonstrated with a vaccine of this type (34). Reduction in corneal ulceration following ocular inoculation of herpes simplex virus in sub-unit vaccinated rabbits has also been reported (8).

All of these examples illustrate that detergent treatment of inactivated herpes simplex virus yields an immunogenic preparation which can be freed from viral nucleic acid.

Incorporation of solubilized glycoprotein sub-units, prepared from herpes simplex virus by deoxycholate treatment, into liposomes in a manner previously described for vesicular stomatitis virus and for measles virus has also been reported (23). These preparations were used to induce cytotoxic T lymphocyte responses in splenocyte cultures from virus-primed mice. This was successful provided both viral and H-2 antigens were present in the liposomes. The ability of *Herpesvirus saimiri* to induce viral antigen on the plasma membrane of the cells that it infects has offered another means of antigen preparation (33). Treatment of infected cells with dithiothreitol and formaldehyde yields vesicles of plasma membrane which contain no detectable live virus. These vesicles are antigenic and produce neutralizing antibodies in guinea pigs and protect mice against intravaginal challenge with herpes simplex type 2.

A method of preparing a sub-unit herpes simplex vaccine by extracting infected cells with 4 M urea has been the subject of a patent application. The evidence for the efficacy of this preparation lies in its ability to protect mice (26).

3.5 Other Herpesviruses

Sub-units have been prepared from two herpesviruses of veterinary importance with a view to making vaccines. Cells infected with Aujeszky's disease virus (porcine herpesvirus 1) treated with Triton X-100 in tris buffer at pH 9.0 yield sub-units which are antigenic and which can be separated from viral DNA by centrifugation. Neutralizing antibodies and a cell-mediated immune response were demonstrated in mice after a single inoculation with this preparation. The treatment also afforded an 87% level of protection from challenge with a lethal dose of virus (35).

A similar method of extracting antigenic sub-units has been applied to cell cultures infected with infectious bovine rhinotracheitis virus. When administered with Freund's complete adjuvant to calves, high levels of neutralizing antibodies were produced. Two doses of this vaccine were sufficient to protect calves exposed to intranasal challenge with infectious bovine rhinotracheitis virus (25). Less predictable results have been observed for a sub-unit vaccine prepared by extracting infected cells with an acidified chloroform and methanol mixture. Although this preparation was highly immunogenic in adult cattle, it failed to protect calves (15).

3.6. Semliki Forest Virus

The importance of obtaining the right state of aggregation in preparations of viral glycoprotein sub-units intended for use in vaccines has been demonstrated with Semliki Forest virus (29, 39). Glycoprotein monomeric sub-units prepared by treatment with Triton X-100 were found to be poorly immunogenic. If the detergent was removed by density gradient centrifugation, an octamer with a sedimentation coefficient of 29S was formed which had much greater immunogenicity. Preparations of liposomes, prepared by octylglucoside dialysis using egg lecithin, were similarly much more immunogenic than the monomeric sub-units. It seems clear that the effectiveness of these preparations as immunogens is due to their multimeric structure.

3.7. Rubella Virus

Sub-unit preparations of rubella virus will produce both humoral antibody and cell-mediated immunity when inoculated into rabbits. Treatment of purified virus with Nonidet P40 in a tris–saline buffer followed by sucrose density gradient centrifugation and dialysis of the active fractions has been used to make the vaccine. The fractions containing the envelope sub-units have been found to be almost as effective as live attenuated vaccine in protection studies in rabbits (6).

3.8. Adenovirus

The production of sub-units from a non-enveloped virus and the use of these preparations to vaccinate human volunteers represent an interesting extension to the examples described above. Crystalline sub-unit vaccines containing either purified hexons or purified fibres of adenovirus type 5 are antigenic and non-reactogenic in man (12).

Adenovirus type 5 was cultured in human embryo kidney cells and the cells centrifuged and treated with fluorocarbon. Soluble antigens were separated from whole virions by caesium chloride density gradient centrifugation and hexons and fibres were separated by column chromatography with DEAE–Sephadex A-50. The separated fractions were further purified by rechromatographing followed by sucrose density gradient separation and crystallization. Although it was shown that the fibre sub-units produced what appeared to be a more potent vaccine than the hexon sub-units, both afforded protection against subsequent infection.

4. BENEFITS AND ADVANTAGES OF SUB-UNIT VACCINES

4.1 Efficacy

The preparative processes used for making viral sub-units entail several steps which increase the purity of the final product. Vaccines prepared in this way will consist almost entirely of specific immunizing protein and are thus highly specific antigens. It would be expected that this specificity would bring with it a high degree of efficacy.

From the examples described above it can be seen that sub-unit preparations are highly immunogenic only if prepared in the right way. The state of aggregation of the purified sub-units (29, 39), or their assembly on liposomes (10, 23, 36) or the retention of part of the cell membrane (19) all improve antigenicity, presumably because the necessary spatial configuration is retained.

However, improved efficacy does not necessarily result from a sub-unit vaccine. The immunogenicity frequently requires enhancing by incorporating an adjuvant into the final product (4). Conventional culture methods may not yield sufficient quantities of virus from which to prepare sub-units at a high enough concentration to be effective. If extensive concentration techniques have to be employed to achieve this, then it is unlikely that the benefits of sub-unit preparation will outweigh the additional costs involved.

4.2. Reactogenicity

Notwithstanding any benefits in efficacy which may or may not result from using sub-unit preparations as a basis for a vaccine, there are good reasons for expecting a reduction in unwanted reactions to the vaccine. The highly purified sub-unit vaccine is almost completely free from extraneous protein of either cellular or viral origin. It contains nothing that is not needed to produce the intended immunological response and non-specific local or systemic reactions will not be induced.

Similarly, the real or imagined dangers of inoculating the viral DNA from viruses of the herpes group are simply avoided by using sub-unit preparations. Numerous experimental vaccines have been used to demonstrate the feasibility of this technique and sub-unit vaccination is probably the only acceptable approach to the prophylaxis of infection with herpesviruses (18, 43).

4.3. Standardisation

The high degree of purification implicit in a sub-unit vaccine provides opportunities for measuring the vaccine potency by chemical rather than biological means. Sub-unit influenza vaccines can be standardised in terms of weight of protein per dose, safe in the knowledge that almost all the protein measured will be specific immunizing glycoprotein (44). Such techniques make for a much more consistent product, which in turn should make it more reliable and predictable in its performance. They allow biological assays, animal potency tests and indirect measurements of immunogenicity to be discarded.

REFERENCES

1. Appleyard, G., Oram, J. D., and Stanley, J. L. (1970). Dissociation of Semliki Forest virus into biologically active components. *J. Gen. Virol.* **9,** 179–189.
2. Atanasiu, P., and Tsiang, H. (1979). Connaisances actuelles sur les propriétés et le pouvoir vaccinant de la glycoprotéine rabique. *Comp. Immun. Microbiol. Infect. Dis.* **1,** 179–184.
3. Balcarova, J., Helenius, A., and Simms, K. (1981). Antibody response to spike protein vaccines prepared from Semliki Forest virus. *J. Gen. Virol.* **53,** 85–92.
4. Brady, M., and Furminger, I. G. S. (1976). A surface antigen influenza vaccine. 1. Purification of haemagglutinin and neuraminidase proteins. *J. Hyg.* **77,** 161–172.
5. Brand, C. M., and Skehel, J. J. (1972). Crystalline antigen from the influenza virus envelope. *Nature (London), New Biol.* **238,** 145–147.
6. Cappel, R., and de Cuyper, F. (1976). Efficacy and immune response to rubella sub-unit vaccines. *Arch. Virol.* **50,** 207–213.
7. Cappel, R., de Cuyper, F., and Rickaert, F. (1980). Efficacy of a nucleic acid free herpetic sub-unit vaccine. *Arch. Virol.* **65,** 15–23.
8. Carter, C. A., Hartley, C. E., Skinner, G. R. B., Turner, S. P., and Easty, D. L. (1981). Experimental ulcerative herpetic keratitis: Preliminary observations on the efficacy of a herpes simplex sub-unit vaccine. *Br. J. Ophthalmol.* **65,** 679–682.
9. Cartwright, B., Talbot, P., and Brown, F. (1970). The proteins of biologically active sub-units of vesicular stomatitis virus. *J. Gen. Virol.* **7,** 267–272.
10. Casali, P., Sissons, J. G. P., Fujinami, R. S., and Oldstone, M. B. A. (1981). Purification of measles virus glycoproteins and their integration into artificial lipid membranes. *J. Gen. Virol.* **54,** 161–171.
11. Corbel, M. J., Rondle, C. J. M., and Bird, R. G. (1970). Degradation of influenza virus by non ionic detergent. *J. Hyg.* **68,** 77–80.
12. Couch, R. B., Kasel, J. A., Pereira, H., Haase, A. T., and Knight, V. (1973). Induction of immunity in man by crystalline adenovirus type 5 capsid antigens. *Proc. Soc. Exp. Biol. Med.* **143,** 905–910.
13. Cowan, K. M. (1968). Immunological studies of foot and mouth disease. *J. Immunol.* **101,** 1183–1191.
14. Crick, J., and Brown, F. (1976). Rabies vaccines for animals and man. *Vet. Rec.* **99,** 162–167.
15. Darcel, C. le Q., and Jericho, K. W. F. (1981). Failure of a sub-unit bovine herpesvirus 1 vaccine to protect against experimental respiratory disease in calves. *Can. J. Comp. Med.* **45,** 87–91.
16. Dyke, S. F., Williams, W., and Seto, J. T. (1978). Glycoproteins of representative paramyxoviruses: Isolation and antigenic analysis using a zwitterionic surfactant. *J. Med. Virol.* **2,** 143–152.
17. Fontanges, R. (1977). Process for obtaining virus proteins. Patent Specification GB 1, 469, 901.
18. Hillman, M. (1976). Herpes simplex vaccines. *Cancer Res.* **36,** 857–858.
19. Hoyle, L., Horne, R. W., and Waterson, A. P. (1961). The structure and composition of the myxoviruses. II. Components released from the influenza virus particle by ether. *Virology* **13,** 448–459.
20. Institut für Angewandte Virologie (1981). Viral vaccine cleavage using detergents in phosphate buffered saline at haemagglutinin concentrations of 160–1600 microgram per litre. Patent Specification DD 146, 386.
21. Kitces, E. N., Morahan, P. S., Tew, J. G., and Murray, B. K. (1977). Protection from oral herpes simplex virus infection by a nucleic acid-free virus vaccine. *Infect. Immun.* **16,** 955–960.

22. Kutinová, L., Šlichtová, V., and Vonka, V. (1979). Immunogenicity of subviral herpes simplex virus preparations. Formation of neutralizing antibodies in different animal species after administration of herpes simplex virus solubilized antigens. *Arch. Virol.* **61**, 141–147.

23. Lawman, M. J. P., Naylor, P. T., Huang, L., Courtney, R. J., and Rouse, B. T. (1981). Cell mediated immunity to herpes simplex virus: Induction of cytotoxic T lymphocyte responses by viral antigens incorporated into liposomes. *J. Immunol.* **126**, 304–308.

24. Lešnik, F., Chudý, Vrtiak, O. J., Konrád, V., Danihel, M., Ragač, P., and Poláček, M. (1980). Immunization trials with virus free antigens against Marek's disease. *Comp. Immunol. Microbiol. Infect. Dis.* **2**, 491–500.

25. Lupton, H. W., and Reed, D. E. (1980). Evaluation of experimental sub-unit vaccines for infectious bovine rhinotracheitis. *Am. J. Vet. Res.* **41**, 383–390.

26. Merck and Co. Inc. (1982). Herpes simplex type 1 sub-unit vaccine and process for its preparation. Patent Application EP 0,048,201.

27. Miles, R. N., Potter, C. W., Clark, A., and Jennings, R. (1981). Reactogenicity and immunogenicity of three inactivated influenza vaccines in children. *J. Biol. Stand.* **9**, 379–392.

28. Miles, R. N., Potter, C. W., Clark, A., and Jennings, R. (1982). A comparative study of the reactogenicity and immunogenicity of two inactivated influenza vaccines in children. *J. Biol. Stand.* **10**, 59–68.

29. Morein, B. Helenius, A., Simons, K., Pettersson, R., Kääriäinen, L., and Schirrmacher, V. (1978). Effective sub-unit vaccines against enveloped animal virus. *Nature (London)* **276**, 715–718.

29a. Morein, B., Sundquist, B., Hoglund, S., Dalsgaard, K., and Osterhaus, A. (1984), Iscom, a novel structure for antigenic presentation of membrane proteins from enveloped viruses. *Nature* **308**, 457–460.

30. Mussgay, M., and Weiland, E. (1973). Preparation of inactivated vaccines against alpha-viruses using Semliki Forest virus—white mouse as a model. *Intervirology* **1**, 259–268.

31. Neurath, A. R., Rubin, B. A., Sillaman, J., and Tint, H. (1971). The effects of non-aqueous solvents on the quaternary structure of viruses: A procedure for the simultaneous concentration, purification and disruption of influenza viruses. *Microbios* **4**, 145–50.

32. Neurath, A. R., Vernon, S. K., Dobkin, M. B., and Rubin, B. A. (1972). Characterization of subviral components resulting from treatment of rabies virus with tri(n-butyl) phosphate *J. Gen. Virol.* **14**, 33–48.

33. Pearson, G. R., and Scott, R. E. (1977). Isolation of virus-free herpes saimiri antigen positive plasma membrane vesicles *Proc. Natl. Acad. Sci. U.S.A.* **74**, 2546–2550.

34. Rajčáni, J., Kutinová, L., and Vonka, V. (1980). Restriction of latent herpesvirus infection in rabbits immunized with subviral herpes simplex virus vaccine. *Acta Virol.* (Engl. Ed.) **24**, 183–193.

35. Rock, D. L., and Reed, D. E. (1980). The evaluation of an experimental porcine herpesvirus 1 (Aujeszky's disease virus) sub-unit vaccine in mice. *Vet. Microbiol.* **5**, 291–299.

36. Ruebush, M. J., Hale, A. H., and Harris, D. T. (1981). Elicitation of anti vesicular stomatitis virus cytotoxic T lymphocytes by using purified viral and cellular antigens incorporated into phospholipid vesicles. *Infect. Immun.* **32**, 513–517.

37. Sandoz (1979). Isolating antigenic glycoprotein from paramyxovirus by selective solubilisation with cationic detergents, and derived vaccines. Patent Specification BE 868 916.

38. Schneider, L., and Diringer, H. (1976). Structure and molecular biology of rabies virus. *Cur. Top. Microbiol. Immunol.* **75**, 153–180.

39. Simons, K., Helenius, A., Morein, B., Balcorova, J., and Sharp, M. (1980). Development of effective sub-unit vaccines against enveloped viruses. *In* "New Developments with Human and Veterinary Vaccines" (A. Mizrahi, I. Hertman, M. A. Klingberg and A. Kohn, eds.), pp. 217–228. Alan R. Liss, Inc., New York.

40. Skinner, G. R. B., Williams, D. R., Buchan, A., Whitney, J., Harding, M., and Bodfish, K. (1978). Preparation and efficacy of an inactivated sub-unit vaccine (NFU. BHK) against type 2 herpes simplex virus infection. *Med. Microbiol. Immunol.* **166**, 119–132.
41. Šlichtová, V., Kutinová, L., and Vonka, V. (1980). Immunogenicity of subviral herpes simplex virus type 1 preparation: Protection of mice against intradermal challenge with type 1 and type 2 viruses. *Arch. Virol.* **66**, 207–214.
42. Webster, R. G., and Laver, W. G. (1966). Influenza sub-unit vaccines; immunogenicity and lack of toxicity for rabbits of ether and detergent disrupted virus. *J. Immunol.* **96**, 596–604.
43. Wise, T. G., Pavan, P. R., and Ennis, F. A. (1977). Herpes simplex virus vaccines. *J. Infect. Dis.* **136**, 706–710.
44. Wood, J. M., Schild, G. C., Newman, R. W., and Seagroatt, V. (1977). An improved single-radial-immunodiffusion technique for the assay of influenza haemagglutinin antigen: Application for potency determinations of inactivated whole virus and sub-unit vaccines. *J. Biol. Stand.* **5**, 237–247.
45. Zaia, J. A., Palmer, E. L., and Feorino, P. M. (1975). Humoral and cellular responses to an envelope-associated antigen of herpes simplex virus. *J. Infect. Dis.* **132**, 660–666.

10

Adjuvants

R. BOMFORD

Department of Experimental Immunobiology
The Wellcome Research Laboratories
Beckenham, Kent, United Kingdom

1. INTRODUCTION

The ideal vaccine should provide life-long protection after a single or a few injections. This is usually the case with live attenuated viral vaccines. When killed bacteria or viruses, or materials isolated from them such as bacterial toxoids, are used to make the vaccine, a rather poor antibody response may be obtained. This deficiency can be corrected by adding an additional component to the vaccine, the adjuvant, which has the property of stimulating the immune response.

Experimental studies in animals have uncovered an astonishing array of diverse materials which function as adjuvants [reviewed by Jolles and Paraf (49); Dukor *et al.* (23); Waksman (70); Edelman (24)]. Only a few of these

Animal Cell Biotechnology, Vol. 2

TABLE I   The Adjuvant Requirements of Different Types of Vaccines

Type of vaccine	Examples	Adjuvants
Live viral	Measles	None
	Polio (Sabin type)	None
	Rubella (German measles)	None
	Yellow fever	
Killed viral	Foot-and-mouth disease[a]	$Al(OH)_3$ plus saponin, or DEAE–dextran, or oil emulsion
Virus subunits	Influenza (purified haemagglutinin)	$Al(OH)_3$
Killed bacterial	Cholera	None
	Pertussis (whooping cough)	None
	Typhoid	None
	Sheep footrot[a]	$Al(OH)_3$ plus oil emulsion
Bacterial components	Diphtheria toxoid	$Al(OH)_3$
	Tetanus toxoid	$Al(OH)_3$

[a] Veterinary vaccine.

have found their way into practical application in human or veterinary vaccines. The only adjuvants in common use in human vaccines are gels of salts of aluminium or calcium [reviewed by Aprile and Wardlaw (6); Joo (50)], although for veterinary vaccines, where the toxicological criteria are less stringent, the choice is greater (43). Table I lists the adjuvant requirements of some representative vaccines.

This chapter reviews the animal studies on adjuvants which are relevant to understanding, as far as that is presently possible, the general principles governing their use in vaccines. It will not be possible to provide details of the manufacturing techniques for individual vaccines, which are, in any case, not usually available in published form except in manufacturers' patents. A useful summary of the U.S. patents covering vaccine preparation techniques in the years 1976 to 1980 has been prepared by Duffy (22).

2. ADJUVANTS IN CURRENT USE

2.1. Aluminium Salts

2.1.1. Mechanism of Action

The adjuvant action of aluminium salts was discovered by Glenny and co-workers (36), who were purifying diphthria toxoid by precipitation with alum

[$KAl(SO_4)_2 \cdot 12H_2O$]. The complex elicited a much better immune response than the free toxoid. Glenny proposed that the precipitated toxoid is retained as a depot at the site of injection, from which it is slowly released, providing a prolonged stimulation of the immune system and building up a heightened immune response of the secondary type, which usually requires repeated injections of antigen on its own (35).

The depot theory was disputed (45) on the grounds that surgical removal of the injected precipitate as early as 4 days after injection did not prevent an augmentation of the immune response, and that multiple immunization with free toxoid did not generate an immune response equivalent to that appearing after a single dose of precipitated toxoid. However, the excision experiment, although demonstrating that the persistence of material at the site of injection is unnecessary for the adjuvant effect, does not entirely refute the depot hypothesis, since there is histological evidence that aluminium salts are transported to the draining lymph node after injection (71), and slow release could still occur in this location. The failure of multiple doses of free antigen to provoke a good response does indicate that slow release alone is insufficient to account for adjuvant action, and that effects on the immune system of the host must be involved as well.

The mechanism of action of adjuvants at the cellular level is beginning to be understood, and the voluminous data have been reviewed by Waksman (70). For the purposes of this chapter it is sufficient to note that the initiation of an immune response requires a series of cellular interactions (Fig. 1), beginning with the uptake of antigen by macrophages, or cells related to macrophages, which then deliver a signal to a particular type of thymus-derived lymphocyte, the T helper cell, which in turn interacts with a bone-marrow-derived lymphocyte, the B cell, which secretes antibody. At present most of the evidence on adjuvant action is consistent with the view that they act on macrophages, promoting their ability to trigger T lymphocytes (Fig. 1). There is rather little data suggesting that adjuvants act directly on lymphocytes, although there is no reason why such adjuvants should not be discovered in the future.

The evidence that the adjuvant action of aluminium salts is mediated via macrophages is circumstantial. They cause a granulomatous reaction (an accumulation of macrophages) at the site of injection (71). This may be connected with their ability to activate complement, a term covering a series of blood proteins involved in the generation of inflammation (57).

As well as attracting macrophages, it has been found that aluminium salts, in common with many other adjuvants, engender an accumulation of lymphocytes in the lymph node draining the site of injection (21, 61). This phenomenon, known as lymphocyte trapping, may also be an indirect consequence of the action of adjuvants on macrophages (31).

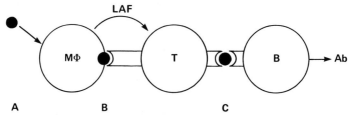

Fig. 1. Stages in antibody synthesis at which adjuvants exert their effect. (A) Free antigen (●) becomes associated with histocompatibility antigen on macrophage (MΦ) membrane. This process may be facilitated if antigen is adsorbed to Al(OH)$_3$ or enclosed in oil droplet. (B) Antigen on macrophage membrane is recognized by receptor on thymus-derived lymphocyte (T). Macrophage secretes soluble lymphocyte-stimulating factor (LAF or interleukin 1). The adjuvant MDP stimulates LAF secretion. (C) T lymphocyte (helper cell) induces B lymphocyte to secrete antibody (Ab).

In summary, it is likely that the adjuvant action of aluminium salts can be attributed to the additive effect of several mechanisms: the prevention of rapid antigen dispersal from the site of injection and draining lymph node; an improved association of antigen with macrophages; the activation of macrophages to stimulate T lymphocytes; and the prolonged retention of lymphocytes in the draining lymph node.

2.1.2. Method of Use

Two methods of adsorbing protein antigens onto aluminium salt adjuvants have been widely used. The first involves the mixing of the antigen with a solution of alum, followed by the rapid addition of NaOH, which leads to the appearance of a precipitate of Al(OH)$_3$:

$$2KAl(SO_4)_2 + 6NaOH \rightarrow 2Al(OH)_3 + 3Na_2SO$$

The antigen adsorbs to the Al(OH)$_3$ by ionic interaction. This procedure yields a rather undefined product (74), since aluminium salts of phosphate or carbonate may be formed if these anions are present in the antigen solution, and also the aluminium cation can interact with the protein to form a protein aluminate. Vaccines prepared in this way are sometimes referred to as alum-precipitated.

The alternative and preferable procedure is to mix the antigen with pre-formed gels of Al(OH)$_3$ or AlPO$_4$. These are available commercially (e.g. Alhydrogel, Superfos, Denmark). Vaccines of this sort are referred to as being adsorbed.

Since the initial interaction between Al(OH)$_3$ and protein is ionic, it will be affected by the net charge difference between the two components, and this in turn will depend on the pH. Aqueous suspensions of Al(OH)$_3$ have a pH of about 6.0, and the $Al(OH)_3$ is positively charged. At this pH most proteins are negatively charged, and will adsorb well.

When attempting to adsorb a new protein it is advisable to determine the degree of adsorption at pH intervals of 0.5 over the range which can be tolerated by the antigen. For instance, foot-and-mouth disease (FMD) virus rapidly loses its immunogenicity (power to stimulate an immune response) below pH 7.0, and so a pH just above this must be used for adsorption. Bacterial toxoids adsorb better when the pH is lowered to between 4.5 and 5.5 with HCl. After 48 h the lowered pH can be restored to 6.5 with NaOH, to avoid pain at injection.

It is frequently found that the potency of adsorbed bacterial vaccines tends to increase on storage (45). Therefore it is unnecessary to be too hasty with quality control testing.

Another factor affecting the efficiency of adsorption is the concentration of salts and buffering ions. Monovalent ions at isotonic concentrations are not inhibitory, but divalent anions do interfere with adsorption, and so such ions as phosphates, borates or sulphates should be avoided, or the lowest possible concentrations used.

The potency of complexes of aluminium salts and antigens has been shown to depend on the ratio between the two components. In experimental models using influenza virus or diphtheria toxoid in the mouse or guinea pig the immune response, with a given dose of antigen, at first rose with the concentration of $Al(OH)_3$, and then declined over quite a narrow concentration range (39, 59). The point at which all the antigen was adsorbed was on the ascending side of the curve. In testing the efficacy of $Al(OH)_3$ as an adjuvant for a new antigen it is advisable to titrate the $Al(OH)_3$.

2.2. Other Adsorbants

A great variety of other adsorbants possess adjuvant properties, including solid particles of latex (52), acrylate (28), and bentonite (33) or a variety of polyelectrolytes (32). In the latter category, DEAE–dextran has been used in porcine FMD vaccines (5).

2.3. Oil Emulsions

2.3.1. Mechanism of Action

Water-in-oil (w/o) emulsions, in which the antigen is entrapped in drops of water in a continuous phase of oil (Fig. 2), were introduced by Freund [30; reviewed by Freund (29)]; who also discovered that their potency can be further increased by adding killed mycobacteria to the oil. In experimental

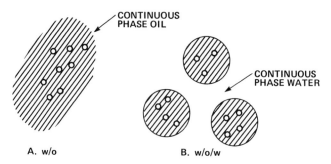

Fig. 2.   Oil emulsion types used as adjuvants. (A) Water-in-oil (w/o). Water droplets, plus antigen, are enclosed in a continuous oil phase. (B) Water-in-oil-in-water (w/o/w). Re-emulsification of the w/o emulsion yields oil droplets, still enclosing the water droplets of the primary aqueous phase, in a secondary aqueous continuous phase.

immunology, emulsions without bacteria are frequently referred to as Freund's incomplete adjuvant, and those with bacteria as Freund's complete adjuvant.

The early work on oil adjuvants led to the conclusion that they work in an analogous way to $Al(OH)_3$, i.e. by slow release of antigen and by stimulation of the host immune system [reviewed by Freud (29)]. The relevant observations may be summarized as follows: (1) w/o emulsions, in which the antigen is enclosed by the oil continuous phase, are much more effective that oil-in-water (o/w) emulsions, in which the antigen is free in a continuous aqueous phase. (2) Particles of emulsion are carried from the site of injection to the draining lymph node, and also to more distant lymph nodes. (3) Excision of the material at the site of injection 1 day after injection only partially reduces the adjuvant effect, and at 2 weeks is without effect. (4) An intense granulomatous reaction develops at the site of injection and in the draining lymph node.

It was later claimed by Herbert (41, 42) that the adjuvant action of oil emulsions can be completely attributed to slow release. Multiple injections of the antigen being used (ovalbumin) did give rise to an immune response quantitatively similar to that seen after a single injection with oil adjuvant. However, as has been noted above while discussing the role of slow release in the adjuvant action of $Al(OH)_3$ for diphtheria toxoid, not all antigens behave in this way. For instance, multiple injections of bovine serum albumin, which is very well adjuvantized by oil emulsions, produce a modest immune response, or even tolerance (55). It is likely that a hypothetical adjuvant working only by slow release would only promote the immune response to intrinsically strong antigens like ovalbumin.

2.3.2. Mode of Use

The components of classical Freund's incomplete adjuvant are mineral oil (e.g., Marcol 52, produced by Esso) in which is dissolved 10% (v/v) of the lipophilic emulsifying agent mannide monooleate (e.g. Arlacel A, ICI or Montanide, Seppic, France). This is emulsified with an aqueous solution of antigen, with an oil : water ratio of about 70 : 30 to 50 : 50.

Although excellent for experimental use, this formulation can give rise to emulsions which are too viscous and insufficiently stable for vaccine manufacture. These difficulties have been overcome by including a hydrophilic emulsifying agent, Tween 80, dissolved in the antigen solution at 1–5% (v/v) (64). This combination of components has been used in vaccines against clostridia (65), FMD (4) and sheep foot rot (66). Another emulsion system containing Markol 52, with Span 85 as the lipophilic emulsifier and Tween 85 as the hydrophilic emulsifier, has recently been described (12). It is too early to know whether this formulation is superior to the older one in terms of reduced toxicity, the major problem with oil emulsion (see below).

Herbert (40) introduced the use of water-in-oil-in-water (w/o/w) emulsion (Fig. 2) as a means of avoiding the viscosity of w/o emulsions. They are produced by re-emulsifying a w/o emulsion in an aqueous phase containing a hydrophilic emulsifier such as Tween 80.

2.3.3. Toxicity Problems

Anti-viral oil emulsion vaccines have been extensively tested in man, but the incidence of local reactions, although low, was sufficient to discourage their continuation [reviewed by Edelman (24)]. Oil-adjuvanted anti-viral and anti-bacterial veterinary vaccines are being produced, but there is always a danger that local reactions will be unacceptable to farmers, or will cause carcass blemish. Since oil emulsions are outstandingly potent adjuvants, there is a strong incentive to devise means of avoiding the local reactions.

One approach to the problem is to replace the mineral oil with a biodegradable alternative, such as vegetable oils (44, 58) or fatty acid esters (14, 46). These emulsions perform quite well in certain experimental systems, but they do not sustain the immune response as well as those employing mineral oil. The author is not aware that any of them have yet found a practical application.

It was hoped that w/o/w emulsions (40) would be less inflammatory, because the oil droplets, being dispersible in water, should not form a concentrated depot at the site of injection. This expectation was not fulfilled with an influenza vaccine in man, where the w/o/w emulsion provoked somewhat more local reaction than the w/o (62). With a foot-and-mouth disease vaccine in pigs, the w/o/w emulsion caused as much inflammation as the w/o emulsion when injected subcutaneously, but very small volumes (0.1 ml) of double emulsion injected intradermally raised no reaction, and generated acceptable neutralizing antibody levels (4).

Investigations into the causes of reactivity of oil emulsions have led to conflicting conclusions. Hardegree and Pittman (38), following up the regular observation that emulsions prepared with antigen are more inflammatory than those without, suggested that the reactivity is caused by the release of oleic acid from the Arlacel A emulsifying agent, which may be accelerated in the presence of antigen. Goto (37), in an extremely thorough study of the factors controlling abscess formation by emulsions of different types in mice and rabbits, found no correlation with the fatty acid content of the emulsions, but rather with their dispersibility. The most compact emulsions, i.e., the perfect w/o emulsions, were the most innocuous, and the most dispersible, o/w, emulsions the most deleterious. The conclusion was that abscess formation is favoured by the infiltration of oil droplets amongst muscle fibres.

In summary, the problem of reactogenicity of oil emulsions has not been solved, and it may be necessary to explore completely new directions (Section 3 below).

2.4. Saponin

2.4.1. Mechanism of Action

Saponins are a diverse collection of steroid or triterpene glycosides found in many plant species (67). The saponin used as an adjuvant is extracted from the bark of the South American tree *Quillaia saponaria*. The structure of the triterpene part of the molecule is known (Fig. 3), but the arrangement of the sugars in the glycosides remains to be worked out (17).

The adjuvant activity of *Quillaia* saponin for tetanus toxoid was demonstrated by Thibault and Richou (63). It has since been widely used as an adjuvant for FMD vaccines (17, 27, 47, 60), and has also turned out to be a very good adjuvant for experimental vaccines against protozoal diseases (19, 48, 53, 54).

Quillaia saponin is a surface-active material, and is used as a foaming agent in soft drinks. It is strongly haemolytic, and causes acute, but not chronic, inflammation at the site of injection. Both these properties are probably linked to the ability of saponin to bind to cholesterol in biological cell membranes (11, 34), creating circular lesions visible under the electron microscope (Fig. 4). Cholesterol binding is probably also involved in adjuvant activity, since this can be removed from preparations of saponin by adsorption with cholesterol (13). When sheep red blood cells, which contain choles-

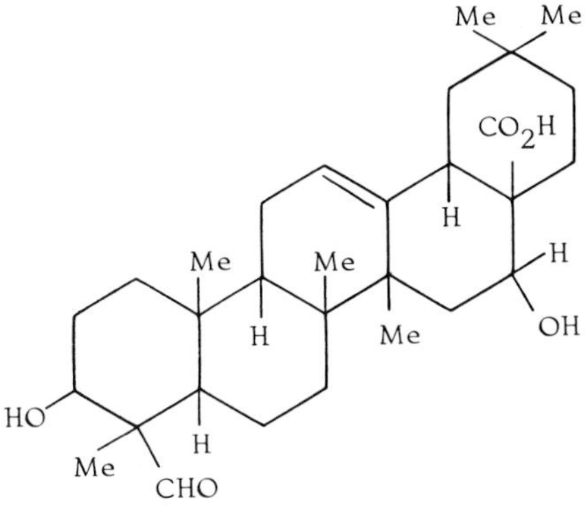

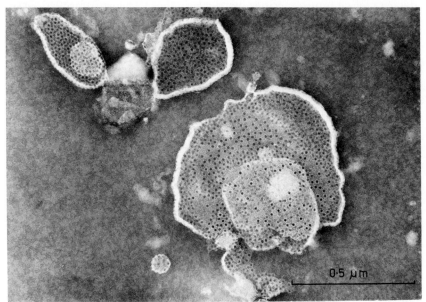

Fig. 3. Quillaic acid, the triterpene of *Quillaia* saponin. Sugars are attached to the hydroxyl group on the lower left-hand side.

terol in their cell membranes, are used as antigen, the adjuvant activity of saponin remains associated with the membrane fragments after haemolysis (*15*).

2.4.2. Mode of Use

Saponin is readily soluble in water, and may be mixed with aqueous solutions of antigen. In the case of FMD vaccine, it is used together with

Fig. 4. Liposomes composed of lecithin and cholesterol after exposure to saponin. Note large numbers of circular lesions.

antigen that has already been adsorbed onto $Al(OH)_3$ and, at the concentrations sufficient for adjuvant activity, it does not elute the virus from the $Al(OH)_3$ (60). Saponized FMD vaccine is not effective in pigs, which respond better to those containing DEAE–dextran or oil (4, 5).

When selecting a supply of saponin for adjuvant purposes it is necessary to ensure that it originates from *Quillaia saponaria,* since some manufacturers purvey saponin of unstated provenance, which may be extracted from other plants. PPF International Ltd., Kent, England, produce a *Quillaia* saponin suitable for adjuvant use (P3 saponin), and a preparation from which all non-triterpene glycoside material has been removed (Quil A Saponin) is offered by Superfos, Denmark.

3. POSSIBLE FUTURE ADJUVANTS

The adjuvants now used in vaccine manufacture have been around for a long time. As noted above, aluminium salts were introduced in the 1920s (36), saponin in the 1930s (63) and oil emulsions in the 1940s (30). There is need for new adjuvants for the following reasons. First, as regards human vaccines, we are confined to aluminium or calcium salts. These are very effective adjuvants for bacterial toxoids, but they may not necessarily be adequate for all future vaccines, particularly those against parasitic diseases such as malaria, which are now in prospect (20, 75). Secondly, in veterinary vaccines, there is the problem of the toxicity of oil emulsions and saponin.

There is a pessimistic school of thought which holds that adjuvanticity and inflammation are causally linked. With the existing adjuvants, there is certainly a degree of correlation between the two. $Al(OH)_3$ and oil emulsions attract and activate macrophages to produce a granulomatous response, which is a form of chronic inflammation. Saponin causes acute inflammation, presumably because it lyses cells at the site of infection by complexing with cholesterol in their cell membranes. However, the macrophage functions which are important in the immune response, the presentation of antigen to helper T lymphocytes and the release of soluble lymphocyte-stimulating factors (Fig. 1), need not induce inflammation. In principle it should be possible to discover adjuvants which selectively control these macrophage functions, or which act directly on lymphocytes, in both cases by binding to cell membrane receptors not present on other cell types. This should avoid the inflammation associated with conventional adjuvants.

New developments in vaccine adjuvants at the experimental level have recently been reviewed by Edelman *et al.* (25) and in Mizrahi *et al.* (56) and Chanock and Lerner (15a). Many new compounds are being tested, and it is too early to assess whether most of them offer any advantages over existing adjuvants. A selection of the most promising will be mentioned here.

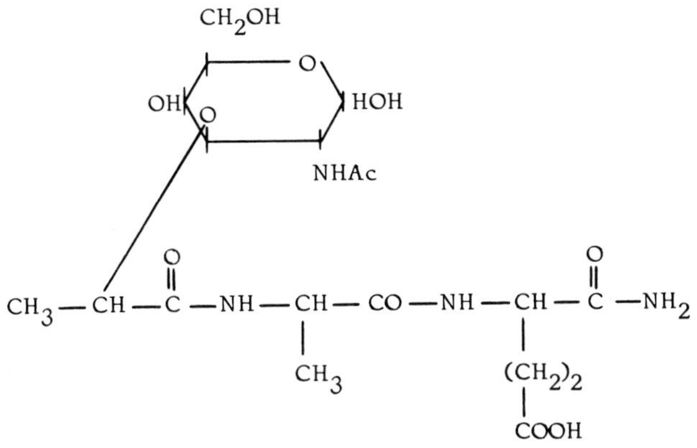

CH$_2$OH

Fig. 5.   Muramyl dipeptide (MDP).

3.1. Bacterial Derivatives and Synthetic Analogues

It was noted above that the potency of oil emulsions can be increased by mycobacteria. Many other microbial materials, of bacterial or fungal origin, exert an adjuvant effect [reviewed by White, (72)].

The only example of the exploitation of the adjuvant effect of bacteria in a human vaccine is provided by *Bordetella pertussis*, the causative agent of whooping cough. Killed *B. pertussis* cells are a good adjuvant, partly, but not completely, because of their constituent endotoxin (10, 73). Usually vaccination against *B. pertussis* is combined with that against diphtheria and tetanus in the tripartite DPT vaccine. The *B. pertussis* acts as an adjuvant for the two bacterial toxoids. The use of *B. pertussis* as an adjuvant in human vaccines will probably be confined to multiple vaccines of this sort, where it is intended to vaccinate against whooping cough.

The component of mycobacteria responsible for their adjuvant activity in Freund's complete adjuvant has been identified as the peptidoglycan (a polymer of sugars and amino acids) of the cell wall (1). A monosaccharide tripeptide isolated from this was the minimum structure sufficient for activity, and a synthetic analogue, muramyl dipeptide or MDP (Fig. 5), was equally effective (26). Most important, MDP not only replaces mycobacteria in oil emulsions, but also works in aqueous solution (8).

MDP has several important advantages as an adjuvant. First, since it is of low molecular weight, it is convenient for studies on the mechanism of adjuvant action. The multiplicity of effects encountered with whole bacteria, many of which may be irrelevant for immunostimulation, are avoided. At present, the evidence is pointing in the direction of an effect via macrophages, the stimulation of the secretion of soluble lymphocyte activating factors (18).

Secondly, MDP is amenable to chemical modification, opening the way to structure–activity relationship studies. From the point of view of its potential use in human vaccines, the most important achievement has been the synthesis of a non-pyrogenic derivative (9b, 16). Also, after the addition of acyl groups, MDP can be incorporated into liposomes (Section 3.2 below).

Finally, MDP can be covalently linked to antigens, thereby reducing the dose required for an adjuvant effect and by using as antigen synthetic peptides corresponding to the antigenic determinants on viral or bacterial proteins, creating completely synthetic, chemically-defined vaccines (7, 9, 9a). The immunogenicity of a synthetic peptide has also been increased by coupling it to a fatty acid carrier (46a).

3.2. Liposomes and Iscoms

Liposomes are vesicles composed of molecular bilayers of lecithin and other lipids. They are useful models for the investigation of the properties of cell membranes [reviewed by Tyrell *et al.* (68)]. Since their properties as adjuvants have recently been reviewed by Alving *et al.* (3) and van Rooijen and van Nieuwmegan (69) a summary of their potential advantages, which have not yet found an application in a commercial vaccine, will suffice here. The lipids used for preparing liposomes, principally lecithin and cholesterol, are non-toxic. Lipophilic adjuvants, such as acylated derivatives of MDP, can be incorporated into liposomes (51), as can lipid antigens (76) or antigens with a lipophilic region in the molecule, such as subunits from enveloped viruses (2). Water-soluble materials can be entrapped in the aqueous compartment inside the vesicle. Thus liposomes offer the possibility of presenting antigen to the immune system in a concentrated form on an artificial membrane, together with an adjuvant.

Iscoms, which are mixed micellar structures of membrane proteins from enveloped viruses and saponin, show promise of combining high immunogenicity with low toxicity (56a).

REFERENCES

1. Adam, A., Ciorbaru, R., Petit, J. F., and Lederer, E. (1972). Isolation and properties of a macromolecular, water soluble, immuno-adjuvant fraction from the cell wall of *Mycobacterium smegmatis. Proc. Natl. Acad. Sci. U.S.A.* **39**, 851–854.
2. Almeida, J. D., Brand, C. M., Edwards, D. C., and Heath, T. D. (1975). Formation of virosomes from influenza subunits and liposomes. *Lancet* **2**, 899.
3. Alving, C. R., Banerji, B., Shiba, T., Kotani, S., Clements, J. D., and Richards, R. L. (1980). Liposomes as vehicles for vaccines. In "New Developments with Human and Veterinary Vaccines" (A. Mizrahi, I. Hertman, M. A. Klingberg, and A. Kohn, eds.), pp. 339–355. Alan R. Liss, Inc., New York.
4. Anderson, E. C., Masters, R. C., and Mowat, G. N. (1971). Immune responses of pigs to

inactivated foot-and-mouth disease vaccines. Response to emulsion vaccines. *Res. Vet. Sci.* **12**, 342–350.

5. Anderson, E. C., Masters, R. C., and Mowat, G. N. (1971). Immune response of pigs to inactivated foot-and-mouth disease vaccines. Response to DEAE–dextran and saponin adjuvanted vaccines. *Res. Vet. Sci.* **12**, 351–357.

6. Aprile, M. A., and Wardlaw, A. C. (1966). Aluminium compounds as adjuvants for vaccines and toxoids in man: A Review. *Can. J. Public Health* **57**, 343–360.

7. Arnon, R., Sela, M., Parant, M., and Chedid, L. (1980). Antiviral response elicited by a completely synthetic antigen with built-in adjuvanticity. *Proc. Natl. Acad. Sci. U.S.A.* **77**, 6769–6772.

8. Audibert, F., Chedid, L., le Francier, P., and Choay, J. (1976). Distinctive adjuvanticity of synthetic analogs of mycobacterial water soluble components. *Cell. Immunol.* **21**, 243–249.

9. Audibert, F., Jolivet, M., Chedid, L., Alouf, J. E., Boquet, P., Rivaille, P., and Siffert, O. (1981). Active antitoxic immunization by a diphtheria toxin synthetic oligopeptide. *Nature (London)* **289**, 593–594.

9a. Audibert, F., Jolivet, M., Chedid, L. Arnon, R., and Sela, M. (1982). Successful immunization with a totally synthetic deptheria vaccine. *Proc. Natl. Acad. Sci. U.S.A.* **79**, 5042–5046.

9b. Audibert, F., Przewlocki, G., LeClerc, C. D., Jolivet, M. E., Gras-Masse, H. S., Tartar, A. L., and Chedid, L. A. (1984). Enhancement by murabutide of the immune response to natural and synthetic hepatitis B surface antigens *Infect. Immun.* **45**, 261–266.

10. Ayme, G., Caroff, M., Chaby, R., Haeffner-Caraillon, N., LeDur, A., Moreau, M., Muset, M., Mynard, M.-C., Roumiantzeff, M., Schulz, D., and Szabo, L. (1980). Biological activities of fragments derived from *Bordetella pertussis* endotoxin: Isolation of a nontoxic, Schwartzman-negative lipid A possessing high adjuvant properties. *Infect. Immun.* **27**, 739–745.

11. Bangham, A. D., and Horne, B. W. (1962). Action of saponin on biological cell membranes. *Nature (London)* **196**, 952.

12. Bokhout, B. A., van Gaalen, C., and van der Heijden, P. J. (1981). A selected water-in-oil emulsion: Composition and usefulness as an immunological adjuvant. *Vet. Immunol. Immunopathol.* **1**, 491–500.

13. Bomford, R. (1980). Saponin and other haemolysins (vitamin A, aliphatic amines, polyene antibiotics) as adjuvants for SRBC in the mouse. Evidence for a role for cholesterol-binding in saponin adjuvanticity. *Int. Arch. Allergy Appl. Immunol.* **63**, 170–177.

14. Bomford, R. (1981). The adjuvant activity of fatty acid esters. The role of acyl chain length and degree of saturation. *Immunology* **44**, 187–192.

15. Bomford, R. (1982). Studies on the cellular site of action of the adjuvant activity of saponin for sheep erythrocytes. *Int. Arch. Allergy Appl. Immunol.* **67**, 127–131.

15a. Chanock, R. M., and Lerner, R. A., eds. (1984). "Modern Approaches to Vaccines. Molecular and Chemical Basis of Virus Virulence and Immunogenicity". Cold Spring Harbor Laboratory.

16. Chedid, L. A., Parant, M. A., Audibert, F. M., Riveau, G. J., Parant, F. J., Lederer, E., Choay, J. P., and Le Francier, P. L. (1982). Biological activity of a new synthetic muramyl peptide adjuvant devoid of pyrogenicity. *Infect. Immun.* **35**, 417–424.

17. Dalsgaard, K. (1978). A study of the isolation and characterization of the saponin Quil A. Evaluation of its adjuvant activity, with a special reference to the application in the vaccination of cattle against foot-and-mouth disease. *Acta Vet. Scan., Suppl.* **69**, 1–40.

18. Damais, C., Riveau, G., Parant, M., Gerota, J., and Chedid, L. (1982). Production of lymphocyte activating factor in the absence of endogenons pyrogen by rabbit or human leukocytes stimulated by a muramyl dipeptide derivative. *Int. J. Immunopharmacol.* **4**, 451–462.

19. Desowitz, R. S. (1975). *Plasmodium berghei:* Immunogenic enhancement of antigen by adjuvant addition. *Exp. Parasitol.* **38,** 6–13.
20. Desowitz, R. S., and Miller, L. H. (1980). A perspective on malaria vaccines. *Bull. W. H. O.* **58,** 897–908.
21. Dresser, D. W., Taub, R. N., and Krantz, A. R. (1970). The effect of localized injection of adjuvant material on the draining lymph node. II. Circulating lymphocytes. *Immunology* **18,** 663–670.
22. Duffy, J. I., ed. (1980). "Vaccine Preparation Techniques." Noyes Data Corp., Park Ridge, New Jersey.
23. Dukor, P., Tarcsay, L., and Baschang, G. (1979). Immunostimulants. *Annu. Rep. Med. Chem.* **14,** 146–167.
24. Edelman, R. (1980). Vaccine adjuvants. *Rev. Infect. Dis.* **2,** 370–383.
25. Edelman, R., Hardegree, M. C., and Chedid, L. (1980). Summary of an international symposium on potentiation of the immune response to vaccines. *J. Infect. Dis.* **141,** 103–112.
26. Ellouz, F., Adam, A., Ciorbaru, R., and Lederer, E. (1974). Minimal structural requirements for adjuvant activity of bacterial peptidoglycan derivatives. *Biochem. Biophys. Res. Commun.* **59,** 1317–1325.
27. Espinet, R. G. (1951). Nouveau vaccin antiaphteux á complexe glucoviral. *Gac. Vet.* **13,** 265.
28. Freeman, M. J. (1968). Heterogeneity of the antibody response of rabbits immunized with acrylic particle–bovine serum albumin complexes. *Immunology* **15,** 481–492.
29. Freund, J. (1956). The mode of action of immunological adjuvants. *Adv. Tuberc. Res.* **7,** 130–148.
30. Freund, J., and McDermott, K. (1942). Sensitization to horse serum by means of adjuvants. *Proc. Soc. Exp. Biol. Med.* **49,** 548–553.
31. Frost, P., and Lance, E. M. (1973). The relation of lymphocyte trapping to the mode of action of adjuvants. *Ciba Found. Symp.* **18,** 24–45.
32. Gall, D., Knight, P. A., and Hampson, F. (1972). Adjuvant activity of polyelectrolytes. *Immunology* **23,** 569–575.
33. Gallily, R., and Garvey, J. S. (1968). Primary stimulation of rats and mice with hemocyanin in solution and adsorbed on bentonite. *J. Immunol.* **101,** 924–929.
34. Glauert, A. M., Dingle, J. T., and Lucy, J. A. (1962). Action of saponin on biological cell membranes. *Nature (London)* **196,** 953.
35. Glenny, A. T., Buttle, G. A. H., and Stevens, M. F. (1931). Rate of disappearance of diptheria toxoid injected into rabbits and guinea-pigs: Toxoid precipitated with alum. *J. Pathol. Bacteriol.* **34,** 267–275.
36. Glenny, A. T., Pope, C. G., Waddington, H., and Wallace, U. (1926). The antigenic value of the toxin–antitoxin precipitate of Ramon. *J. Pathol. Bacteriol.* **29,** 31–40.
37. Goto, N. (1978). Comparative studies on effects of incomplete oil adjuvants with different physical properties. *Jpn. J. Med. Sci. Biol.* **31,** 53–79.
38. Hardegree, M. C., and Pittman, M. (1967). Influence of antigens on release of free fatty acids from Arlacel A (mannide monooleate). *Proc. Soc. Exp. Biol. Med.* **123,** 179–184.
39. Hennessen, W. (1965). The mode of action of mineral adjuvants. *Prog. Immunobiol. Stand.* **2,** 71–79.
40. Herbert, W. J. (1965). Multiple emulsions: A new form of mineral-oil antigen adjuvant. *Lancet* **2,** 771.
41. Herbert, W. J. (1966). Antigenicity of soluble protein in the presence of high levels of antibody: A possible mode of action of the antigen adjuvants. *Nature (London)* **210,** 747.
42. Herbert, W. J. (1968). The mode of action of mineral-oil emulsion adjuvants on antibody production in mice. *Immunology* **14,** 301–318.

43. Herbert, W. J. (1970). "Veterinary Immunology." Blackwell Scientific Publications.
44. Hilleman, M. R. (1966). Critical appraisal of emulsified oil adjuvants applied to viral vaccines. *Prog. Med. Virol.* **8**, 131–182.
45. Holt, L. B. (1950). "Developments in Diphtheria Prophylaxis." Heinemann, London.
46. Holt, L. B. (1967). Oily adjuvants. *Symp. Gen. Immunobiol. Stand.* **6**, 131–136.
46a. Hopp, T. P. (1984). Immunogenicity of a synthetic HBsAg peptide: Enhancement by conjugation to a fatty acid carrier. *Mol. Immunol.* **21**, 13–16.
47. Hyslop, N. St. G., and Morrow, A. W. (1969). The influence of aluminium hydroxide content, dose volume and the inclusion of saponin on the efficacy of inactivated foot-and-mouth disease vaccines. *Res. Vet. Sci.* **10**, 109–120.
48. Johnson, P., Neal, R. A., and Gall, D. (1963). Protective effect of killed trypanosome vaccines with incorporated adjuvants. *Nature (London)* **200**, 83.
49. Jolles, P., and Paraf, A. (1973). "Chemical and Biological Basis of Adjuvants." Chapman & Hall, London.
50. Joo, I. (1974). Mineral carriers as adjuvants. *Symp. Ser. Immunobiol. Stand.* **22**, 123–130.
51. Kotani, S., Kinoshita, F., Morisaki, I., Shimono, T., Okunaga, T., Takada, H., Tsujimoto, M., Watanabe, Y., Kato, K., Shiba, T., Kusumoto, S., and Okada, S. (1977). Immunoadjuvant activities of synthetic 6-O-acyl-N-acetylmuramyl-L-alanyl-D-isoglutamine with special reference to the effect of its administration with liposomes. *Biken J.* **20**, 95–103.
52. Litwin, S. D., and Singer, J. M. (1965). The adjuvant action of latex particulate carriers. *J. Immunol.* **95**, 1147–1152.
53. McColm, A. A., Bomford, R., and Dalton, L. (1982). A comparison of saponin with other adjuvants for the potentiation of protective immunity by a killed *Plasmodium yoelii* vaccine in the mouse. *Parasite Immunol.* **4**, 337–347.
54. Mitchell, G. H., Richards, W. H. G., Coller, A., Kietrich, F. M., and Dukor, P. (1979). Nor-MDP, saponin, corynebacteria and pertussis organisms as immunological adjuvants in experimental malaria vaccination of macaques. *Bull. W. H. O.* **57**, Suppl. 1, 189–197.
55. Mitchison, N. A. (1964). Induction of immunological paralysis in two zones of dosage. *Proc. R. Soc. London, Ser. B* **161**, 275–292.
56. Mizrahi, A., Hertman, I., Klingberg, M. A., and Kohᵢn, A., eds. (1980). "New Developments with Human and Veterinary Vaccines." Alan R. Liss, Inc., New York.
56a. Morein, B., Sundquist, B., Hoglund, S., Dalsgaard, K., and Osterhaus, A. (1984). Iscom, a novel structure for antigenic presentation of membrane proteins from enveloped viruses. *Nature (London)* **308**, 457–460.
57. Ramanathan, V. D., Badenoch-Jones, P., and Turk, J. L. (1979). Complement activation by aluminium and zirconium compounds. *Immunology* **37**, 881–888.
58. Reynolds, J. A., Harrington, D. G., Crabbs, C. L., Peters, C. J., and DiLuzio, N. R. (1980). Adjuvant activity of a novel metabolizable lipid emulsion with inactivated viral vaccines. *Infect. Immun.* **29**, 937–943.
59. Schmidt, G. (1967). The adjuvant effect of aluminium hydroxide in influenza vaccine. *Symp. Ser. Immunobiol. Stand.* **6**, 275–282.
60. Strobbe, R., Leunen, J., Mammerickx, M., and Debecq, J. (1964). Valeur á longue échéance du vaccin anti-aphteux adsorbé saponiné trivalent. *Bull. Off. Int. Epizoot.* **61**, 1059–1078.
61. Taub, R. N., Krantz, A. R., and Dresser, D. W. (1970). The effect of localized injection of adjuvant material on the draining lymph node. I. Histology. *Immunology* **18**, 171–186.
62. Taylor, P. J., Miller, C. L., Pollock, T. M., Perkins, F. T., and Westwood, M. A. (1969). Antibody response and reactions to aqueous influenza vaccine, simple emulsion vaccine and multiple emulsion vaccine. A report to the Medical Research Council committee on influenza and other respiratory virus vaccines. *J. Hyg.* **67**, 485–490.
63. Thibault, P., and Richou, R. (1936). Sur l'accroissement de l'immunité anti-toxique sous

influence de l'addition de diverses substances á l'antigéne (anatoxines dipthérique et tétanique). *C. R. Seances Soc. Biol.* **121,** 718–721.

64. Thomson, R. O., and Batty, I. (1967). Experimental clostridial oil emulsion vaccines. *Bull. Off. Int. Epizoot.* **67,** 1569–1581.

65. Thomson, R. O., Batty, I., Thomson, A., Kerry, J. B., Epps, H. B. G. E., and Foster, W. H. (1969). The immunogenicity of a multi-component clostridial oil emulsion vaccine in sheep. *Vet. Rec.* **85,** 81–85.

66. Thorley, C. M., and Egerton, J. R. (1981). Comparison of alum-absorbed or non-alum-absorbed oil emulsion vaccines containing either pilate or non-pilate *Bacteroides nodosus* cells in inducing and maintaining resistance of sheep to experimental foot rot. *Res. Vet. Sci.* **30,** 32–37.

67. Tschesche, R., and Wulff, G. (1973). Chemie und Biologie der Saponine. *In* "Fortschritte der Chemie den organischen Naturstoffe" (W. Herz, H. Griesenbach, and G. W. Korby, eds.), pp. 462–606. Springer-Verlag, Berlin and New York.

68. Tyrrell, D. A., Heath, T. D., Colley, C. M., and Ryman, B. E. (1976). New aspects of liposomes. *Biochim. Biophys. Acta* **457,** 259–302.

69. van Rooijen, N., and van Nieuwmegen, R. (1982). Immunoadjuvant properties of liposomes. *In* "Targeting of Drugs" (A. G. Gregoriadis, ed.), NATO ASI Ser. pp. 301–326. Plenum, New York.

70. Waksman, B. H. (1979). Adjuvants and immune regulation by lymphoid cells. *Springer Semin. Immunopathol.* **2,** 5–33.

71. White, R. G., Coons, A. H., and Connolly, J. M. (1955). Studies on antibody production. III. The alum granuloma., *J. Exp. Med.* **102,** 73–82.

72. White, R. G. (1976). The adjuvant effect of microbial products on the immune response. *Annu. Rev. Microbiol.* **30,** 579–600.

73. Wirsing von Koenig, C. H., Heymer , B., Hoff, H., Finger, H., and Emmerling, P. (1981). Biological activity of *Bordetella pertussis* in lipopolysaccharide-resistant mice. *Infect. Immun.* **33,** 223–230.

74. World Health Organization (1953). Diphtheria and pertussis vaccination. *WHO Tech. Rep. Ser.* **61.**

75. World Health Organization (1976). Immunological adjuvants. *WHO Tech. Rep. Ser.* **959.**

76. Yasuda, T., Dancey, G. F., and Kinsky, S. C. (1977). Immunogenicity of liposomal model membranes in mice: Dependence on phospholipid composition. *Proc. Natl. Acad. Sci. U.S.A.* **74,** 1234–1236.

11

Formulation and Storage

K. A. CAMMACK
G. D. J. ADAMS
Therapeutic Products Laboratory
Public Health Laboratory Service
Centre for Applied Microbiology and Research
Salisbury, Wiltshire
United Kingdom

Animal Cell Biotechnology, Vol. 2

1. INTRODUCTION

1.1. Objectives

Our primary objective in this contribution is to draw together many aspects of the theory and practice of formulation, and then to discuss them in relation to optimisation of product stability during storage over many months or even years. It should become evident from these discussions that formulation and storage are totally inter-dependent concepts when designing a strategy for storage and distribution.

The second objective must therefore be the identification of the stabilising factors and the presentation of experimental work to demonstrate the efficacy of stabilising or protective agents for the freezing and thawing and freeze-drying of well-characterised enzymes and virus preparations.

1.2. Definition of Formulation

By the term formulation is meant a full quantitative and qualitative description of the chemical constituents that make up the final product. It should include stabilisers, buffer composition, any additional electrolytes and all the relevant information on the processing immediately preceding storage and distribution.

In practice formulation may refer to nothing more than the composition of a buffer solution used in the final dialysis after elution in column chromatography. Then the likelihood is that no further thought will be given to the buffer formula other than some reassurance that the nominal pH is well within the pH stability range of the product and parenterally acceptable.

There are, of course, many who have wide experience in pharmaceutical formulation, especially in the production of well-established attenuated or killed vaccines and standard antisera, who could claim quite justifiably that over the past 50 years or more perfectly satisfactory empirical procedures have been available. But in anticipation of future needs in modern biotechnology, with the reports of greater difficulty in stabilising highly purified biopolymers like monoclonal antibodies, interferons, viruses and viral subunit vaccines, it has to be recognised that many impurities in the past were fortuitously acting as stabilising agents.

2. PHYSICOCHEMICAL CHARACTER OF BIOPOLYMERS

2.1. General

Most of the bioproducts described by the contributors are biopolymers or biopolymer assemblies as in viruses or membrane material derived from animal cell cultures. Proteins are the dominant class of biopolymer although the lipoproteins in membranes, various glycoproteins, and nucleic acids or nucleohistone fractions are just as important.

An understanding of their macromolecular structure is invaluable in anticipating the stability of these products in relation to their optimal storage conditions.

2.2. Proteins

Clearly it is inappropriate and impractical to review the outstanding achievements in structural protein chemistry over the past 30 years. Since the pioneer work of Kendrew and Perutz (44) on myoglobin and haemoglobin, more than 80 proteins, mostly enzymes, have been resolved to at least 2.5 Å by means of x-ray crystallographic methods.

Nevertheless, there are aspects of this work (16) that are well worth noting as they help us to gain an appreciation of the the thermodynamic restraints on macromolecular conformation in aqueous media.

2.2.1. Primary Structure—Covalent Bonding

Polypeptide chains are characterised by the linear sequence of more than 23 species of amino acids linked by the covalent peptide bond. Their intrinsic chemical stability is evident from the severe treatment at 110°C required to achieve a complete hydrolysis into free amino acids. A spontaneous degradation of a polypeptide is therefore most unlikely at room temperature. On the other hand, catalytic degradation by proteases is a common occurrence once the host cells are disrupted.

2.2.2. Secondary Tertiary and Quaternary Structure—Non-covalent Interactions

More important in the context of the results to be discussed later are the comparatively weak non-covalent interactions, which, by acting cooperatively, are the prime factor in stabilising the native molecular conformation of a protein and conserving the structural micro-environment of a biologically active site.

Typical are hydrogen bonds, for example the inter-peptide hydrogen bonding in the α-helix, generally referred to as secondary structure. Of equal importance are hydrogen bonds formed by water molecules (acting as proton donor or acceptor) interacting with polar groups on the surface of the protein molecule.

The tertiary structure of a protein is extremely vulnerable during processing and storage. Polypeptide chains fold into a three-dimensional matrix, which is the native conformation and the biologically active form. Hydrogen bonds link groups that formerly were some distance apart in the amino acid sequence. Other weak interactions include electrostatic repulsive or attractive forces between protonated or deprotonated groups, van der Waals forces (43) and the very important concept of hydrophobic bonding (43, 66, 83, 84), where various apolar groups are able to come into close proximity, excluding water molecules, which in turn produces a favourable decrease in thermodynamic free energy.

Equally vulnerable is the quaternary structure of multi-subunit proteins (48) especially in the case of allosteric or regulatory enzymes. Hydrophobic interactions at the subunit interfaces probably stabilise the quaternary structure, and by simple analogy with the hydrophobic interior of a globular protein water molecules are excluded. In understanding the dissociation of multi-subunit proteins it is important to note that hydrophobic bonds are weakened as the temperature of the system is reduced (9).

2.2.3. Bonding Energy

The weak non-covalent interactions defined above are those most likely to be perturbed by freezing/thawing, dehydration and changes in water activity in general (7, 8). The magnitudes of the thermodynamic changes in enthalpy and entropy are not compatible with the energy required to sever a covalent bond of bond energy 80–100 kcal mole^{-1} but are totally compatible with bonding energies of 0.5–10 kcal mole $^{-1}$ for the range of non-covalent interactions that operate in stabilising secondary, tertiary and quaternary structure.

2.3. Nucleic Acids

Some of the comments made on protein structure may usefully be extended to RNA and DNA. The equivalent of the highly stable peptide bond is the covalent phosphodiester bond. Polynucleotides are moderately resistant to chemical hydrolysis but very susceptible to degradation by nucleases. Single- or double-stranded helices are stabilised by base pair hydrogen bonding and hydrophobic interactions in base stacking.

2.4. Membranes—Lipoproteins

Far less is known about the detailed molecular structure of cell membranes; however, the 'bilayer model' (54, 79) is the one conventionally used. Apolar hydrocarbon chains of saturated or unsaturated fatty acids form the hydrophobic interior while hydrophilic carboxyl or phosphate groups are exposed on the membrane surface to which proteins or water molecules are attached. Protein molecules rich in hydrophobic residues have been shown to be embedded in the lipid interior.

As the integrity of the membrane is supported entirely by non-covalent interactions it is very susceptible to freezing and thawing.

3. WATER

3.1. Hydration and Solvation

Strong emphasis must be given to the complementarity of the noncovalent interactions and solvent mediation in defining the optimal conditions for conformational stability (7, 8, 17). It is essential to consider the system as a whole—that is the macromolecule, other ionic and non-ionic constituents, bound, ordered and random water molecules. All of these must be regarded as a complex energy matrix maintained in thermodynamic equilibrium (9, 45, 75).

A useful example of the importance of solvent mediation in protein structure is the reversible dissociation and unfolding of aldolase subunits in aqueous urea (77). Upon removal of the urea by exhaustive dialysis the disordered polypeptide chains refold, regaining their native tertiary structure, and the ordered subunits reassemble into active enzyme.

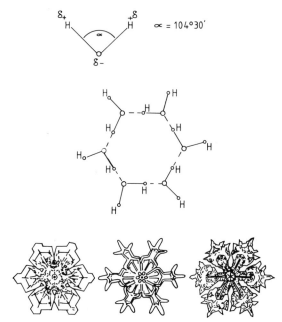

Fig. 1. Hydrogen-bonded water molecules. Hexagonal symmetry and ice crystal structures.

3.2. Water Structure

Because solvation effects are so fundamental to the structure and function of a biopolymer, the structure of water itself cannot be neglected (*22, 46, 66*), although there is a marked tendency to overlook its importance in formulation.

Water molecules in Fig. 1 have a dihedral configuration. This is clearly an oversimplification of the structure of water in the liquid or solid ice phases. The extensive hydrogen bonding is greatly facilitated by water being a proton donor and proton acceptor, and even in the vapour phase dimers will exist (*78*).

A striking and beautiful manifestation of hexagonal hydrogen-bonded structures (*56*) is the extension of the symmetry in crystals of ice, frost and snow, as shown in Fig. 1 (*56*).

Static hexagonal structures do not exist in the liquid phase but the popular model is one based on "mini-icebergs' or 'flickering-clusters' of hydrogen-bonded water in rapidly reversible equilibrium with randomly distributed and more mobile dimers and trimers (*21, 78*).

With this type of structure kept in mind the anomalous physical properties of liquid water can be readily accepted. The average molecular weight is much higher than the nominal figure of 18; consequently the vapour pressure is depressed and the boiling and melting temperatures considerably raised. The cohesion forces directly attributable to hydrogen bonding produce a high surface tension, low compressibility and an inversion in the dependence of density on temperature in the region of 2–4°C. The high latent heats of fusion and vapourisation indicate the amount of energy required to disrupt the hydrogen bonding.

The bonding forces stabilising macromolecular conformation and water structure are evidently similar so that dehydration caused by freezing or sublimation represents a distinct physical stress which can upset the balance of complementarity between solute and solvent.

4. SALT SOLUTIONS AND FREEZING

4.1. General

Omitting for the present all impractical aspects of distributing a product in the frozen state including the cost of distribution, there are obvious advantages in freezing pharmaceutical and diagnostic bioproducts. Apart from limiting the further growth of microbial contaminants, a major effect is on the chemical kinetics of the system; for practical purposes all chemical reaction rates are zero at −196°C in liquid nitrogen.

What has to be considered very carefully however, is the potential damaging effect of increasing ionic strength of electrolytes and possible pH changes as the solution freezes (5, 6, 53).

4.2. Freezing Point Depression and Eutectic Temperatures

Since the depression of the freezing point of water is a colligative property, it follows that relatively high molar concentrations of electrolyte ions may produce freezing points as low as −20° to −45°C. Proteins in solution do not significantly depress the freezing point.

Cell-derived material and other biopolymers are almost invariably suspended or dissolved in saline or some other weak electrolyte solution. Examination of the simple eutectic phase diagram for NaCl solution in Fig. 2 shows how even very dilute salt solutions produce dangerously high concentrations on cooling.

Curve CB is a solubility plot for NaCl which is almost independent of the temperature from below 0° to +10°C. A saturated salt solution, when cooled slowly, will precipitate salt crystals, but ice does not form until the eutectic temperature at −21.8°C is reached.

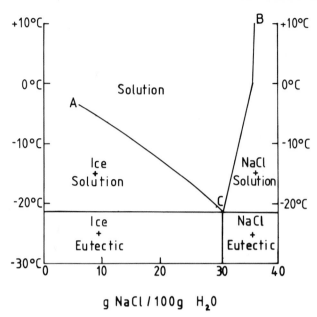

Fig. 2.   Eutectic phase diagram of sodium chloride solution.

Similarly, the phase diagram along curve AC, starting at 100% water and proceeding to increasing concentrations of NaCl, shows that any temperature–NaCl combination above this curve, on cooling, will intercept AC, at which point ice crystals precipitate and not salt, and continuing along AC the salt concentration increases until the eutectic temperature at −21.8°C is again reached.

The eutectic composition coming from either direction is 4.6 M NaCl (23% w/v or 30% w/w NaCl) and the transition from 'liquidus' to 'solidus' is indistinguishable from a normal freezing point transition since the composition of the two phases is identical. Thus at this stage, when the eutectic temperature is reached a matrix of both ice and salt crystals equivalent to 4.6 M NaCl is formed.

No matter how dilute the original solution may be initially, the cooling curve must travel along AC to the eutectic. The effective liquid volume decreases significantly because of ice formation but its concentration increases steadily to greater than 2–3 M NaCl.

A common error in using deep-freeze cabinets at −18° to −21°C is to accept a visual freezing of the sample. The eutectic diagram in Fig. 2 shows how dangerously close these temperature conditions are to the saline eutectic. Most of the samples stored in this way would, by careful microscopic examination [e.g. Luyet (56)], reveal micro-pockets of interstitial fluid of approximately 4.0 M NaCl 'liquidus'. There is positive evidence that biopolymers are often inactivated by exposure to such high salt concentrations within the micro-pockets mentioned above.

In order to avoid a prolonged and potentially damaging exposure to high salt concentration in the liquidus, some workers prefer to freeze as rapidly as possible to −78°C in solid CO_2 and acetone or even liquid nitrogen. However, this is known to be unsuitable for some cell-derived products, and, as a preliminary to freeze-drying commercially, is impractical.

4.3. Measurement of Eutectic Freezing

In order to establish a true freezing point, the marked increase in resistance as liquidus is transformed to solidus can be measured with a conductivity meter. In practice, for pharmaceutical products sterilised by filtration, and therefore lacking nuclei for crystallisation, supercooling becomes a greater problem than usual, so that the freezing point may be depressed by more than 20°C. Standard procedure is to freeze well below the eutectic in solid CO_2 and acetone at approximately −78°C and then to measure the abrupt decrease in resistance during melting and thus avoid supercooling.

A superior method for the detection of multiple eutectics was introduced some years ago by Rey (69), principally as an aid to freeze-drying research and development. Differential thermal analysis, or DTA, is designed to pick up the very small exothermic and endothermic changes during a phase transition.

4.4 Implications of Eutectic in Buffer Formulations

Van den Berg (6) and van den Berg and Rose (5) published a unique and invaluable account of the effect of cooling mixtures of sodium and potassium phosphate salts in buffer formulations in the pH range 5.0 to 8.0 Two examples are shown in Table I. The first covers a temperature reduction from −4° to −12.7°C for sodium and potassium phosphate and NaCl. As a result of the co-precipitation of ice and the hydrated $Na_2HPO_4 \cdot 12H_2O$ the pH decreased from 5.5 to 3.7, nearly a hundred-fold increase in the hydrogen-ion concentration, while the molarity increased from 0.36 to 3.25 M. The second example covers a temperature reduction from −0.6° to −6.6°C for pH 8.1 phosphate buffer where the pH decreased to 6.7 and the molarity increased from 0.14 to 2.38 M.

As phosphate buffers are very widely used in the biological field, the results in Table I provide a definitive case for a thorough scrutiny of all types of buffer formulations before cooling and freezing, and indeed there is at least one report on interferon where this behaviour affects activity (see Section 12).

TABLE I Changes in pH and Molarity of Phosphate Buffer Systems as a Result of Freezing
and Co-precipitation of Ice and $Na_2HPO_2 \cdot 12H_2O$

A.[a]

Freezing point (°C)	$pH^{25°C}$	$KH_2PO_4{}^b$	$Na_2HPO_4{}^b$	$Na_2HPO_4{}^b$	$NaCl^b$
−4.0	5.5	0.17	0.06	0.06	0.7
−6.0	5.1	0.25	0.09	0.10	1.19
−8.3	4.4	0.35	0.12	0.14	1.69
−12.7	3.7	0.50	0.15	0.20	2.40

B.[c]

Freezing point (°C)	$pH^{25°C}$	$K_2HPO_4{}^b$	$KH_2PO_4{}^b$	$Na_2HPO_4{}^b$
−0.6	8.1	0.02	0.01	0.11
−0.7	7.3	0.05	0.03	0.12
−1.7	7.0	0.15	0.11	0.12
−2.4	7.0	0.20	0.13	0.13
−3.9	6.8	0.47	0.41	0.16
−6.6	6.7	1.13	1.02	0.23

[a] From van den Berg (6).
[b] Data are measured in moles per litre.
[c] From van den Berg and Rose (5).

5. BIOPOLYMER MODIFICATION BY FREEZING AND FREEZE-DRYING

5.1. Dehydration Effect

Whether the material is frozen and thawed or freeze-dried and reconstituted, both processes involve a dehydration and rehydration step. A recapitulation of the issues we have discussed under structure, hydration, solvent mediation and eutectics should lead us to expect some modifications of the biopolymer.

Freezing of a novel biopolymer may be abandoned if total or partial biological inactivation occurs initially, while in routine freeze-drying, losses in activity may be alleged to be due to the incompetence of some hapless operator.

The point one is trying to make here is that rarely does the formulation itself become identified as being the intrinsic problem, and a problem meriting further investigation. In the pharmaceutical industry pilot investigations are sometimes limited to a few trial runs, which may be insufficient to reveal a metastable physical condition (see Section 5.5).

TABLE II   Modification of Proteins Caused
by Freezing and Thawing and Freeze-drying

Protein	References
Haemocyanin	52
Catalase	74
	85
	15
	35
Myosin	74
	35
	91
Lactate dehydrogenase (LDH)	60
	12
	80
Lysozyme	92
	25
Invertase	87
α-Lactalbumin	93
Actomyosin	88
L-Asparaginase	71
	61
	37,38
Chromoprotein phycoerythrin	49

5.2. Examples of Protein Modification

Since 1958, when haemocyanin was first reported to be unstable in solution after freezing and thawing (52), there has been a succession of papers, mainly on enzymes, describing the denaturation of proteins caused by a single cycle or multiple cycles of freeze/thaw and by freeze-drying (Table II). Of special interest are measurements of laser Raman spectra with crystals and freeze-dried samples of lysozyme (92) and α-lactalbumin (93) which provide undisputed evidence of conformational change induced by freeze-drying.

5.3. Basis for the Improvement of Protein Formulation

Prior to 1950 the early cryobiologists and microbiologists with the responsibility of maintaining culture collections were able to specify chemical compounds collectively referred to as 'cryoprotectants', which, by trial and error, and later by more scientific experimentation (62, 64, 67), were chosen as useful protective additives in freezing plant or animal tissue and freeze-drying microorganisms.

Most encouraging is to find that some of these cryoprotectants are proving to be just as effective in the preservation of the quaternary structure of multi-subunit enzymes. In other words, their protective action can be demonstrated and partly understood at a biomolecular level. Anticipating the results of these experiments, it would appear that hydrophobic interactions are a dominant feature of the quaternary structure of multi-subunit enzymes and also of their compartmentalisation and inter-dependent functions in a complex three-dimensional cell matrix. The weakening of hydrophobic interactions at low temperature predicted by thermodynamic calculations in papers by Brandts (7, 8) is consistent with the denaturing effect of freezing, which protective agents seem to inhibit.

5.4. Review of Protective Agents

Hanafusa (35) has identified a number of protective agents which preserve the molecular integrity of the catalase tetramer and its enzyme activity after freeze-drying and reconstitution (see Table III). Loss of quaternary structure denoted by dissociation into monomers coincided with enzyme inactivation. Electrolytes like NaCl and KCl were, in fact, a slight improvement as protective additives over no additive at all when there was an 87% loss in activity. Most impressive, however, was the 70–90% recovery of catalase activity in the presence of sucrose, inositol, glycine and potassium glutamate. These results were very similar to those reported for myosin (35, 91) and lactate dehydrogenase (12, 80) in Table II.

TABLE III   Effect of Additives on the Dissociation of Catalase by Freeze-drying[a,b]

Additive	$S_{20,W}$	Catalase activity	Molecular state[c]
None[d]	11.8	100	Tetramer
None[e]	6.2	13	Dimer, monomer
NaCl	12.2, 8.0	34	Tetramer, dimer
KCl	11.4, 5.4	38	Tetramer, dimer, monomer?
Na pyrophosphate	5.7	10	Monomer
Sucrose	12.0	85	Tetramer
Inositol	12.3	71	Tetramer
Glycine	12.4	91	Tetramer
K glutamate	11.9	90	Tetramer

[a] Data from Hanafusa (35) and Tanford and Louvrien (85).

[b] Material: 0.5% catalase (beef liver), MW 260,000 (tetramer), in 10 mM phosphate buffer, pH 7.0, plus 30 mM additive.

[c] Tetramer (MW 260,000, $S_{20,W}$ = 11.6), dimer (MW 130,000, $S_{20,W}$ = 7.6), monomer (MW 65,000, $S_{20,W}$ = 4.2).

[d] Control.

[e] Freeze-dried.

TABLE IV Effect of Additives on the Dissociation of L-Asparaginase by Freeze-drying[a,b]

Additive		Enzyme activity (%)	Tetramer (%)	Residual moisture (%)
Reconstituted at pH 7.5	None	100	100	2.02 ± 0.36
Reconstituted at pH 10.0	None	18	20	
	D-Glucose	96	100	1.88 ± 0.01
	D-Mannose	100	100	0.88 ± 0.03
	D-Sorbitol	100	100	1.01 ± 0.07
	D-Ribose	100	90	2.07 ± 0.11
	D-Sucrose	93	100	1.20 ± 0.03
	PVP	99	90	
	Inositol	85	80	0.76 ± 0.06
	Erythritol	65	60	0.68 ± 0.09
	Ammonium sulphate	68	70	0.69 ± 0.04
0.15% w/v	L-Glutamine	80	70	
	D-Mannitol	49	50	0.69 ± 0.11
	D-Galactitol	37	40	0.80 ± 0.11
0.15% w/v	L-Aspartic acid	39	40	1.79 ± 0.19
	Sodium chloride	9	10	
	Lithium bromide	0	—	
	Urea	6	—	

[a] From Hellman *et al.* (38).

[b] Material: 0.8% L-asparaginase (*Erwinia carotovora*), M.Wt 140,000 (tetramer). All samples freeze-dried in unbuffered 10 m*M* NaCl with 2% w/v additive, reconstituted at pH 10 in 0.1*M* glycine and dialysed.

Hellman, Miller, and Cammack (37, 38) have carried out a study of a very wide range of additives with the enzyme L-asparaginase isolated from *Erwinia carotovora* previously characterised by Cammack, Miller, and Marlborough (11, 61). This enzyme tetramer proved to be an ideal model protein in that cycles of freezing and thawing did not modify the biopolymer while freeze-drying did. Marlborough *et al.* (61) found that when asparaginase was freeze-dried in unbuffered 10 mM NaCl it became unstable and began to dissociate into inactive monomers above pH 8, whereas the native enzyme was very stable up to pH 11.8.

A list of some of the additives investigated by Hellman and his co-workers in Table IV confirms that glucose and sucrose are very effective as protective agents in freeze-drying. However, the sugar alcohols mannitol, galactitol and erythritol were somewhat less effective, providing 35–55% protection, although clearly this was a marked improvement over 10 mM NaCl in the absence of additive. It is interesting to comment here that mannitol is probably one of the most widely used protective agents in freeze-drying by the pharmaceutical industry in spite of its indifferent protective properties in Table IV; however, mannitol does have the useful additional property of improving the quality of the ice matrix so that the 'collapse' phenomenon is prevented during sublimation (58); also see Section 13.2.3.

5.5. Metastable Physical State after Freeze-drying

A careful study of the paper by Hellman and his co-workers (38) leads to a very important conclusion which needs to be emphasised. From the model they presented it is evident that any freeze-dried enzyme may in fact be unwittingly stored in a metastable physical state. This metastable state may not be recognised simply because reconstitution (at neutral pH in the case of *Erwinia* asparaginase) sometimes effectively reassembles the monomers even in the absence of protective additive. It must be more logical to store the enzyme as the intact native tetramer and not in a metastable dissociated state which precedes denaturation. With more complex virus assemblies, where reassembly becomes correspondingly more difficult, this would certainly be true.

6. PROTECTIVE AGENTS AND WATER STRUCTURE

Clearly some progress has been made in the past decade in the classification of protective agents, although because of the wide diversity of bioproducts, no universally acceptable formulation is likely to emerge. As indicated in the introduction, stabilisation of a biopolymer is a complementary procedure where the properties of the solvent and solute are optimised to achieve a satisfactory result.

Glucose, sucrose, lactose, mannitol, glycine and sodium or potassium glutamate are used routinely as cryoprotectants during freeze-drying. Glycerol and dimethyl sulphoxide (DMSO), which are the nearest to being universal cryoprotectants in cell biology and would be essential in any practical freezing protocol prior to freeze-drying (86), are for several reasons unsuitable for freeze-drying.

Stabilisation of biopolymers in aqueous solution is dependent on the structure of water. Klotz (46) and von Hippel and Wong (39) were able to make a useful distinction between additives that were chaotropic or caused a disordering of water structure, and those known to have a tendency towards the ordering of water structure, the latter being more suited to stabilisation of biopolymers. Meryman (64) lists the chaotropic cations and anions as being Na^+, Li^+, Ca^{2+}, Zn^{2+}, I^-, NO_3^-, HCO_3^-, and these should be excluded from any formulation if possible; the water-ordering ions were NH_4^+, K^+, Mg^{2+}, Cl^-, N_3^-, CO_3^{2-}, CH_3COO^-, and generally they do seem to produce better results than the chaotropic series.

Referring again to Table IV, it is very interesting to find that $(NH_4)_2SO_4$ is a surprisingly good protective agent whereas most electrolytes are much less effective than sugars. Before the introduction of superior purification methods such as molecular sieving and affinity chromatography, fractional precipitation methods were almost entirely dependent on this salt or ethanol, and indeed immune globulin fractions were stored as pastes at 4°C in saturated solutions of the salt without any evidence of physical or immunological change.

Clathrate water structures described by Klotz (46, 47) are hydrogen-bonded polyhedra. They are open, cage-like structures, into which guest apolar groups or residues can be fitted, and thermodynamically the system is stable. Not all the hydrophobic amino acid residues—valine, leucine and phenylalanine, for example—are confined to the interior of a globular protein, and where they are exposed on the surface ordered clathrate water structures have been postulated to accommodate these hydrophobic groups. Chaotropic cations and anions therefore interact unfavourably with these structures.

In the past many workers have commented on the possible role of hydrophilic sugars as hydrating agents, thus providing a buffer against excessive dehydration of the product, which is universally accepted as being undesirable. The results in Table IV do not reveal any correlation between the relative efficacies of sugars as protective agents and the moisture contents of asparaginase after freeze-drying, which covered the range 0.6–2.5% (w/v).

Parker (68) has discussed models of hexose sugars as structural analogues for hexagonal ordered water structure bound to protein, replacing the essential bound water where necessary. A good example was inositol, although in Table IV this is classified as being only a middle-grade protective agent when compared with sucrose and glucose.

7. INTRODUCTION TO PRACTICAL ASPECTS OF PRESERVING COMPLEX BIOLOGICAL PRODUCTS

All the systems reviewed in this chapter have a common feature, namely that decay processes can be minimised by either dehydration or immobilising (freezing) the free water in the system. Having examined the effects of formulation on storage and stability of highly purified systems such as enzymes, we must now consider more complex and less well defined systems such as virus suspensions.

Many of the problems associated with the freezing of biological products have been discussed above. Three serious practical disadvantages of freeze preservation, namely the high cost of maintaining stocks of frozen products, the associated high cost of transporting these stocks and the potential for total loss of material due to failure of the freezer plant, have encouraged the adoption of methods of dehydrating products. The high costs of storing and transporting frozen materials are especially important factors to be considered when mass vaccination programmes are undertaken in under-developed countries.

8. METHODS OF DEHYDRATION AT ELEVATED TEMPERATURES

The authors consider that methods of dehydration at elevated temperatures for drying biological materials are outside of the scope of this chapter. For a more detailed description of such methods, the reader is referred to the many standard texts related to biochemical engineering (90).

9. FREEZE-DRYING

9.1. General Considerations and Advantages

There is no doubt that, despite both the high initial capital equipment cost and the high running costs of the method, freeze-drying has become the most important method for dehydrating pharmaceutical materials, diagnostic reagents and similar high-cost thermo-labile bioproducts. The method was developed on a commercial scale in 1935 by Flosdorf and Mudd (20) and is particularly useful where a wide range of different pharmaceutical products have to be dried under conditions which minimise batch-to-batch contamination while retaining high levels of microbial sterility during dehydration.

Some of the more important advantages of freeze-drying are:

1. The water content of the product can be reduced to very low levels. In general the lower the water content, the more stable the product, although over-drying may cause a loss of product stability.

2. Since the product is normally sealed *in vacuo* or in inert gas, oxidative denaturation is greatly reduced.

3. Loss of water equals a loss of product weight and this may be important for products where transport costs are significant.

9.2. Disadvantages of the Method

Freeze-drying is not suitable for materials which supercool to form glasses, for products which produce surface skins upon cooling, thereby inhibiting the evolution of subliming water vapour, and for certain cell types (eukaryotes) which are able to retain viability, when frozen, only in the presence of special additives incompatible with freeze-drying.

Several authors have recently reported the successful freeze-drying of spermatazoa; however, the methods used, particularly in relation to the final residual moisture content, cannot be generally used as methods for preserving mammalian cells (42).

9.3. General Review of Methodology

It is clearly not possible in a review of this length to give more than a very basic idea of the most important principles involved in freeze-drying. For more detailed accounts of the process, the reader is referred to the many reviews and articles on the subject (63, 64, 72).

It is advantageous to consider the process of freeze-drying in four stages: pre-freezing, primary drying (vacuum sublimation), secondary drying (vacuum desorption) and, finally, stoppering and removal.

Freezing on the shelves of the freeze-drier is used almost universally on commercial freeze-driers for processing pharmaceutical products.

Freezing in solid CO_2/acetone or by specialised techniques such as evaporative freeze-drying (26) is of limited commercial value and such methods are normally only found in small laboratory freeze-driers.

9.3.1. Problems Associated with Incomplete Freezing

We have discussed in Section 4 the dangers of storing biological materials in a domestic deep-freeze. This fact has been clearly demonstrated by Davies and Beran (14), using Aujeszky's disease virus. These authors demonstrated that the virus was rapidly inactivated at −13°C and should be stored at −196°C for adequate survival. Suspensions of viruses often contain mixtures of several electrolytes and should be cooled below the lowest eutectic temperature to ensure complete freezing so that free liquid is not present in the apparently frozen mass (Section 4.2, Fig. 2). It is imperative that a vacuum is not applied until measurements of resistivity indicate a true freezing.

Supercooling, mentioned in Section 4.3, presents particular problems in freeze-drying. If a vacuum is applied to an incompletely frozen or supercooled liquid, foaming may occur, resulting in an unacceptable product.

The problem of collapse, which is often confused with a failure to go below the eutectic temperature, will be discussed below (see Section 2.3).

9.3.2. Rates of Freezing Prior to Sublimation

When considering the freezing rates obtained on the shelves of the freeze-drier, one must appreciate that freezing proceeds from the base of the vial progressing upwards through the depth of the liquid.

In this case it is useful to regard freezing rates as depths of liquid frozen, in millimetres per minute. The optimal rate of freezing prior to freeze-drying is 1 mm/min and such a freezing rate will ensure the correct ice crystal formation necessary to permit water vapour to escape freely from the ice matrix. At freezing rates below 0.5 mm/min or much greater than 1.0 mm/min, the final structure of the ice matrix will be such that vapour flow is impeded during sublimation and drying times significantly extended. It is not possible to achieve a cooling rate of approximately 1.0 mm/min when the depth of fill in the vial greatly exceeds 10 mm, and therefore 10 mm should be regarded as the upper limit of fill. In practice, one is often requested to freeze-dry material filled to a depth considerably in excess of 10 mm.

9.4. Primary Drying

Following freezing, the chamber is evacuated to allow free diffusion of water vapour from the frozen mass to the colder condenser surface. Since the migration of water from the product may be regarded basically as a diffusion process, this may help the reader to appreciate why freeze-drying is a relatively slow process.

Heat input into the product is essential to compensate for latent heat losses during sublimation (approximately 640 calories per gram of ice). Without heat, the product temperature would fall, finally approaching that of the condenser so that freeze-drying ceases.

Heat must not be applied until a vacuum of 0.2 torr has been attained or product melting may occur. If the pressure in the chamber should rise significantly above this value, because of vacuum pump failure or excessive heat input, the product may melt.

Throughout most of the primary drying cycle the product will remain at a significantly lower temperature than the shelf; only towards the end of primary drying, when sublimation has virtually ceased, will the product temperature rise rapidly to attain that of the shelf. At the end of primary drying only water adsorbed to the bulk powder and structurally bound to the biopolymer will remain and the product could be removed from the drier at this stage. In practice, since the water content of the powder is usually too high (7–10%) for optimal stability, the drying cycle is extended.

9.5. Secondary Drying

This prolonged drying stage is called secondary drying or desorption. During secondary drying, water contents will fall from approximately 7% to 1–2% [see Section 5.4, Table IV]. Biological materials may be inactivated by overdrying and it is usual to retain some residual moisture in the powdered product (73).

In older accounts of freeze-drying methodology, it was sometimes suggested that primary and secondary drying stages were performed in separate machines. Because of the greater efficiency of vapour-trapping systems on modern machines special secondary driers are no longer commonly used.

9.6. Stoppering and Sealing

At the end of the drying cycle, pharmaceutical products are normally stoppered within the freeze-drier. There has been considerable controversy over the design of container in which dried products should be sealed. The choice available to the processor rests between two basic systems, all-glass sealed ampoules and rubber-sealed glass vials.

Intuitively one would suppose that a hybrid rubber–glass sealed vial would be more prone to leakage than an all-glass sealed ampoule, particularly when the dried product is stored below 0°C and rubber bung shrinkage and lack of resilience must be considered. Indeed Barbaree and Smith (3) have reported that rubber-sealed vials are prone to leakage.

Corresponding reports by Greiff, Melton, and Rowe (33), claim to reveal small capillary imperfections extending through the tips of heat-sealed ampoules. In practical terms, it cannot be denied that reagents and stock cultures of dried microorganisms stored for several years in glass ampoules reveal little evidence of instability. Conversely, there is no doubt that the rubber-sealed vial represents the system most compatible with modern freeze-drying technology, allowing vials to be sealed within the freeze-drier, under predetermined residual moisture contents and conditions of vacuum or gas fill.

Overall, one would recommend a vial for the pharmaceutical product, where short to medium storage life is required, while using ampoules for microbial cultures and reagents which must remain unchanged over the course of several years.

10. PRESERVATION OF VIRUSES AND VACCINES

10.1. Introduction

Many viral vaccines are stored and distributed in the liquid state (for example vaccinia or Foot and Mouth Disease Virus in 50% glycerol) although freeze-drying is becoming an increasingly popular method of preparing a stable vaccine.

Majer *et al.* (59), producing a rabies vaccine grown on human diploid WI-38 Cells and inactivated with β-propiolactone, pointed out that earlier vaccines were unstable in the liquid state and required storage in a refrigerator. By freeze-drying this vaccine in a mixture of tris-EDTA-potassium glutamate-degraded gelatin, stability was greatly improved.

(The subject of adjuvant addition which is clearly an important factor when considering the formulation of liquid vaccines is dealt with elsewhere in the volume.)

10.2. Principles of Medium Formulation Using Influenza A Virus

The medium in which the vaccine or virus culture is to be suspended prior to freeze-drying should have the following properties:

1. The medium should act as a bulking agent, having the capability of retaining the product within the vial, preventing its escape with the subliming water vapour.

2. It should be inert to the product, causing no loss of biological activity in both the liquid and dried states.

3. For pharmaceutical use the medium should be free of harmful reactions, compatible with body fluids and of an ethically acceptable quality.

4. For reagent use, freedom from interfering reactions with the substrate or products is clearly essential.

5. The medium should maintain a high level of product stability during storage.

6. It should be capable of successful processing within the freeze-drier, ideally displaying a eutectic temperature approximately 10°C higher than the shelf temperature.

7. The medium should be formulated to allow a margin of unavoidable operational variability in order to ensure maximum uniformity of successive batches.

Other considerations, not primarily a property of the medium formulation, are:

1. The light-sensitive nature of the product and the need to use ambered or coloured vials.

2. The type of filler gas most appropriate to maintain product stability.

3. The final storage temperature of the product necessary to ensure longevity of shelf-life.

4. Miscellaneous factors, such as the nature of the vial, stopper etc. Often, although these considerations may be regarded as arbitrary requirements, adherence to a non-ideal formulation or freeze-drying procedure may be required to satisfy the demands of the pharmaceutical or reagent marketing departments.

10.3 Experimental Formulation Using Influenza A Virus

It is obviously desirable to arrive at the simplest medium which will satisfy the requirements listed above. It is not uncommon, however, to encounter cases where additives have been incorporated into the medium with little apparent reason or experimental justification for their inclusion. Acceptance of long-standing 'evolved' media formulations should be made only after careful experimental verification of the freeze-drying protocol.

A number of the more important points mentioned above may be illustrated by reference to some of the work in the authors' laboratory using influenza A virus, cultured and assayed as described by Appleyard and Maber (2).

10.3.1. Avoidance of Physical Loss of Virion during Freeze-drying

Table V (experiment I) illustrates how mechanical loss of both medium and virus occurred when influenza virus suspended in physiological saline

TABLE V   To Demonstrate That Mechanical Loss of Suspending Medium Can Occur
During Sublimation of Influenza A Virus and That Such Mechanical Loss Can Be
Prevented by Addition of a High Molecular Weight Compound to the Medium[a]

Experiment number	Suspending medium composition	EEL flame photometer light intensity readings		Percentage loss—Na+
		Pre-freeze-drying readings	Post-freeze-drying readings	
I	0.9% sodium chloride solution, pH 7.0	44	34	22%
II	0.9% sodium chloride + 1% DEAE– dextran, pH 7.0	46	45	2.0%
III	0.9% sodium chloride + 1% calcium lactobionate, pH 7.0	38	36	5.0%

[a] Samples freeze-dried for 24 hr.

was freeze-dried and powder deposit within the chamber was observed. It is
important to realise that these losses occurred only during sublimation and
not as a result of incomplete freezing followed by 'boiling'.

Experiments II and III in Table V demonstrate that mechanical loss can be
prevented by the incorporation of a high-molecular-weight additive into the
medium.

10.3.2. Special Hazards Associated with Sodium Azide Loss During Sublimation

Product loss with the vapour flow during sublimation is a well-recognised
problem and mechanical transport of solid from the vial will also account for
losses of non-volatile or non-subliming hazardous materials such as sodium
azide, which may be incorporated into a reagent formulation to prevent the
growth of microbial contaminants either before freeze-drying or upon
reconstitution.

Sodium azide *should not* be incorporated into freeze-drying media, since
not only can azide denature the virus (18) but also there have recently been a
number of reports of serious explosions associated with azide contamination
of the pumping systems of freeze-driers.

Other fungicides or bacteriocides, such as merthiolate, have been added
to freeze-drying media as alternatives to sodium azide.

All biostats have a potentially damaging effect on cell-derived material,
and if they have to be present in the medium they should be incorporated
into the reconstitution fluid which is added to the dried product.

TABLE VI   Infectivity Titres for Influenza A Virus (Strain WSN) Suspended
in Various Media after Freeze-drying for 24 hr

Purified virus suspended in:	Virus infectivity (plaque-forming units/ml)		
	Initial (pre-freeze-dried) titre	Post-freeze-dried titre	Percentage loss of titre
0.9% sodium chloride solution, pH 7.0	2.0×10^7	2.4×10^6	88
0.9% sodium chloride + 1% DEAE–dextran, pH 7.0	6.2×10^6	1.5×10^3	99.98
0.9% sodium chloride + 1% calcium lactobionate, pH 7.0	7.9×10^4	2.4×10^4	69.0
1% methoxy polyethylene glycol (MW 550)	1.7×10^{5a}	Less than 5×10	100
1% polyvinyl pyrollidone (MW 44,000)	Less than 5×10^a	Less than 5×10	100
Dextran (MW 70,000)	1.0×10^5	1.3×10^4	87
1% bovine serum albumin	4.8×10^6	1.0×10^6	79
1% bovine serum albumin + 1 : 5 dilution allantoic fluid + 1% calcium lactobionate	4.7×10^6	4.0×10^6	15

[a] Loss of infectivity in suspension before freeze-drying.

10.3.3. Assessment of 'Protective Agents'

Table VI demonstrates that, although high-molecular-weight additives proved to be good mechanical binding agents, their inclusion in media for preserving biological activity resulted in wide variations in protective function. Extremely poor viabilities were noted when influenza was suspended in PVP + saline and then freeze-dried, although Suzuki (81) has shown that a mixture of PVP and sodium glutamate will preserve infectivity of vaccinia virus during freeze-drying. Only 1 : 5 allantoic fluid + 1% BSA + 1% calcium lactobionate + physiological saline was totally acceptable as a protective agent for preserving the biological activity of the virus. This formulation is very similar to that suggested by Greiff and Rightsell (30) using Parke-Davis additive (a mixture of serum albumin and calcium lactobionate).

Freeze-drying media formulations used in the authors' laboratory, incorporating BSA or allantoic fluid alone, gave erratic results, which were sometimes acceptable although never reproducible. By combining both BSA and allantoic fluid with 1% calcium lactobionate, viral titres were much more consistent and recovery during freeze-drying was never less than 70% (see Table VII).

TABLE VII   Relative Protective Effects of Calcium Lactobionate Sterilized
by Filtration or Autoclaving on the Infectivity of Influenza A Virus Freeze-dried for 24 hr[a]

	Viral infectivity (plaque-forming units/ml)		
Medium	Pre-freeze-dried titre	Post-freeze-dried titre	Percentage loss of infectivity
0.9% sodium chloride + 1.5 allantoic fluid + 1% bovine serum albumin, pH 7.0	(a) 6.5×10^{7b} (b) 2.0×10^{7b}	8.2×10^{6b} 1.8×10^{7b}	87 10
0.9% sodium chloride + 1.5 allantoic fluid + 1% bovine serum albumin + 1% calcium lactobionate (autoclaved), pH 7.0	5.7×10^7	4.0×10^7	30
0.9% sodium chloride + 1.5 allantoic fluid + 1% bovine serum albumin + 1% calcium lactobionate (filtered)	4.2×10^7	1.8×10^6	96

[a] Material: influenza A virus (WSN) suspended in 1 : 5 dilution allantoic fluid + 0.85% (w/v)
NaCl + 1% bovine serum albumin + 1% calcium lactobionate.
[b] Range of variation in absence of calcium lactobionate (autoclaved).

One interesting observation (Table VII) was that, if the 10% calcium lactobionate stock solution was sterilised by filtration, the lactobionate appeared
to lose its stabilising property when compared to lactobionate solution which
had been sterilised by autoclaving. This observation underlines the need for
careful experimental evaluation of all the steps in the medium formulation.

10.3.4. Formulations Used for Parenteral Products

The allantoic fluid–BSA–calcium lactobionate–physiological saline medium described above was a perfectly acceptable medium for preserving virus
stability, but because the medium contained foreign proteins, it was considered unsuitable for inclusion into the formulation of a vaccine. For this
reason, suspensions of influenza A (strains $BELA_2$ and WSN) virus, purified
by ultracentrifugation, were suspended in 2% Dextraven 110 (average MW
110,000; Fisons Ltd.) + physiological saline (pH 6.8–7,0) (Table X). Dextraven 110 was more suitable than allantoic fluid + BSA for parenteral use.
Calcium lactobionate did not appear to enhance stability when added to
formulations of saline and Dextraven 110.

10.3.5 Rates of Freezing

It has been clearly demonstrated that for mammalian cell suspensions, the rate of freezing is an important factor affecting survival (50). When less complex biological systems such as viruses and other cell-derived components are considered, the freezing rate may assume a less important role. Greaves, Davies, and Steele (27) have demonstrated that freezing rates are important when T4 bacteriophage is frozen. These workers illustrated that suspensions of T4 became more stable as the freezing rate was increased from 1°C/min to 450°C/min, although at rates of freezing in excess of 900°C/min there was a sharp reduction in survival. Damage occurred between $-10°$ and $-20°C$ and 10% sugar or 20% peptone incorporated into the medium acted as a protectant, reducing damage by depressing the temperature for onset of injury.

Steeves and Grant (76), using Friend spleen focus-forming virus, perhaps a more realistic model than the structurally complex T-even phages, showed that viability of the frozen virus was influenced by the rate of freezing prior to storage and the rate of subsequent thawing.

In practice, when virus suspensions are freeze-dried on an industrial scale, it may be impossible to use an experimentally optimised freezing rate. Only two freezing rates are available for pre-freezing material within the freeze-drier:

1. The cooling rate obtained when the product is cooled on the freeze-drier shelf, starting at ambient temperature ('slow' freezing rate)

2. The cooling rate obtained when the material is placed onto pre-cooled shelves ('fast' freezing)

Table VIII shows how freezing rates obtained in the Edwards EF6 shelf freeze-drier had little effect on the viability of influenza A virus, even when fast or slow thawing rates were combined with these cooling rates.

TABLE VIII   Sensitivity of Influenza A Virus Infectivity to Fast and Slow Cooling and Fast and Slow Thawing Rates When Virus Is Frozen to $-36°C$ for 4 hr[a]

Group	Cooling rate	Thawing rate	Initial virus titre	Post-freezing titre
A	Fast (82°C/min)	Fast (46°C/min)	1.4×10^8	1.7×10^8
B	Fast	Slow (2.9°C/min)	1.4×10^8	1.5×10^8
C	Slow (2.5°C/min)	Fast	1.4×10^8	1.4×10^8
D	Slow	Slow	1.4×10^8	1.7×10^8

[a] Material: influenza A virus (strain WSN) in 1 : 5 dilution allantoic fluid + 0.85% NaCl + 1% bovine serum albumin + 1% calcium lactobionate.

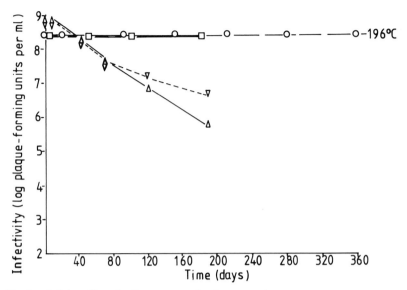

Fig. 3.   Stability of liquid influenza virus after freezing and storage at various temperatures. Material: influenza A virus (strain WSN) suspended in 1 : 5 dilution allantoic fluid + 0.85% (w/v) NaCl + 1% bovine serum albumin + 1% calcium lactobionate. (○) Storage at −196°C; (□) storage at −70°C; (▽) storage at −20°C; (△) storage at +4°C.

10.3.6. Optimal Storage Temperature of Frozen Virus Suspension

While the freezing rate may be an important factor to be considered when preserving virus suspensions, of fundamental importance is the terminal temperature at which frozen suspensions are to be stored (*14, 29*).

Figure 3 indicates that liquid suspensions of influenza A virus are extremely stable when stored at −70° or −196°C, when compared with similar material stored at +4° or −20°C.

Greiff and Rightsel (*29*) suggested that only at −130°C could physicochemical changes be regarded as negligible. In practice, these authors demonstrated that measles virus remained stable when stored at −65°C in Parke-Davis Additive or DMSO although significant reductions in titre were shown to occur on storage at −20°C. Greiff and Rightsel suggested that stability was related partly to the stability of the viral RNA and predicted that Mg^{2+} ions would protect RNA viruses from thermal inactivation. Experimentally they were able to demonstrate that incorporation of $MgCl_2$ into the suspending medium protected freeze-dried poliovirus (an RNA virus) from denaturation during storage.

The relationship of Mg^{2+} ions and the stability of RNA has been well established. Cammack, Miller and Grinstead (10) demonstrated clearly how the conformational stability of RNA, isolated from bacterial ribosomes, was extremely dependent on the concentration of Mg^{2+} ions in the medium.

10.3.7. Factors Affecting Storage Stability of Freeze-dried Virus

Long-term stability is in some ways more important than immediate post-freeze-drying viability and improved long-term stability might well compensate for a slight loss of viability during freeze-drying.

In an extensive experimental appraisal of the factors related to freeze-dried viral stability, Greiff and Rightsel (30) suggested a number of factors which could influence stability and concluded that influenza (suspended in an allantoic fluid–serum albumin–calcium lactobionate–saline solution) should only be dried to an optimal moisture content of 1.7%; under-drying or overdrying both reduced the long-term stability of the material. Many of these observations have been confirmed by other workers (4) and have been confirmed in the authors' laboratory, extending the study to include other strains of influenza A virus and protective media more acceptable for inclusion in a vaccine (see Table IX).

Earlier, Fry and Greaves (23) had demonstrated the danger of over-drying bacteria, although as Greiff and Rightsel (30) state, there is often a legal requirement by licensing authorities to produce vaccines with an absolute minimum water content.

Although Greiff and Rightsel's influenza suspensions, freeze-dried to 1.7% moisture, exhibited excellent stability when stored at $-20°C$ and sealed *in vacuo*, clinically the dried suspension was more acceptable if the vial was filled with an 'inert' gas prior to sealing.

Dry, oxygen-free nitrogen is normally used for back-filling. Greiff (31), however, pointed out that nitrogen should not be regarded as a chemically inert element and experimentally showed that the rare gases argon and helium were better choices as filler gases.

10.3.8. Problems Associated with Accelerated Storage Tests (AST)

A reasonably good correlation was found between predicted storage stabilities (estimated by AST) (28) and actual storage data obtained by periodic estimates of viral titres (long-term storage, LTS) for material optimally dried or under-dried. However, a discrepancy was observed between the two methods for results obtained for over-dried material (0.6% residual moisture). The data obtained by LTS testing indicated that the virus was more stable than had been predicted from AST data.

TABLE IX Stability of Influenza A Virus Freeze-dried to Various Moisture Contents and Sealed Under Various Gases[a]

		Time (days) to lose 1 log infectivity			
	Residual moisture content (%)	Estimated by accelerated storage test[b]		Estimated by long-term storage test[b]	
Filler gas		+4°C	−20°C	+4°C	−20°C
Helium (500 mbar)	3.6	31	790	49	670
Nitrogen (500 mbar)	3.8	14	160	40	475
Vacuum (10^{-1} mbar)	5.3	14	230	37	305
Helium (500 mbar)	1.8	260	>2000	320	0.36 log loss in 926 days
Helium (500 mbar)	1.7	190	>2000	225	N/A
Nitrogen (500 mbar)	1.7	84	720	113	926
Vacuum (10^{-1} mbar)	1.9	148	1600	140	0.71 log loss in 926 days
Vacuum (10^{-1} mbar)	1.4	35	580	40	418
Helium (500 mbar)	0.6	26	100	300	543
Nitrogen (500 mbar)	0.6	11	54	240	524
Vacuum (10^1 mbar)	0.6	20	90	265	650

[a] Material: influenza A virus (strain WSN) suspended in 1 : 5 dilution allantoic fluid + 0.85% NaCl + 1% bovine serum albumin + 1% calcium lactobionate.

[b] +4°C and −20°C indicate temperatures at which virus was stored.

No explanation can be offered for the discrepancies observed between the two tests when virus was stored in the over-dried state (0.6% residual moisture). Only by a more extensive appraisal of the factors involved in stability can a hypothesis be advanced.

Accelerated storage tests are, of course, widely used to estimate product stability (32, 82) and the convenience of such tests is discussed by Griffin et al. (34). Nevertheless, the use of any accelerated storage test for estimating the stability of a biological product does require caution. Indeed Damjanovic and Radulovic (13) have pointed out that accelerated storage data should be used only to indicate stability and not used in an attempt to interpret mechanisms of death or damage.

Because of this lack of confidence in the AST, results summarised in Table X, showing the stability of influenza virus when suspended in 2% Dextraven

TABLE X Stability of Purified Influenza A Virus Freeze-dried
to Different Moisture Contents and Stored Under Various Gases[a]

Virus strain	Filler gas	Residual moisture content (%)	Time to lose 1 log infectivity (days)[b]	
			+4°C	−20°C
WSN	Helium (500 mbar)	2.6	175	440
WSN	Nitrogen (500 mbar)	2.6	175	382
WSN	Vacuum (10^{-1} mbar)	2.6	210	390
WSN	Helium (500 mbar)	1.6	125	200
WSN	Nitrogen (500 mbar)	1.6	94	175
WSN	Vacuum (10^{-1} mbar)	1.6	102	210
BEL A_2	Helium (500 mbar)	2.6	190	353
BEL A_2	Helium (500 mbar)	1.6	108	250

[a] Material: influenza A virus (strain WSN or BEL A_2) suspended in 0.85% NaCl + 2%
Dextraven (dextran MW 110,000).
[b] +4°C and −20°C indicate temperature at which the virus was stored.

110, are based only on data obtained from long term storage tests. Again, it
was apparent that over-drying the virus, suspended in 2% Dextraven, from
2.6 to 1.6% residual moisture content reduced its stability.

There was much less clear evidence that helium improved the stability of
the dried virus compared with nitrogen when these gases were used for
back-filling the vials.

10.3.9. Differential Thermal Stability of Infectivity and Haemagglutination Activity

Apart from an overall increase in the virion stability, individual viral components may exhibit increased resistance to physical stresses after freeze-
drying.

Figure 4 illustrates that while infectivity of freeze-dried influenza fell
rapidly when the virus was heated to +90°C the haemagglutinin activity
remained unimpaired even after heating for 4 hr at this temperature.

Apostolov and Damjanovic (1) reported a similar observation with Sendai
virus and suggested that selective heat inactivation of a freeze-dried virus
might be of some value in producing a killed viral vaccine.

10.4. Theoretical Aspects of Media Formulation

Throughout this chapter, continual emphasis has been given to the role of
water and this role would seem to be equally valid when applied to the
preservation of viruses.

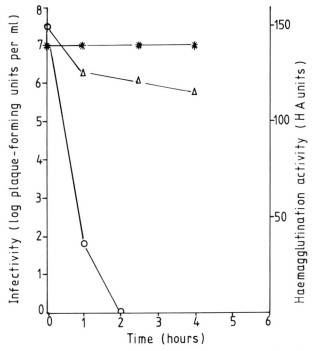

Fig. 4. Comparative heat stabilities of haemagglutination activity and total infectivity for freeze-dried influenza virus after heating dried virus to +70° and +90°C. Material: influenza A virus (strain WSN) suspended in 1 : 5 dilution allantoic fluid + 0.85% (w/v) Nacl + 1% bovine serum albumin + 1% calcium lactobionate, dried to residual moisture content 3.6%, filler gas helium (500 mbar). (★) HA activity virus heated to +70° or +90°C; (△) infectivity virus heated to +70°C; (○) infectivity virus heated to +90°C.

Fellowes (19) demonstrated that two physically quite dissimilar viruses, foot-and-mouth disease virus (FMDV) and vesicular stomatitis virus (VSV), both showed a similar response to freezing and thawing or freeze-drying and reconstitution. Rightsel and Greiff (70) had earlier attempted a more comprehensive classification of viruses based on their physical stabilities, particularly in relation to the stresses of freezing and dehydration. These authors demonstrated most clearly that the structure of an infective viral particle cannot be regarded simply as the sum of its biochemical components but that the water content must be regarded as an integral part of the virus structure. In influenza virus, for example, the central ribonucleic acid core was shown to be surrounded by a structured layer of water with spokes of water radiating from this central core region and extending to the outer lipid membrane.

Greiff and Rightsel (30), using influenza virus freeze-dried to give different residual moisture contents, re-stated a general view that 'ordered water' in close proximity to the surface of a protein molecule is an essential component of conformational stability. This ordered water is generally regarded as being more difficult to sublime than the 'free water' of the system.

The authors postulated two mechanisms for the degradation of freeze-dried virus, depending on the residual moisture content:

1. With under-dried virus (i.e. 3.2% residual moisture) there is sufficient free water present to allow conformational changes in the viral coat proteins to occur, similar to those normally experienced in solution or suspension.

2. With over-dried virus (i.e. 0.4% residual moisture) saturation or blocking or hydrophilic sites of proteins by oxidation is the dominant factor.

Hence, the optimal residual moisture content of 1.7% reflects a compromise, balancing degradation through conformation change and oxidation during storage.

11. FREEZE-DRYING OF BLOOD AND BLOOD PRODUCTS

Because of the sensitivity of individual cells, suspensions of whole blood cannot be freeze-dried, although erythrocytes, white blood cells and platelets can be successfully frozen for subsequent therapeutic use.

Sera, plasma and extracted serum proteins have been successfully freeze-dried for a number of years. Indeed sera and sera-derived products are often incorporated into freezing or freeze-drying media formulations to improve stability.

In addition to their therapeutic value, freeze-dried blood fractions and products may be used as diagnostic reagents.

In the authors' laboratory human anti-rubella serum is being routinely freeze-dried. Earlier preparations, in which serum was diluted with peptone water, showed clear evidence that 'boiling' had occurred resulting in a hard, brittle product. Although the reagent reconstituted satisfactorily and retained full biological activity, the appearance of the material was considered commercially unacceptable. A change to 2% Lomodex dextran (Fisons Ltd.) (average MW 70,000) + saline for diluting the serum resulted in a light, amorphous, white plug which reconstituted readily and had undiminished biological activity, satisfactory shelf-life and commercially acceptable appearance.

12. FREEZE-DRYING OF INTERFERONS

The interferons represent a group of cell-derived proteins which have certain basic physical and chemical properties in common, the principal one being the ability to induce antiviral resistance (89).

Lindemann, Burke, and Isaacs (51), in one of the earlier papers describing the activity, purification and preservation of chick cell interferon, noted that reconstituted freeze-dried interferon retained complete biological activity. However, when these authors used freeze-drying to concentrate the interferon, an insoluble precipitate was formed and some activity lost. Later, Galasso (24) reported variable and unpredictable losses in the potency of interferon preparations during storage and transport. Jameson *et al.* (41) conducted a detailed study of the stability of murine interferon and described how pre-treatment of interferon to inactivate the inducer virus, methods of purification, suspending medium composition and final moisture content all influenced the stability of the preparation. The authors predicted that interferon, suspended in 0.1 M sodium phosphate, pH 7.0, augmented with 0.5% bovine serum albumin and dried to 3.0% residual moisture, would lose 1000 units of activity in 110 years when stored at +4°C. Substitution of potassium ions for sodium ions in the buffer reduced stability. Jameson *et al.* suggested that the enhanced activity of interferon freeze-dried in a medium with a high sodium ion concentration may have been due to changes in the pH of the solution upon cooling (6), these pH changes being more marked in media with high sodium ion rather than high potassium ion concentrations.

13. PRESERVATION OF HORMONES

13.1 Insulin

Insulin is perhaps the best known example of a hormone in therapeutic use and at present insulin is marketed as a liquid solution or suspension of the crystallised hormone in sodium acetate or a formulation designed to improve its *in vivo* activity. There is no requirement to distribute the hormone in the frozen state, the liquid product remaining stable for a considerable time.

13.2 Pituitary Gland-derived Hormones

13.2.1 General Remarks

With other hormones, stability in the liquid state may be poor and many of these hormones are distributed in the freeze-dried state.

In this laboratory human growth hormone (hGH) has been successfully freeze-dried, and experimental determination of the medium formulation illustrates clearly the need for close cooperation between the production and freeze-drying departments.

13.2.2. Purification and Stability

At present, for clinical use, the hGH is extracted from human pituitary glands, which are often stored in acetone for some months prior to purification (36). Both prolonged storage in acetone and subsequent extraction procedures may cause denaturation or aggregation of the hormone prior to freeze-drying (40, 65).

Earlier hormone extracts were freeze-dried from water after purification (55), mannitol being added to improve the quality of the product. Polymerisation has always been of concern with preparations of hGH since polymerised material may react in an immunologically adverse manner in children treated by frequent injection over a long period of time.

Because of changes in the extraction methodology, hGH has been freeze-dried in 50 mM tris–saline plus 0.5% mannitol. Although retaining good biological activity, this preparation was considered commercially unacceptable after freeze-drying, extensive plug shrinkage and boiling being observed.

Figure 5 shows that removal of the mannitol (curve B) and reduction of tris from 50 to 25 mM (curve C) had a marked influence on measurements of conductivity versus temperature, observed when solutions were warmed

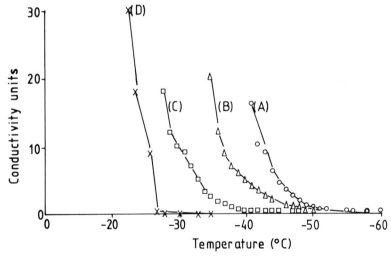

Fig. 5. Profiles illustrating the relationship between conductivity and temperature for various aqueous solutions frozen to −60°C and warmed until melted. Curve A:50 mM tris + 100 mM sodium chloride + 0.5% mannitol + 1.5 mg/ml bovine serum albumin, pH 8.0. Curve B: 50 mM tris + 100 mM sodium chloride + 1.5 mg/ml bovine serum albumin, pH 8.0. Curve C: 25 mM tris + 125 mM sodium chloride + 1.5 mg/ml bovine serum albumin, pH 8.0. Curve D: 100 mM sodium chloride.

from −70°C to various melting temperatures. One is somewhat reluctant to describe these measurements as eutectic measurements, because of the asymptotic curves produced, compared with the rapid increase in conductivity recorded for a saline solution (curve D).

Subsequent preparations of hGH in 25 mM tris solution without the addition of mannitol resulted in an acceptable product.

13.2.3. Collapse Phenomenon

Mackenzie (58) has suggested that the problem of plug shrinkage or boiling, which has often been suggested as indicating incomplete cooling below the eutectic, is due to a much more complex phenomenon termed collapse. Collapse occurs when a solute system fails to crystallise completely even when the solution has been cooled below its apparent eutectic, so that minute pockets of unfreezable moisture remain within the apparently frozen mass. It is a particular problem where the solute or solute mixture persists in an amorphous state. Collapse temperatures are characteristic of the nature of the solute (57) and should not be confused with eutectic temperatures, which are appropriate in instances where the solute does crystallise, and collapse temperatures are independent of freezing rate or solute concentration.

In order to avoid the phenomenon of collapse it is necessary to freeze-dry the solution below its collapse temperature (for example −41°C for glucose and −48°C for fructose solutions). In practice, it is difficult to estimate the collapse temperature of a solution and usually one avoids gross collapse phenomena by pragmatic alterations in the medium formulation, possibly as we observed with the hGH or rubella reagent.

REFERENCES

1. Apostolov, K., and Damjanovic, V. (1973). Selective inactivation of infectivity of freeze-dried Sendai virus by heat. *Cryobiology* **10**, 255–259.
2. Appleyard, G. A., and Maber, H. (1974). Plaque formation by influenza viruses in the presence of trypsin. *J. Gen. Virol.* **25**, 351–357.
3. Barbaree, J. M., and Smith, S. J. (1981). Loss of vacuum in rubber stoppered vials stored in a liquid nitrogen vapour phase freezer. *Cryobiology* **18**, 528–531.
4. Beardmore, W. B., Clark, T. D., and Jones, K. (1968). Preservation of influenza virus infectivity by lyophilisation. *App Microbiol.* **16**, 362–365.
5. Berg, L. van den, and Rose, D. (1959). Effect of freezing on the pH and composition of sodium and potassium phosphate solutions to the reciprocal system $KH_2PO_4-Na_2HPO_4-H_2O$. *Arch. Biochim. Biophys.* **81**, 319–329.
6. Berg, L. van den (1959). The effect of addition of sodium and potassium chloride to the reciprocal system: $KH_2PO_4-H_2O$ on pH and composition during freezing. *Arch. Biochim. Biophys.* **84**, 305–315.
7. Brandts, J. F. (1964). The thermodynamics of protein denaturation. I. The denaturation of chymotrypsinogen. *J. Am. Chem. Soc.* **86**, 4291–4301.

8. Brandts, J. F. (1964). The thermodynamics of protein denaturation. II. A model of reversible denaturation and interpretations regarding the stability of chymotrypsinogen. *J. Am. Chem. Soc.* **86**, 4302–4314.

9. Brandts, J. F., Joan, F., and Nordin, J. H. (1970). Low temperature denaturation of chymotrypsinogen in aqueous and frozen solution. *In* "The Frozen Cell" (G. E. W. Wolstenholme and M. O'Connor, eds.), pp. 189–213. Churchill, London.

10. Cammack, K. A., Miller, D. S., and Grinstead, K. H. (1970). Physical properties of ribosomal ribonucleic acid isolated from bacteria deficient in ribonuclease I. *Biochem. J.* **117**, 745–755.

11. Cammack, K. A., Marlborough, D. I., and Miller, D. S. (1972). Physical properties and subunit structure of L-asparaginase isolated from *Erwinia carotovora*. *Biochem. J.* **126**, 361–379.

12. Chilson, O. P., Costello, L. A., and Kaplan, N. O. (1965). The effect of freezing on enzymes. *Fed. Proc. Fed. Am. Soc. Exp. Biol.* **24**, Suppl. 15, 55–65.

13. Damjanovic, V., and Radulovic, D. (1968). Predicting the stability of freeze-dried *Lactobacillus bifidus* by the accelerated storage test. *Cryobiology*, **5**, 101–104.

14. Davies, E. B., and Beran, G. W. (1981). Influence of environmental factors upon survival of Aujeszky's disease virus. *Res. Vet. Sci.* **31** (1), 32–36.

15. Deisseroth, A., and Dounce, A. L. (1967). Nature of the change produced in catalase by lyophilisation. *Arch. Biochem. Biophys.* **120**, 671–692.

16. Dickerson, R. E., and Geiss, I. (1969). "The Structure and Actions of Proteins." Benjamin, Menlo Park, California.

17. Eagland, D. (1975). Nucleic acids, peptides and proteins. *In* "Water: A Comprehensive Treatise." (F. Franks, ed.), Vol. 4, pp. 305–516. Plenum, New York.

18. Easterbrook, K. B. (1967). Analysis of the early stages in vaccinia virus infection in KB cells using sodium azide. *Virology* **15**, 417–427.

19. Fellowes, O. N. (1968). Comparison of cryobiological and freeze-drying characteristics of foot and mouth disease virus and of vesicular stomatitis virus. *Cryobiology* **4**, 223–231.

20. Flosdorf, E. W., and Mudd, S. (1935). Procedure and apparatus for preservation in lyophile form of serum and other biological substances. *J. Immunol.* **29**, 389–425.

21. Franks, F. (1972). Introduction—Water, the unique chemical. *In* "Water: A Comprehensive Treatise" (F. Franks, ed.), Vol. 1, pp. 1–13. Plenum, New York.

22. Franks, F. (1975). The hydrophobic interaction. *In* "Water: A Comprehensive Treatise" (F. Franks, ed.), Vol. 4, pp. 1–94. Plenum, New York.

23. Fry, R. M., and Greaves, R. I. N. (1951). The survival of bacteria during and after drying. *J. Hyg.* **49**, 220–246.

24. Galasso, G. J. (1970). Standard reagents in interferon research. *Symp. Ser. Immunobiol. Stand.* **14**, 272–276.

25. Goteni, S., Raymond, J., Ducastaing, A., Robin, J.-M., and Creach, P. (1974). Effets de la lyophilisation sur la solubilité et l'activité enzymatique du lysozyme du blanc d'oeuf de Poulet. *C. R. Seances Soc. Biol.* **168**, 280–285.

26. Greaves, R. I. N. (1946). The preservation of proteins by drying. *Med. Res. Counc. (G.B.), Spec. Rep. Ser.* **258**, 37.

27. Greaves, R. I. N., Davies, J. D., and Steele, P. R. M. (1966). The freeze-drying of frost sensitive organisms. *Cryobiology* **3**, 283–287.

28. Greiff, D., and Rightsel, W. (1965). An accelerated storage test for predicting the stability of suspensions of measles virus dried by sublimation in vacuo. *J. Immunol.* **94**, No. 3, 395–400.

29. Greiff, D., and Rightsel, W. (1967). Stabilities of suspension of viruses after vacuum sublimation and storage. *Cryobiology* **3**, No. 6, 435–443.

30. Greiff, D., and Rightsel, W. (1968). Stability of suspensions of influenza virus dried to different contents of residual moisture by sublimation in vacuo. *Appl. Microbiol.* 16, No. 6, 835–840.

31. Greiff, D. (1970). Stabilities of suspensions of influenza virus dried by sublimation of ice in vacuo to different contents of residual moisture and sealed under different gases. *Appl. Microbiol.* 20, No. 6, 935–938.

32. Greiff, D., and Greiff, C. (1972). Linear non-isothermal, single step stability studies of dried preparations of influenza virus. *Cryobiology* 9, 34–37.

33. Greiff, D., Melton, H., and Rowe, T. W. G. (1975). On the sealing of gas-filled glass ampoules. *Cryobiology* 12, 1–14.

34. Griffin, C., Cook, E., and Mehaffey, M. (1981). Predicting the stability of freeze-dried *Fusobacterium mortiferum* proficiency testing samples by accelerated storage tests. *Cryobiology* 18, 420–425.

35. Hanafusa, N. (1969). Denaturation of enzyme protein by freeze-thawing and freeze-drying. *In* "Freezing and Drying of Microorganisms" (T. Nei, ed.), pp. 117–129. Univ. of Tokyo Press, Tokyo.

36. Hartree, A. S. (1966). Separation and partial purification of the protein hormones from human pituitary glands. *Biochem. J.* 100, 754–761.

37. Hellman, K. H., Miller, D. S., and Cammack, K. A. (1979). Dissociation of L-asparaginase by freeze-drying. *Congr. Biochem., 11th,* 1979, p. 219.

38. Hellman, K. H., Miller, D. S., and Cammack, K. A. (1983). The effect of freeze-drying on the quaternary structure of L-asparaginase from *Erwinia carotovora. Biochim. Biophys. Acta* 749, 133–142.

39. Hippel, P. H. von, and Wong, K.-W. (1964). Neutral salts: The generality of their effects on the stability of macromolecular conformations. *Science* 145, 577–580.

40. Holmstrom, B., and Fholenhag, K. (1975). Characterisations of human growth hormone preparations used for the treatment of pituitary dwarfism: A comparison of concurrently used batches. *J. Clin. Endocrinol. Metab.* 40, No. 5, 856–862.

41. Jameson, P., Greiff, D., and Grossberg, S. E. (1979). Thermal stability of freeze-dried mammalian interferons. Analysis of freeze-drying condition and accelerated storage tests for murine interferon. *Cryobiology* 16, 301–314.

42. Jeyendran, R. S., Graham, E. F., and Schmel, M. K. L. (1981). Fertility of dehydrated bull semen. *Cryobiology* 18, 292–300.

43. Kauzmann, W. (1959). Some factors in the interpretation of protein denaturation. *Adv. Protein. Chem.* 14, 1–57.

44. Kendrew, J. C., and Perutz, M. F. (1957). X-ray studies of compounds of biological interest. *Annu. Rev. Biochem.* 26, 327.

45. Klotz, I. M. (1958). Protein hydration and behaviour. *Science* 128, 815–822.

46. Klotz, I. M. (1965). Role of water structure in macromolecules. *Fed. Proc., Fed. Am. Soc. Exp. Biol.* 24, No. 2, 24–33.

47. Klotz, I. M. (1970). Polyhedral clathrate hydrates. *In* "The Frozen Cell" (G. E. W. Wolstenholme and M. O'Connor, eds), pp. 5–26. Churchill, London.

48. Klotz, I. M., Darnall, D. W., and Langerman, N. B. (1975). Quaternary structure of proteins. *In* "The Proteins" (H. Neurath, R. L. Hill, and C. Boeder, eds.), 3rd ed., Vol. 1, pp. 293–411. Academic Press, New York.

49. Leibo, S. P., and Jones, R. F. (1964). Freezing of the chromoprotein phycoerythrin from the red alga *Porhydridium cruentum. Arch. Biochem. Biophys.* 106, 78–88.

50. Leibo, S. P., and Mazur, P. (1971). The role of cooling rates in low temperature preservation. *Cryobiology* 8, 447–452.

51. Lindemann, J., Burke, D. C., and Isaacs, A. (1957). Studies on the production, mode of action and properties of interferon. *Bri. J. Exp. Pathol.* 38, 551–562.

52. Litt, M. (1958). Preservation of haemocyanin. *Nature (London)* 181, 1075.

53. Lovelock, J. E. (1953). Haemolysis of human red blood cells by freezing and thawing. *Biochim. Biophys. Acta* **10**, 414–426.
54. Lucy, J. A. (1968). Theoretical and experimental models for biological membranes. In "Biological Membranes Physical Fact and Fiction," (D. Chapman, ed.), pp. 233–285. Academic Press, New York.
55. Lumley-Jones, R., Benker, G., Salacinski, P., Lloyd, T., and Lowry, P. (1979). Large scale preparation of highly purified pyrogen-free human growth hormone for clinical use. *J. Endocrinol.* **82**, 77–86.
56. Luyet, B. (1960). On various phase transitions occurring in aqueous solutions at low temperature. *Ann. N. Y. Acad. Sci.* **85**, 549–569.
57. Mackenzie, A. P. (1967). The collapse phenomenon in the freeze-drying process. *Cryobiology* **3**, 387.
58. Mackenzie, A. P. (1977). The physico-chemical basis for the freeze-drying process. *Dev. Biol. Stand.* **36**, 51–67.
59. Majer, M., Herrmann, A., Hilfenhaus, J. H., Reichert, E., Mauler, Z., and Hennessen, W. (1976). Freeze-drying of a purified human diploid cell rabies vaccine. *Dev. Biol. Stand.* **36**, 285–289.
60. Markert, C. L. (1963). Lactate dehydrogenase isoenzymes: Dissociation and recombination of subunits. *Science* **140**, 1329–1330.
61. Marlborough, D. I., Miller, D. S., and Cammack, K. A. (1975). Comparative study on conformational stability and subunit interactions of two bacterial asparaginases. *Biochim. Biophys. Acta* **386**, 576–589.
62. Mazur, P. (1970). Cryobiology: The freezing of biological systems. *Science* **168**, 939–949.
63. Mellor, J. D. (1978). "Fundamentals of Freeze Drying." Academic Press, London.
64. Meryman, H. T. (1966). Freeze-drying in cryobiology. In "Cryobiology" (H. T. Meryman, ed.), pp. 609–663. Academic Press, New York.
65. Moore, W. V. (1978). The role of aggregated hGH in the therapy of hGH-deficient children. *J. Clin. Endocrinol. Metab.* **46**, No. 1, 20–27.
66. Nemethy, G., and Scheraga, H. (1962). The structure of water and hydrophobic bonding in proteins. *J. Chem. Phys.* **36**, No. 12, 3401–3417.
67. Orndorf, G. R., and Mackenzie, A. P. (1973). The functioning of the suspending medium during the freeze-drying preservation of *Escherichia coli*. *Cryobiology* **10**, 475–487.
68. Parker, J. (1972). Spatial arrangement of some cryoprotective compounds in ice lattices. *Cryobiology* **9**, 247–250.
69. Rey, L. (1960). Thermal analysis of eutectics in freezing solutions. *Ann. N. Y. Acad. Sci.* **85**, 510–534.
70. Rightsel, W., and Greiff, D. (1967). Freezing and freeze-drying of viruses. *Cryobiology* **3**, No. 6, 423–431.
71. Rosenkrantz, H., and Scholtan, W. (1971). Circular Dichroismus and Konformation der L-asparaginase. *Hoppe-Seyler's Z. Physiol. Chem.* **352**, 1081–1090.
72. Rowe, T. W. G. (1971). Machinery and methods in freeze-drying. *Cryobiology* **8**, 153–172.
73. Seligman, E. B., and Farber, J. F. (1971). Freeze-drying and residual moisture. *Cryobiology* **8**, 138–144.
74. Shikama, K., and Yamazaki, I. (1961). Denaturation of catalase by freezing and thawing. *Nature (London)* **190**, 83–84.
75. Sinanoglu, O., and Abdulnur, S. (1965). Effect of water and other solvents on the structure of biopolymers. *Fed. Proc. Fed. Am. Soc. Exp. Biol.* **24**, No. 2, 5–12 to 5–13.
76. Steeves, R. A., and Grant, V. R. (1978). Biological activity of Friend spleen focus-forming virus after cold storage. *Cryobiology* **15**, 109–112.
77. Stellwagen, E., and Schachman, H. K. (1962). The dissociation and reconstitution of aldolase. *Biochemistry* **1**, No. 6, 1056–1069.

78. Stillinger, F. H. (1980). Water revisited. *Science* **209**, 451–457.
79. Stoeckenius, W., and Engelman, D. M. (1969). Current models for the structure of biological membranes. *J. Cell Biol.* **42**, 613–646.
80. Sudi, J., and Khan, N. G. (1970). Factors affecting freeze-drying induced transient dissociation of lactate dehydrogenase tetramers. *Acta Biochim. Biophys. Acad. Sci. Hung.* **5** (2), 159–175.
81. Suzuki, M. (1973). Protectants in the freeze-drying and the preservation of vaccinia virus. *Cryobiology* **10**, 435–439.
82. Suzuki, M. (1973). Stability and residual moisture content of dried vaccinia virus. *Cryobiology* **10**, 432–434.
83. Tanford, C. (1962). Contribution of hydrophobic interactions to the stability of the globular conformation of protein. *J. Am. Chem. Soc.* **84**, 4220–4247.
84. Tanford, C. (1973). "The Hydrophobic Effect: Formation of Micelles and Biological Membranes." Wiley, New York.
85. Tanford, C., and Louvrien, R. (1962). Dissociation of catalase into subunits. *J. Am. Chem. Soc.* **84**, 1892–1896.
86. Taylor, R., Adams, G. D. J., Boardman, C. E. B., and Wallis, R. G. (1974). Cryoprotection—permeant versus non-permeant additives. *Cryobiology* **11**, 430–438.
87. Tong, M.-M., and Pincock, R. E. (1969). Denaturation and reactivity of invertase in frozen solutions. *Biochemistry* **8**, 908–913.
88. Tsuchiya, T., Tsuchiya, Y., Nonomura, Y., and Matsumoto, J. J. (1975). Prevention of freeze denaturation of carp actomyosin by sodium glutamate. J. Biochem. (Tokyo) **77**, 853–862.
89. Vilcek, J. (1969). "Interferon," Viro. Monog. No. 6. Springer-Verlag, New York.
90. Webb, F. C. (1964). Dehydration. *In* "Biochemical Engineering," Chapter 15, pp. 398–426. Van Nostrand-Reinhold, Princeton, New Jersey.
91. Yasui, T., and Hashimoto, Y. (1966). Effect of freeze-drying and denaturation of myosin from rabbit skeletal muscle. *J. Food Sci.* **31**, 293–299.
92. Yu, N. T., and Jo, B. H. (1973). Comparison of protein structure in crystals and in solution by laser Raman scattering. I. Lysozyme. *Arch. Biochem. Biophys.* **156**, 469–474.
93. Yu, N. T. (1974). Comparison of protein structure in crystals, in lyophilised state, and in solution by laser Raman scattering. III. α-Lactalbumin. *J. Am. Chem. Soc.* **96** (14), 4664–4668.

PART III

PRODUCT TESTING

The acceptance of a medical product depends upon the balance in the risks versus gains equation (Chapter 15, this volume). There is no set formula—it depends upon the severity of the disease, the geographical location, the age and health of the recipient and, in the case of low-mortality diseases, economic considerations such as the cost of hospitalisation, or loss of productivity of the workforce. Some of the risks are known and can be recognised, even quantified, and the quality control procedures described in this part are based on this knowledge. The controversial part of quality control is trying to protect against the unknown, or the factors about which we know only a little. This particularly applies to cancerogenicity where, because we even now still do not know the fundamental cause of cancer, we have to protect recipients of medical products from any factor which theoretically could be involved in malignancy. The use of primary and diploid cells from normal tissues is a consequence of this concern. Whilst the possibility exists, however remote, that a cell derived from malignant tissue, or a cell showing malignant characteristics (i.e. transformed cells), could be contaminated with an oncogenic agent, or pass on transforming nucleic acids, proteins or cancer genes, then it cannot be used to produce a human medicinal product.

Before a biological product can be used it has to be shown to be safe and effective. To ensure that these criteria are met, many countries have regulatory agencies that investigate the product themselves and assess the available data on the product. These data (Chapters 12 and 13) are generated from (1) the quality control procedures that check for microbial contamination of the product and items used in its production, and product identity; (2) safety tests that check whether the virus is completely inactivated, or has retained its attenuated state; and (3) efficacy tests which check that the correct dose of immunogen is present, correctly formulated and capable of giving the desired clinical effect. Efficacy is tested during the development of a medicinal product by using it on human volunteers. *In vitro* testing is useful but can never completely replace animal tests. Animal tests, although very meaningful and essential, cannot with any certainty be extrapolated to efficacy in humans. Thus, at some stage a new product must be assessed in humans, initially in a few volunteers (on the named patient basis by medically qualified personnel) and eventually in a large clinical, or field, trial (Chapter 14).

In this part the efficacy tests, assessment by clinical trial and the quality control procedures that are imposed are described. The role of the regulatory agencies as public watchdogs is also reviewed. This is all factual information on what is done to a cell product before it is considered fit for clinical use. Behind the scenes, however, many controversies rage. In the face of ignorance over either the long-term effects of a drug, or the mechanisms involved in the molecular biology of disease, regulatory agencies have to err on the side of safety. This is understandable in view of historical events, when newly discovered agents, once recognised, have been found in medical products, e.g. SV40, an oncogenic virus from monkey cells. A recent warning is the discovery of acquired immuno-deficiency syndrome (AIDS), which has a consequence on the way blood-derived products have to be tested. Luckily, human serum is not normally used for cell culture, but human bovine serum albumin is used as a stabilising agent in some human vaccines. However, the degree of the swing implemented merely to take safety precautions against hypothetical dangers has two undesirable effects. Firstly, it increases the cost of the product enormously and, secondly, it increases the time and costs of developing new products. Obviously regulatory agencies are criticised for being over-cautious, and the way out of this dilemma is (1) to gain more knowledge of the fundamentals of disease so that risks can be quantified and (2) to use increasingly pure products so that the probability of an agent being present can be minimised. Also, the development of more sophisticated assay techniques will be a valuable help. For instance, the minimum quantity of DNA in a gene is in the order of 10^{-15} g; thus if DNA can be measured to this degree of sensitivity, and a product is shown to have less than this amount, it cannot be oncogenic. (At present, techniques of measuring down to 10^{-12} g/ml DNA are available.)

If products are to be generated from cells more cheaply and in larger amounts than at present, then the quality control requirements must be reduced and a free choice of cell types allowed. For this to happen depends largely upon the results of basic research. It is possible (as shown by the production of interferon from lymphoblastoid cells), and needs the right sort of data to be generated to convince the regulatory authorities that the product is safe and effective.

12

Product Testing: An Introduction

J. FONTAINE, A. BRUN, R. CAMAND, G. CHAPPUIS,
C. DURET, M. LOMBARD, Y. MOREAU, P. PRECAUSTA,
M. RIVIERE, C. ROULET, J. P. SOULEBOT, and
C.STELLMANN
Rhône Mérieux
Laboratoire IFFA
Lyon, France

1. INTRODUCTION

Ongoing progress in microbiological technology and the ever-increasing quality-to-price ratio demanded by consumers of biologicals have caused microbiologists not only to perfect final quality control testing techniques, but also to implement a system of routine in-process tests over the entire production process, with the result that the product is totally monitored from the raw material stage to the retail product stage.

Animal Cell Biotechnology, Vol. 2

Risks of negative results in the final quality control test have been reduced and practically eliminated by the rigour of such in-process controls and consequently the rejection of an entire batch of vaccine potentially avoided.

2. BASIC CONCEPTS

Biological products (biologicals) present a certain number of specific problems both in process and at final quality control. The origin of such problems can be due to the following:

1. That they have no chemically defined composition, which could lead to a series of errors throughout the production process.

2. That the only means of qualitatively and quantitatively evaluating them is by measuring their effect in animals or man; this effect is a definable, non-specific appearance or increase of a biological property (antibodies, resistance) in the target individual, specifically characteristic of the biological in question, which can only be revealed by specially adapted techniques.

3. That it is currently possible to guarantee the quality of such products not from a final test result on a random sample but from an amalgam of several in-process tests which continuously monitor all production stages and conditions.

2.1. Good Manufacturing Practice

A certain number of guidelines have been drawn up and a booklet entitled "Good Manufacturing Practice" (1–3) published in several countries. It is a set of regulations for the manufacturers of human and veterinary biologicals, and if the instructions therein are followed, the safety of the final product is virtually assured. The major principles of the booklet can be summarised as follows:

1. The laboratory should be in good working order and large enough to correctly assume those functions required for production purposes: basic services (electricity, gas, steam, water etc.), liquid and solid waste disposal, air-conditioning, temperature and humidity control etc. Individual and separate testing/packaging/storage units to be strictly used for specific individual functions should also be provided for. The design of such units should be such that cleaning is easy, risk of contamination low and that products can in no way be confused (by mislabelling, miscoding etc.).

2. Equipment (culture vessels, centrifuges, etc.) should be easy to sterilise, check and maintain routinely, and should be designed to prevent contamination of the product.

3. Strict control on goods received should ensure that demands made on product quality are fully met at all times. Such goods should be properly stored and certified as being fit for use.

4. Personnel should be competent, well trained and sufficient in number to accomplish the different production stages; precise instructions (hygiene, safety precautions etc.) should be respected to minimise risks of contamination and confusion between products. Different levels of responsibility should be clearly defined. Records kept by personnel should be clear and kept up to date. In-process controls by qualified staff guarantee product reliability at all stages of manufacture.

2.2. Certified Quality

Within the framework of the "Good Manufacturing Practice" and organisation, *quality* can be guaranteed only if the highest of standards are maintained all along the line, from the testing of raw materials right up to the final quality control on the retail product, i.e. the more the tests, the higher the cost, the lower the risk, the higher the price (Fig. 1). But will the consumer be able or want to pay for such a guarantee? In human biology the answer is obvious, but unfortunately the picture is not always the same in veterinary biology.

A suitable balance should be established between these elements, taking into account the imperatives of both consumer and manufacturer. This is a very important concept which requires a certain dialogue between the parties involved—manufacturer, quality controller and consumer—with a very real centre of exchange of information and decisions. Decisions on the products are taken with reference to standards established by the relevant parties themselves (Fig. 2).

In order to be successful, the above-mentioned system must be backed by a clearly organised production process, which in turn depends on three intrinsic elements (4):

2.2.1. Master Cell Stock (MCS)

The MCS corresponds to a stock of line cells stored in liquid nitrogen and originating from the same culture; the cells of the MCS are evenly distributed in ampoules or other suitable containers, which are then used to

TESTS   COST   RISK   PRICE   CONSUMER ? ?

Fig. 1.   An increase in tests, in order to guarantee the quality of a product, raises the cost, lowers the risk and raises the price. The consumer may not be able or want to pay for such a guarantee.

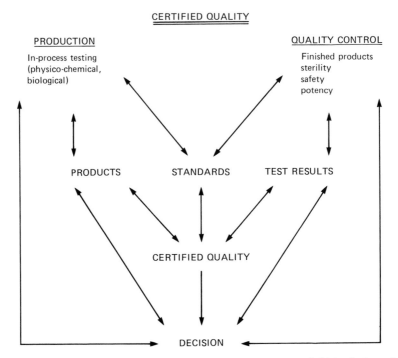

Fig. 2. Quality control should include (a) in-process testing and (b) finished product testing.

start off a series of industrial production cultures. The MCS is tested for the following: chromosomes (number, shape), tumorigenicity and oncogenicity, specificity of the species and that it is free of contamination by bacteria, mycoplasmas, virus etc. Whenever possible, primary cells are prepared from specific pathogen-free (SPF) animals bred under strictly controlled conditions.

2.2.2. Master Seed Virus (MSV)

The idea of the MSV is similar to that of the MCS in that it is used to designate the stock of virus used to inoculate the cell culture. It is thoroughly tested for absence of bacteria, mycoplasmas and viruses, identity and immunogenicity.

2.2.3. Batch and Series

Along with the MCS and the MSV, the idea of batch and series aims to attain maximum homogeneity at a given level of production in order to have as consistent a product as possible; this same batch of product is then used to produce a maximum number of units at following stages. The result of this concern for homogeneity is a reduction in the number of certain tests (if it is accepted that tests done on one batch are valid for others in the series) and a supplementary assurance of the accuracy of the tests if these tests are repeated for all batches of a series. For live vaccines, simple titration of the vaccine virus can replace a potency test when taken in conjunction with results of preceding tests on the MSV and on the industrial production virus. For inactivated vaccines, a potency test on a single batch is valid for all batches of the series; a simple antigen titration suffices for other batches. On an industrial scale, controls organised in such a way offer certain advantages; i.e. they are *simple* to implement (thus economical) whilst assuring quality, *fast* to implement and give *consistent* results (homogeneous raw materials) (see Fig. 3).

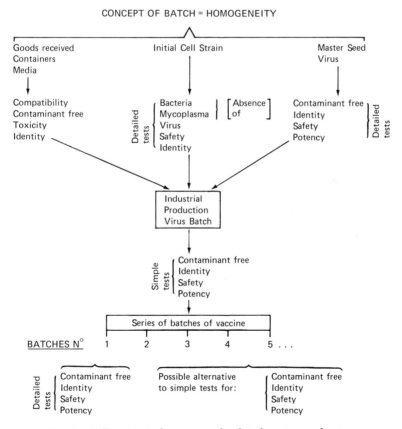

Fig. 3. Different tests done on a cycle of viral vaccine production.

3. PRODUCT TESTING

Biological products of animal origin undergo several specific tests in addition to those done on more conventional pharmaceuticals (e.g. vial testing, stopper testing etc.). Broadly speaking, these biological product-specific tests (5) use three types of techniques: physical, chemical and biological, between which it is often difficult to distinguish. For the purposes of this study, the tests have been divided as follows: (1) physicochemical techniques, (2) biological techniques and (3) recently applied techniques.

The object of individual tests within their respective category will be developed after the definition of certain terms employed in the text:

Inactivated virus: Virus which, under the effect of physicochemical treatment, has lost its capacity to multiply whilst retaining its antigenicity and immunogenicity.

Killed vaccine: Vaccine prepared from inactivated virus to which an immunity adjuvant is often incorporated.

Live vaccine: Vaccine prepared from virus whose pathogenicity has been more or less modified or attenuated for the animal species for which it is intended; it does, however, retain its capacity to multiply and its antigenicity and immunogenicity.

3.1. Physicochemical Tests

At present, physicochemical tests used to assay biologicals are not as important as those done on chemotherapeutic products since they usually lead neither to the identification of the product nor to an accurate evaluation of the contents, either of the bulk material or of potential impurities. However, they are used to reveal physical or chemical alterations in the composition of the products which obviously reflect anomalies in the manufacturing process. Virus produced industrially and finished products are tested using such techniques as presented below.

3.1.1. Examination of Appearance

This examination is valid for all products and at all stages. It draws the attention of the controller to any visible alterations in the quality of the product which could be due to a defect in the manufacturing process, e.g. microbial growth, precipitation, specks of dust, discoloration, state of freeze-dried pellets etc.

3.1.2. Density and Viscosity

Special apparatus (pycnometer, viscosimetric pipette) is used for these routine tests. A full description can be found in the European Pharmacopoeia (6, 7).

3.1.3. pH

Wide variations in the pH value can lead to the destruction of the virus or of the antigen; the pH should thus be closely monitored during the production process. Measurement of pH has become a routine technique in ordinary laboratories and is done by using electronic voltmeters and glass electrodes (8, 9). However, where biological products are concerned, this apparently simple operation can prove to be quite difficult with this type of ordinary apparatus, as the thiols (from the cysteine residues of the proteins) react with the Ag–AgCl electrodes and rapidly cause them to deteriorate. It is thus preferable to use double-junction calomel electrodes as reference electrodes.

3.1.4. Emulsions

Certain vaccines have oily emulsions incorporated to potentiate immunity (e.g. foot-and-mouth disease pig-specific vaccine), and for such products, it is important to be sure of the following: (1) the type of emulsion (water-in-oil or oil-in-water) and (2) that the correct formulation has been used, by measuring density and viscosity (6, 7).

TECHNIQUE: Put a drop of the product to be tested on the surface of some water held in a beaker. If the drop forms a globule without mixing with the water, it is a water-in-oil type emulsion; if it forms an odd shape and after gently shaking the water becomes milky, it is an oil-in-water type emulsion.

3.1.5. Ultraviolet Spectrum Absorption

This test determines the concentration of virus present in a virus suspension. The liquid to be tested is ultracentrifuged in a sucrose gradient and its absorbance along the gradient is recorded. This reading reveals a peak whose surface area is proportional to the quantity of virions present. This test can be applied to all viruses so long as suitable gradients and conditions are defined for each group.

TECHNIQUE. (foot-and-mouth disease virus) (10). 1 to 3 ml of virus suspension is put onto a sucrose gradient (15–45% or 10–35%) and is ultracentrifuged at low speed (16,000 rpm for 16 hr) or at high speed (40,000 rpm for 3 hr) on a rotor SW41 and the absorbance of the material discharged from the gradient measured at a wavelength of 260 nm.

3.1.6. Residual Moisture

This test is very important and specific to freeze-dried products as the stability of numerous viruses depends on the residual water content of the freeze-dried pellet; it is thus of utmost importance that this value be within very accurate limits. The method used derives from Karl Fischer's technique described in the European Pharmacopoeia and is based on the fact that iodine (*only* in the presence of water) oxidises sulphur dioxide to give into sulphuric acid

$$I_2 + 2H_2O + SO_2 \rightarrow H_2SO_4 + 2HI$$

TECHNIQUE. The freeze-dried pellet is reconstituted with anhydrous methanol; then iodine and sulphur dioxide dissolved in pyridine are added with a burette. Results can be read by colorimetry or even with the naked eye.

3.1.7. Inactivation Agent Contents

Inactivation agents such as formaldehyde, glycidaldehyde, aziridine etc. are not products which are chemically very stable, so attention should be paid to the fact that for as yet ill-defined reasons, the quantity of inactivating agent actually added can sometimes be lower than the quantity required to inactivate the virus. In order to ensure total inactivation at finished product stage, specific colorimetry is routinely used to evaluate the amount of inactivating substance actually present. The measurement of aziridine contents can be given as an example (see Fig. 4) (*11*).

TECHNIQUE. 2 ml of acetate 25 mM (pH 4.4) and 1 ml of 4(4-nitrobenzyl) pyridine in a 2.5% acetone solution is added to a 2-ml solution containing at most 0.5 micromoles of aziridine. This solution is put into tubes, shaken, put into a boiling-water bath for 15 min and cooled; 5 ml of acetone is then added and 1 ml of 0.1 M potassium hydroxide. The absorbance is immediately measured at 600 nm. A comparison is done with a well-defined reference product such as bromoethylamine which gives the same chromogen as aziridine with the same yield.

3.1.8. Measurement of Adjuvant Content

Adjuvants are substances used to enhance the activity of a vaccine and the ones most commonly used are aluminium hydroxide, saponin and oily excipients. At finished product level, it is important to check that the initial

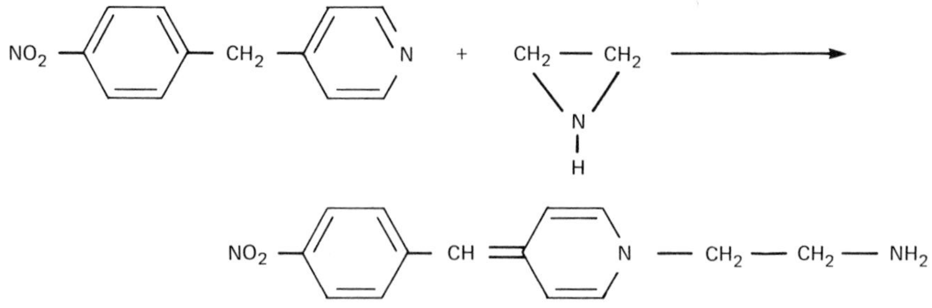

Fig. 4. Reaction of aziridine with 4(4-nitrobenzyl) pyridine.

quantity of adjuvant introduced has not varied and that it has been uniformly distributed in the batches of vaccine.

The amount of aluminium hydroxide present in foot-and-mouth disease vaccine is measured using the technique which is described in the European Pharmacopoeia (12).

TECHNIQUE. (foot-and-mouth disease). The vaccine is reduced to an ash and the aluminium hydroxide complexed by a known quantity of ethylenediamine tetraacetic acid used in excess. The excess of EDTA is back-titrated using a copper solution in the presence of pyridyl-azonaphthol as indicator.

3.1.9. Biochemical Evaluation

Little can be said about biochemical measurement as a testing technique, as it is only rarely applied. This is mainly due to the fact that the biologically active part of a vaccine represents a relatively small percentage of the total material. Research, however, is being done into this measuring method, which will undoubtedly gain importance in the future.

3.2. Biological Tests

Broadly speaking, biological tests can be divided into the following categories: (1) freedom from contaminants test, (2) identity test, (3) safety test, (4) potency test.

The principle of such tests is to prove whether a specific or non-specific element or effect, which can be detected by the naked eye or by an appropriate laboratory method using either an inert medium (agar) or a living medium (cell culture, living animal), is *present* or *absent*. Certain testing techniques can be used for several different categories of tests. When testing freeze-dried products, the pellet should first be dissolved in a suitable diluent.

The above-mentioned categories of biological tests can be regrouped under two major headings: (1) tests to prove the *absence* of an effect or element (contamination, safety), (2) tests to prove the *presence* of an effect or element (identity, potency).

3.2.1. Tests to Prove the Absence of an Effect or Element

Biological products should not contain any micro-organisms which could interfere with the activity of the bulk material of the product. It is important that the finished product be proved safe for the man or animal into which it is to be inoculated. Tests are done systematically on every batch of vaccine to prove the following: (1) that it does *not* contain any undesirable micro-organisms such as bacteria, mycoplasma or virus: *test for contaminants;* (2) that the product is safe, i.e. that it does *not* cause any pathological incident post-inoculation: *safety test.*

Test for Contaminants. This test is normally done at many selected stages of the manufacturing process, but is done exhaustively on the finished product. The object of the test is to prove that the product is free of any organisms of bacterial, mycoplasmic or viral origin which could result in the subsequent infection of the target organism.

TEST FOR BACTERIA. Each country has well-established standards regarding tests for bacterial contamination set out in its own individual pharmacopoeia (*13*):

TECHNIQUE. (European Pharmacopoeia). The test should be done on between 4 and 10 vials of product depending on the size of the batch produced. A certain number of stages can be necessary before a batch is accepted, schematised in Fig. 5.

TEST FOR ABSENCE OF MYCOPLASMAS (*14*). Mycoplasma are organisms which are difficult to culture and require the use of special solid or liquid culture media.

TECHNIQUE. (solid medium). Six Petri dishes containing 10 ml of suitable agar (heart infusion gel) are inoculated with 0.2 ml of a vaccine suspension consisting of a blend of at least five bottles from the batch being tested. Half of the dishes are incubated aerobically and the other half anaerobically at +36°C for 14 days. At the end of the incubation period, the cultures are

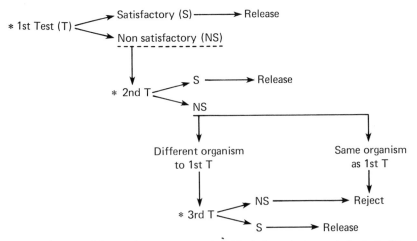

Fig. 5. Absence of bacterial contaminants in biological products: control of sterility conditions.

examined microscopically for the presence of mycoplasmas. The preparation passes the test if growth of mycoplasmas does not occur in any of the inoculated media.

TECHNIQUE. (liquid medium). 9 ml of culture medium (heart infusion broth) is put into four tubes. One millilitre of a vaccine suspension is put into the first tube, from which a further 1 ml is taken, and so thereby making four dilutions: 1 : 100, 1 : 1000, 1 : 10,000 and 1 : 100,000. The tubes are incubated at +36°C for 14 days and then inoculated onto solid medium as above. For both liquid and solid media control cultures should be done: negative (tubes inoculated with the culture medium used in the test) and positive (tubes of the same culture medium inoculated with known cultures of mycoplasmas). The preparation passes the test if growth of mycoplasmas does not occur in any of the inoculated media and does occur in the positive control.

TEST FOR ABSENCE OF VIRUSES.   The object of this test is to check that the vaccine virus is the only one really present in the product; it particularly concerns live vaccines, as for killed vaccines the test is the same as the safety test (see below). Two techniques can be applied, one which consists of neutralising the vaccine virus with a monospecific hyperimmune serum and inoculating it to cell or animal cell cultures selected for the specific receptivity; and the other, which is to test the potential presence of neuropathogenic agents.

TECHNIQUE 1 (canine distemper virus). Vero cells of monkey origin are used. 0.15 ml of a suspension containing 100,000 cells/ml is put into each well of a 96-well Falcon Microtest II plate. 0.1 ml of a serum/virus mixture is then added to each well. On the 7th day post-inoculation, the medium is removed and a haemadsorption test done on the cells. Neither haemadsorption nor cytopathic effect should be observed.

TECHNIQUE 2. (veterinary rabies vaccine). The potential presence of neuropathogenic agents is tested by inoculating the specific antibody-neutralised vaccine being examined, intracerebrally to laboratory animals (generally mice); the animals should remain in good health. Six immature mice are generally used and are inoculated with 0.03 ml of neutralised vaccine; they are then observed for 21 days.

Safety Test.   There are two aspects to safety testing: (1) to check that the vaccine has been properly inactivated (i.e. that there are no residual living virus particles), a test done *before* the addition of the immunity adjuvant and on the finished product: *inactivation test;* (2) to check that the finished product is not toxic: *abnormal toxicity test.*

INACTIVATION TEST.   The chemically or physically inactivated virus is inoculated to animals or added to cell cultures which are sensitive to the type of virus in question; no pathological symptom or cytopathic effect should occur which could be attributed to this same virus.

TECHNIQUE. (veterinary rabies vaccine) (*15*).

1. 0.03 ml of inactivated virus is injected intracerebrally to 10 mice weighing between 14 and 16 g, which are observed for 21 days. They should remain in good health. If one mouse dies after the 5th day without any clinical sign of rabies, its brain is removed, ground up and put into suspension; it is then inoculated to other mice.

2. Another way to check inactivation is to see whether the virus can still multiply in susceptible monolayer cells (*16*). If so, mice inoculated later with the supernatant of these cultures will present rabies symptoms.

TECHNIQUE. (foot-and-mouth disease vaccine) (*17–21*). 60 ml of inactivated virus is inoculated onto a lamb, pig or calf kidney cell monolayer or cattle thyroid cells, grown in roller bottles at the rate of 1 ml/cm² surface area, and left in contact for 30 min at +37°C. After adsorption, a sufficient quantity of minimum essential medium containing lactalbumin is added. 72 hr later, the bottles' contents (which should display no CPE) are frozen, thawed and new cell cultures inoculated with the culture liquid in the same manner. If after a further 72 hr of observation no CPE has been noted on the subculture, it can be concluded that the product has been correctly inactivated. This method is very highly sensitive and can be used for finished vaccines by inoculating the eluted antigen.

ABNORMAL TOXICITY TEXT.   The object of this test is to check the potential toxicity of the product which is demonstrated by a local or generalised reaction in animals (*22, 23*).

TECHNIQUE. Two guinea pigs are each inoculated with 2 ml of vaccine intraperitoneally (or subcutaneously if the vaccine has an adjuvant incorporated) and five mice each subcutaneously with 0.5 ml of vaccine. The animals are observed for 7 days, and if one of the animals dies or shows symptoms of disease, the test is repeated. The test is considered as being satisfactory if there are no local or general reactions further to this second series of tests.

SPECIFIC SAFETY TEST.   The specific safety test is a means of checking that the virus has really lost its virulence for the species concerned. For both types of vaccine (live attenuated and killed), susceptible animals are inoculated with at least one dose of the vaccine being examined.

TECHNIQUE. (foot-and-mouth disease vaccine). The vaccine is injected intradermolingually into 20 sites in the tongues of cows at a rate of 0.1 ml per site. If no signs of FMD are observed at the inoculation sites 4 days later, three doses of vaccine are injected via the subcutaneous route. The animals are then observed for a further 6 days and their temperature taken daily. The vaccine is considered safe if abnormal reactions or local lesions on the tongue have not been noted.

TECHNIQUE. (canine distemper) (24). Two ferrets or two puppies are inoculated subcutaneously with two doses of vaccine. The animals are observed for 21 days and their temperature monitored. The vaccine is considered safe if no symptoms of canine distemper occur.

3.2.2. Tests to Prove the Presence of an Effect or Element

A ready-to-use biological product should effectively contain the bulk material with which it has been prepared and also cause the desired effect in the target species. Tests are done on every batch of vaccine to prove the following: (1) that the intended bulk material is really present: *identity test;* (2) that the product is efficient and the intended reaction caused in the target organism: *potency test.*

Identity Test. The principle of the identity test consists in: (1) making the virus or the antigen (i.e. the vaccine) react with a corresponding antiserum; (2) revealing the presence (or absence) of a specific antibody–antigen complex in animals, in cell cultures or using serological techniques.

This is straightforward and can be done directly on the finished product in the case of living vaccines; for adjuvanted, inactivated vaccines, the presence of the adjuvant (which can disturb the reaction) can require a preliminary treatment stage (e.g. desorbtion of antigen), which is obviously not necessary when the test is done before the addition of the adjuvant. It should be noted that the potency test (see below) is also an identity test.

TECHNIQUE. (canine distemper vaccine) (25). The seroneutralisation index technique, which expresses the difference in titre between the non-neutralised virus and the virus put into the presence of an immune serum, is used. The virus is diluted fourfold into tubes and each dilution is distributed into six wells (0.05 ml) to which 0.05 ml of serum is added and into another six wells without serum. After incubation for 1 hr at +20° to +22°C, 0.15 ml of cell suspension is added to each well and the plates incubated at +37°C. The CPE is examined after 7 days. The titre of the virus affected or unaffected by the serum is evaluated as well as the neutralisation index (the difference).

Note: The examination of the cells in the wells where the virus has been neutralised can also be a test for contaminants.

TECHNIQUE. (veterinary rabies vaccine). The complement fixation test (Kolmer's method) is used. The inactivated virus is put into contact with the corresponding antiserum in the presence of complement (fresh guinea pig serum); after 30 min incubation at +37°C, sensitized sheep red blood cells are added. If the required virus is present, haemolysis does not occur in the test tube, as all of the complement has been removed by attachment to the antigen–antibody complex.

Potency Test.   The object of the potency test is to demonstrate that the product is fully active. The level of potency can be evaluated by measuring the following: (1) the quantity of antigen per unit of volume present in the vaccine: *infectivity titration;* (2) the resistance to challenge of the vaccinated target animal: *direct testing* (virulent challenge); (3) the antibodies or cell reactions which develop in the target animals after vaccination: *indirect testing* (seroneutralisation etc.)

INFECTIVITY TITRATION.   This technique consists in making a series of successive dilutions with a virulent suspension using a previously established dilution factor and in testing the virulence of the different dilutions for the animal, egg embryo (mortality, symptoms, lesions, haemagglutination capacity) or cell culture (cytopathic effect). For this, a certain number of animals, egg embryos or cell cultures are inoculated with a convenient dose of each of the prepared dilutions (generally six per dilution). The range of dilutions is selected in such a way that effects from 100% to 0% are observed within the test, which is specifically designed for the particular cell–virus system selected. Effects are enumerated and the calculation (using one or other of the statistical methods detailed at the end of the chapter) of an infective titre expressed in a number of 50% doses per volume unit is worked out.

TECHNIQUE. (veterinary rabies vaccine) (26). Five successive dilutions are used and are injected intracranially to each of six mice at a volume rate of 0.03 ml. They are observed for 14 days. Deaths due to rabies are noted. The infective titre is expressed as the $\log_{10}$ of the number of 50% lethal doses (LD_{50}) per millilitre of virus suspension.

TECHNIQUE. (canine distemper vaccine) (25). Fourfold dilutions of virus in MEM medium are used. 0.15 ml of a cell suspension of primary culture containing 100,000 cells/ml is put into the wells of a Falcon Microtest II 96-well plate; at the same time, 0.1 ml of diluted virus is distributed at the rate of six wells per dilution. The titre is calculated using Karber's method and corresponds to the log of the reciprocal of the initial virus dilution which produces a cytopathic effect in 50% of the wells.

TECHNIQUE. (Newcastle disease vaccine) (27, 28). Fourfold dilutions of virus in peptone water are used. They are inoculated at the rate of 0.1 ml into the allantoic cavity of 9-day-old embryonated eggs (six eggs per dilution); after 72 hr at +37°C, the eggs are placed at +4°C for 6 hr before the allantoic fluid is removed for a haemagglutination test with a 2% suspension of chicken red blood cells. The eggs which give a positive reaction are considered as being infected and the 50% titre is calculated as above.

DIRECT TESTING (Virulent Challenge). The principle of the potency test (which is the most direct method) is to inoculate the vaccine under test into animals which are susceptible to the virus and which the vaccine should protect. When immunity has developed, the animals and non-vaccinated controls are challenged with a virulent strain of virus. After an observation period longer than that necessary for normal development of the experimental disease, the controls should prove to have the disease and the vaccinated animals to be protected. Ideally, it is best to work with animals from the species for which the vaccine is intended—target animals such as dogs for canine distemper, cattle for FMD etc. This is often expensive, so for both economical and practical reasons, small laboratory animals can also be used: guinea pigs for FMD, mice for rabies etc. The test is generally qualitative (sometimes quantitative) when the proper animals are used and quantitative when laboratory animals are used.

Qualitative Test. TECHNIQUE. (canine distemper vaccine) (29). Each of five dogs is vaccinated with one dose of vaccine. Fourteen days post-vaccination, these same dogs and two non-vaccinated controls are inoculated suboccipitally with 1000 50% lethal doses of the Snyder Hill strain of canine distemper virus and are observed for 21 days. The two controls should present typical canine distemper symptoms with lesions and the five vaccinated animals should remain in good health.

Quantitative Test. When a quantiative potency test is done, one of two methods is used: (1) virus titration on vaccinated animals and on controls (K index for FMD vaccine) (30); (2) vaccination of animals with varying doses of vaccine in order to determine, after challenge, a 50% protective dose.

TECHNIQUE. (foot-and-mouth disease vaccine) (31–33). The vaccine is diluted fourfold in an immunologically neutral buffer solution; one volume of each dilution (equal to the volume of the normal vaccination dose) is injected subcutaneously to five cattle. 20 to 22 days post-vaccination, the animals are inoculated intradermolingually in two sites with 0.1 ml of a suspension containing $10,000 \text{ ID}_{50}$ of a virulent virus strain of the same strain as that used to prepare the vaccine. 8 days post-challenge, the cattle are slaughtered. The controls should present FMD lesions and the others should show no generalised lesions (i.e. on the feet) at all. The number of 50% protective doses (or bovine potency) contained in one dose of vaccine is calculated using the probits method (see statistical methods at the end of this chapter).

TECHNIQUE. (veterinary rabies vaccine) (34). This is the method advised by the National Institutes of Health (NIH) in the United States for rabies vaccines and is called the NIH test. Fivefold dilutions of vaccine are prepared, and 0.5 ml of each dilution of vaccine is inoculated two times via the intraperitoneal route to 10 mice with an interval of 7 days. Seven days after the second inoculation, the mice are challenged intracerebrally with 0.03 ml of a suspension of CVS virus containing 5–50 50% lethal doses; control mice are also inoculated with the same strain and observed for 14 days (animals that die before the 5th day do not count). The 50% protective dose is calculated using the logits methods. The reciprocal of this vaccinating dose is known as 'mouse potency'; the comparison with the reference vaccine gives the value of the vaccine tested in international units (IU).

It should, however, be pointed out that although direct testing is the most reliable form, it does present a certain number of disadvantages:

1. Economically: The high cost of operations—all animals used are usually sacrificed as the inoculation of a highly contagious virus can form a focus of infection necessitating elaborate and costly protection of premises.

2. Technically: The frequent difficulty in certain cases of obtaining regular easy-to-detect symptoms and lesions, which leads to doubtful interpretation of results. In certain cases the appearance or non-appearance of the expected effect or the length of the test poses such problems that it is not possible to repeat the test on all batches of vaccine. This is the case for chicken infectious bronchitis vaccine, whose efficacy is estimated in relation to the laying graph of hens post-vaccination.

On an industrial scale, direct potency testing cannot always be routinely applied, but it is of great value in the development of experimental vaccines and in the correlation of other testing techniques.

INDIRECT TESTING.   Instead of using direct challenge in animals, this technique uses the antigen–antibody reaction in the laboratory; the amount of antibody which is found in the serum of animals after vaccination is determined. These methods can also be used to titrate the antigen contained in the vaccine (thus its potential activity), without having anything to do with the vaccination of an animal with the vaccine. These methods are indirect methods. They are based on serological reactions which measure or titrate the antibodies or antigens when in the presence of a reference antigen or antibody. The advantage of this type of test is that it can be applied easily to all industrial batches and is readily repeatable. It can give very accurate results *so long as correlations have previously been established with virulent challenge methods*. Although the antigen–antibody reaction can sometimes be *directly* visible (preciptation, flocculation), it is more common to use *indirect* methods for the observations, such as the labelling of the antigen or antibody with fluorochrome, radio-isotopes or enzymes or the exploiting of certain properties possessed by the antigen, the antibody or the antigen–antibody complex (see Fig. 6).

The tests most commonly used for the antigen–antibody reaction are: seroneutralisation, haemagglutination inhibition and immunofluorescence. Other tests of more recent origin are dealt with in Section 3.3.

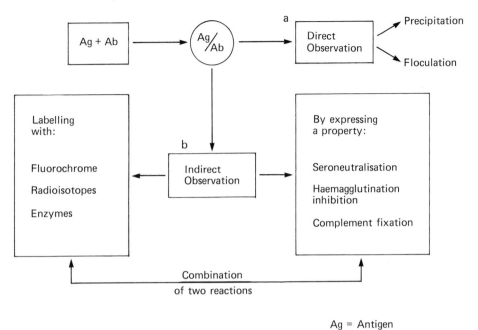

Ag = Antigen

Ab = Antibody

Fig. 6. Revelation of antigen–antibody reactions: (a) direct reading, (b) indirect reading (labelling or expression of a directly visible property: cytopathic effect, hemagglutination, etc.).

Seroneutralisation (35).   The principle of this method consists in demonstrating by using an indicator (susceptible cell or animal) that the antibodies in the serum have or have not neutralised the infective capacity of the virus. In each dilution of a series prepared from the serum under test, an equal volume of virus suspension of known titre, identical to that which was used for the preparation of the vaccine, is added. After an appropriate contact time at a suitable temperature, the antigen–antibody complex is inoculated into a group of laboratory animals (generally five or six animals: mice, guinea pigs etc.) or inoculated into five or six tubes or wells (for plate or tissue culture). When mortality symptoms or cytopathic effects have been noted at the end of the prescribed period, the titre corresponding to the dilution of serum protecting 50% of the animals or inhibiting a cytopathic effect in 50% of the cultures is calculated.

Dilution–titre relationship: If a 1:1000 dilution will give 50% protection then this is equal to 1000 50% protection units or $3 \log_{10}$ or a $-\log_{10}$ titre of 3.

TECHNIQUES. (veterinary rabies vaccine) (36). Five cats or dogs free of rabies antibodies and never vaccinated are each inoculated subcutaneously or intramuscularly with one dose of vaccine. 15 to 30 days later, after the collection of a blood sample, they are revaccinated; 15–20 days later, another blood sample is taken. A series of dilutions of a corresponding serum is then made and 1 ml of each of them is added to 1 ml of virus of a known titre and placed at +37°C for 1 hr, then cooled to +4°C before being inoculated intracerebrally (0.03 ml of each mix) to five mice weighing between 14 and 16 g. The neutralising titre is the reciprocal of the dilution of serum which protects 50% of the mice. This technique is used less and less and is being replaced by the rapid fluorescent focus inhibition test (RFFIT) (see below) (37, 38).

Haemagglutination Inhibition (39, 40).   This test exploits the property possessed by certain viruses of agglutinating the red blood cells of certain animal species and the property of the corresponding immune sera of inhibiting this same reaction.

TECHNIQUE. (live canine parvovirus vaccine). The haemagglutinating titre and the corresponding quantity of virus to be used are first determined in the following way:

A suspension of parvovirus is diluted twofold. 0.05 ml of each dilution is placed in a well into which 0.05 ml of a suspension of pig red blood cells containing 30 million blood cells/ml is added. After 3 to 4 hr at +4°C, the last dilution where total haemagglutination is observed is noted. This is the dilution which, by definition, contains 1 haemagglutination unit per 0.05 ml of volume. For the inhibition test, 4 haemagglutination units are used. The serum to be tested is diluted twofold; 0.05 ml of each of these dilutions is put into the well of a microplate with 0.05 ml of virus suspension expressing 4 haemagglutination units; this mix is left for 1 hr at laboratory temperature for incubation. Then 0.05 ml of the pig red blood cell suspension is added to each well. Results are read after one night at +4°C. The serum titre is expressed as the highest dilution of the serum where haemagglutination did not occur.

Immunofluorescence.   In this technique, either the antigen or the antibody is labelled with a fluorochrome (fluorescein isothiocyanate). When the antigens or antibodies are activated by ultraviolet radiation they emit radiation in the visible spectrum which can be observed with a fluorescence microscope fitted with excitation and barrier filters. For rabies, the viral antigen accumulated in the cytoplasm can be seen if it has previously been attached to its corresponding antibody (so long as this antibody has been labelled with fluorochrome); green-coloured cytoplasmic inclusions then form. This property is exploited directly in the diagnosis of rabies and also in the titration of rabies antibodies in vaccine potency testing. This last method is the RFFIT (37, 38).

TECHNIQUE. (RFFIT). Different dilutions of serum to be titrated are added to a fixed quantity of rabies virus and then incubated. The different mixtures are inoculated into cell cultures and, after incubation at +37°C, the cell sheets are treated with a labelled rabies antibody. The cell sheets are examined under the microscope and the presence or absence of fluorescent cytoplasmic inclusions is noted. The presence of cytoplasmic inclusions means that the rabies antigen is present (virus multiplication), i.e. the virus added to the dilution which corresponded to the serum being titrated was not neutralised, or even that initially this serum dilution did not contain any antibodies. An absence of cytoplasmic inclusions means an absence of the rabies antigen, i.e. that there was no viral multiplication since the dilution corresponding to the serum to be titrated neutralised the infective virus during the first part of the test. In practice, a certain number of microscopic fields are observed and those which demonstrate fluorescent cells are counted. The titre of the serum is given by the reciprocal of the serum dilution which reduces the number of fluorescent fields (compared to cultures of control virus) by 50%.

It should be pointed out that these indirect potency tests are applied not only for the routine quality control of batches of vaccine produced on an industrial scale (prior to commercialisation) but also over long periods of time for vaccines when field trials are done to determine immunity persistence and modes of use which lead to the design of vaccination programmes.

3.3. Recently Applied Techniques

At a given time in the development of scientific techniques, certain techniques (although not systematically applied) are used in research and development activities. This indeed is the case for certain techniques used in the quality control of vaccines produced from animal cell cultures: (1) electron microscopy; (2) antigen or antibody labelling; (3) cell immunology.

3.3.1. Electron Microscopy

Electron microscopy is still mainly used during research, but is being implemented more and more regularly to examine preparations produced from virus cultures at different stages (*41, 42*). The following can be tested by using electron microscopy:

1. That the cell culture is not contaminated by other viruses (purity test)
2. The amount of virus (potency test),
3. That the virus is the one it is supposed to be (identity test)
4. That freezing or any other physicochemical action does not harm the integrity of the virion

3.3.2. Antigen or Antibody Labelling

Labels (L), generally fluorochrome, radio-isotopes (*43*) or enzymes (enzyme-linked immunosorbent assay, ELISA) (*44*), which are easy to detect by radioactivity or the action of an enzyme on a substrate, are used to label either the antigen or the antibody, thereby making it easy to detect the antigen–antibody complex or the antigen or antibody present in excess. In order to do so, the separation of the antigen–antibody complex from the free forms present in excess is done either by precipitation or by fixing onto a solid support. Two types of test can be distinguished: a simple test and a complex test.

Simple Test.

$$Ag^L + Ab \longrightarrow Ag^L\,Ab \;(+ \;Ag^L \text{ if } Ag^L \text{ is in excess}) = \textit{Model 1}$$
$$Ag + Ab^L \longrightarrow AgAb^L \;(+ \;Ab^L \text{ if } Ab^L \text{ is in excess}) = \textit{Model 2}$$

Complex Test. Either antigen or antibody is fixed (F) onto a solid support before adding the other antigen or antibody and then another labelled element (either antigen or antibody from the same system, or anti-antibody antibodies, i.e. anti-species antibody):

$$Ag^F + Ab \longrightarrow Ag^FAb$$
$$+Ag^L \longrightarrow Ag^FAbAg^L \qquad = \textit{Model 3}$$

$$Ab^F + Ag \longrightarrow Ab^FAg$$
$$+Ab^L \longrightarrow Ab^FAgAb^l \qquad = \textit{Model 4}$$

$$Ag^F + Ab \longrightarrow Ag^FAb$$
$$+Ab^L \longrightarrow Ab^FAbAb^L \qquad = \textit{Model 5}$$
$$\text{(anti-ab)} \qquad \text{(anti-ab)}$$

Competition assays

Ag? is identical to Ag^L but in an unknown quantity. Ab^F corresponds to both Ag^L and Ag?

$$Ab^F + Ag? + Ag^l \longrightarrow \text{If Ag? does not exist} \longrightarrow Ab^FAg^L \qquad = \textit{Model 6}$$
$$\text{If Ag? does exist} \longrightarrow Ab^FAg? + Ab^FAb^L$$

Ag? is identical to Ag^F but in an unknown quantity. Ab^L corresponds to both.

$$Ag^F + Ab^L + Ag? \longrightarrow \text{If Ag? does not exist} \longrightarrow Ag^F\,Ab^L \qquad = \textit{Model 7}$$
$$\text{If Ag? does exist} \longrightarrow Ag^F + Ag^FAb^L + Ag?Ab^L$$
$$\text{(washed out)}$$

The amount of Ag^FAb^L detected is inversely proportional to the amount of Ag?

Labelling techniques are relatively recent and several are still at the research stage. Here, however, are a few examples of how they can be applied:

MODEL 1. Meningococcal antibody (globulins) titration: the $Ag^{125I}Ab$ complex is precipitated by ammonium sulphate and separated by centrifugation. By measuring the degree of radioactivity of the supernatant (which gives the quantity of antigen in excess) the amount of Ab used in the reaction can be calculated.

MODEL 2. Rabies antigen identification or titration: cell purity (absence of rabies antigen) can be established by checking the absence of a reaction with the labelled antibody or the amount of virus contained in a preparation.

MODEL 3. Tetanus antibody (globulin) titration—known as the sandwich method; the support is a polymer tube.

MODEL 4. Test for hepatitis B antigen, which might contaminate production.

MODEL 5. Immunoenzymatic test (using enzyme as label). The ELISA is used to detect infectious bovine rhinotracheitis. At the end of the reaction, a substrate change in colour under the effect of the enzyme label means that the antigen–antibody reaction has taken place.

MODEL 6. Titration of alpha-foetoproteins (using enzyme as label). ELISA is used.

MODEL 7. Foot-and-mouth disease sub-type research (using enzyme as label). ELISA is used.

The major advantage of these techniques is their great accuracy, but care should be taken that *specific* antisera are used which do not contain any foreign antibodies which might interfere with the reaction. The use of appropriate controls is essential.

3.3.3. Cell Immunology

The study of cell immunity is currently gaining importance further to the development of techniques devoted to it. It allows the potential immunodepressant effect of certain live virus vaccines (i.e. pox, herpes) or the potency of vaccines to be evaluated. Three major tests are used:

1. Lymphoblast transformation test
2. Macrophage or leucocyte inhibitory factor
3. Cytotoxic cell test

Lymphoblast Transformation Test (45). This test demonstrates the absence of an immunodepressive capacity of a vaccine. It is based on the fact that the function of the lymphocytes of an animal is not altered post-vaccination. The lymphocytes can become lymphoblasts under the influence of various lectins. The transformation process is revealed by the incorporation of labelled DNA precursors (radio-isotopes).

Macrophage or Leucocyte Inhibitory Factor (MIF or LIF) (46). This test is directly correlated to cell immunity. The lymphocytes of animals primed against a defined antigen produce various mediators when inoculated with the antigen in question, *in vitro*. The best known and best documented is the MIF.

TECHNIQUE. Guinea pig peritoneal cells (which contain a majority of macrophages and a minority of lymphocytes) are put into a capillary tube, which is then put into a horizontal migration cabinet and incubated for 24 hr at +37°C. A cell halo forms around the opening of the tube if the peritoneal cells come from an immunised animal; the addition of the antigen responsible for this stage inhibits macrophage migration, as shown by the decrease in the migration index:

$$MI = \frac{\text{migration surface with Ag}}{\text{migration surface without Ag}} \times 100$$

Note that the stimulus causing the production of the MIF is strictly specific to the immunising antigen.

Cytotoxic Cell Test (47).   Certain cells are capable of killing other cells *in vitro*. Most of these cells do it with no relation to vaccination of any sort and without specificity (natural killer cells, macrophages etc.). However, after vaccination, the cytotoxic T cells act specifically against cells bearing the antigen corresponding to the vaccine. So the T lymphocytes of an animal vaccinated against, for example, Marek's disease can kill infected target cells bearing the Marek's disease antigen *in vitro*. The quantity of cells destroyed is evaluated by measuring the isotope liberated in the supernatant of the cultures after destruction of the target cells.

4. STATISTICS IN TESTING

Statistics is now an accepted tool in many fields and the application of mathematical methods to the biological sciences is to be seen in the state-of-the-art techniques in control procedures and research into vaccine production. However, because of the number of variables influencing laboratory tests (i.e. the receptivity of types of cells, nature of diluents, incubation periods, presence of viral aggregates and of cell debris etc.) it seems useful to normalise the standard conditions of these tests, after having studied and then defined a certain number of criteria as follows:

1. The *reliability* of the systems, which should be characterised by repeatability, reproducibility and sensitivity.

2. The *practicability*, which is a compromise between cost and ease of use.

3. *Efficiency* or *Feasibility*. which represent a compromise between pragmatism and the theoretical ability to discern a statistical difference between two similar manipulations. In this way, and thanks to a knowledge of the analysis of variance, control charts can be drawn up based on past data using the average as a base. Thus data are arrayed with two standard deviations. The limits thus set out define confidence limits within which future data should fit. From this basis, data are considered as established on the index card and are termed 'controlled data'. Data cards are made up at each stage of the production line to ensure datum points that are particularly rich and important in terms of a *specified quality* system.

The application of statistical methods to the control of vaccines produced on cell cultures uses two fundamental models, the pharmaceutical hypothesis and the unique particle hypothesis (48–51).

4.1. Pharmaceutical Hypothesis

It is accepted that the indicator acts in an 'all-or-nothing' way as soon as its own receptivity threshold is reached, and that these receptivity thresholds are distributed according to a normal (Gaussian) curve; in this case, a progressive response can be observed from 0 to 100% on a given population in a concentrated area of the product. The dose able to affect 50% of the population tested thus corresponds to the average threshold of receptivity of the indicator (which is the case in animal indicators). This median effective dose, or ED_{50}, was unanimously adopted to express the titre of the biological product concerned and is calculated from the logarithm of the dilutions used, since biological effects are almost always proportional to the logarithm of the doses (52–63).

There are many calculation methods derived from this hypothesis, each having a mathematical application implying limits that are more or less stringent: Reed–Muench, Kärber–Irwin–Cheeseman, Thomas, logit, probit, neoprobit, Lichfield, coordinate transforms etc.

The fullest, most simple and least limited is that of Kärber–Irwin–Cheeseman, in which:

1. The dilution period, expressed in $\log x$, is a constant h.
2. Doses to give 0 and 100% effects must be experimentally determined.
3. The sample of the indicators in terms of dilution is a constant n.

A *priori* knowledge of the biological system enables the definition of dilution limits of the dosage (64–66).

The ED_{50} is calculated from the following formula:

$$ED_{50} = ED_{100-x}$$
$$x = \frac{h}{n}\left(\frac{n}{2} + \Sigma f\right)$$

The variance of the ED_{50} is calculated from the following:

$$s^2_{EDs} = \frac{h^2}{(n-1)n^2} \sum_{ED_0}^{ED_{100}} f \cdot s$$

in which h = log dilution period, Σf = number of positive indicators between ED_0 and ED_{100}, n = population of indicators between ED_0 and ED_{100} and $f \cdot s$ = product of positive and negative indicators for each dilution.

The ED_{50} titre is given by the decimal logarithm of the reciprocal of the dilution causing a 50% effect for the unit volume tested, usually omitting the logarithm base. The potency of a vaccine is expressed by the number of 50% vaccinating dose (VD_{50}) contained in one dose (67).

It is standard practice to use the expressions ID_{50} (50% infective dose) and $CCID_{50}$ (50% cell culture infective dose) for the infective titre in cell cultures. For seroneutralisations, the ED_{50} concept is applicable, the seroneutralisation titre being the logarithm of the reciprocal of the dilution of serum neutralising a normalised quantity of virus. The titre of a serum is sometimes expressed as a seroneutralisation index (Identity Test section above). This technique is, generally speaking, less precise than the preceding one, but more sensitive.

The method of calculation described above should not prevent a study of other methods, the interest in which is a function of the intention, and since the most complex are also the most exact it is necessary to use calculators or computers. There are also graphic methods that are simple, rapid and often sufficient which require the use of special graph paper, whose scales are mathematical transformations (probit, neoprobit, logit, coordinate transforms etc.) which enable a linear representation of the dose–effect relationship. It can also happen that base 2 logarithms or Napierian logarithms are used in particular circumstances for the expression of titres.

4.2. Unique Particle Hypothesis

In this hypothesis, it is admitted that when the receptivity thresholds are equal, the only variability which is observed is linked to the fact that the distribution of virus particles in the suspension follows a Poisson distribution (68–71).

Two methods of analysis merit special attention:

1. Fischer's method (72, 73) in tubes or wells, which gives an ED_{50} titre by using either complex calculations or graph paper or, better still, tables whose application conditions are the following:

a. The h log interval of the dilution is constant and equal to $1:2$, $1:4$ or $1:10$.

b. The population n per dilution is constant.

c. The dilutions tested should cover approximately from 0 to 100% responses.

2. The method of counting on a cell sheet or 'dish equivalent' (74–76), which considers that each plaque of lysis observed on the cell sheet comes from a virus particle or from a group of indissociable particles. A dish equivalent Σ is determined for each dilution by multiplying the number of corresponding dishes by the arithmetical expression of this dilution, having taken as base 1 the highest dilution. The N/E ratio gives a value which corresponds to a number of units per volume unit of the base 1 dilution which was at the

origin of the lysis plaques. The number of plaques N is counted. The final expression of the titre compared to the initial suspension takes into account this base 1 dilution: if for example the N/E ratio is equal to 0.64 for a high dilution (base 1) of $10^{7.4}$, the titre will be (0.64 to $10^{7.4}$) or 7.2 plaque-forming units (PFU) per volume. The accuracy of this titre depends uniquely on the population of plaques counted: N.

It is possible to go from a titre expressed in ED_{50} to a titre in PFU by using the ratio 10 ED50 = 10 PFU. The titre expressed in PFU is then inferior by log 0.15 to the titre expressed in ED_{50}. This method of calculation can be applied to the technique of the plating efficiency test (PET) in the testing of cell culture media for example.

5. CONCLUSION

The quality control tests applied to biological products prepared in animal cell culture are summarised in Fig. 7.

This review of quality control testing is by no means exhaustive but assists in emphasising the rather peculiar nature of biologicals as opposed to chemo-therapeuticals in the pharmaceutical industry. For chemical products, once a new formula has been established, each batch can be quality controlled by

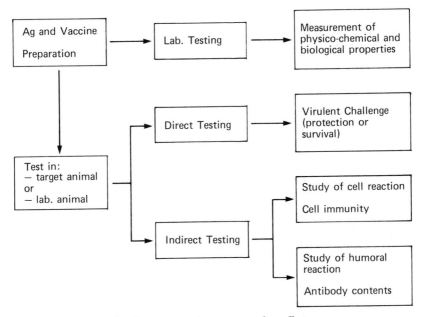

Fig. 7.   Test to prove the presence of an effect.

merely checking that the right number of molecules is present. There is then no doubt about the activity of the product in question.

In biology, however, particularly where vaccines are concerned, such chemical precision is impossible and the verification of the qualitative aspect uncertain; from the quantitative point of view, the only thing that can be measured and that can serve as an assurance is the observed *effect*.

If high quality is required, the indications of the Good Manufacturing Practice Code should be carefully followed, i.e. routine tests at all stages of production. The important elements in achieving this are easily reproducible testing methods and maintained quality for all batches of product.

Present research and new perspectives which have opened up with the advent of genetic engineering and antigen synthesis mean that more accurate and less restrictive technologies can be envisaged for the future. It will, for example, be possible to determine biochemical formulas of immunogen fractions, as is already the case for meningitis vaccine, whose quality is assured by the controlled presence of a minimum quantity of a biochemically defined polyriboside.

It is obvious that in the future other techniques will be developed in immunology. Also, viral and bacterial immunogen peptide charts will be drawn up, thereby revolutionising the entire concept of routine testing and quality control testing of biological products.

ACKNOWLEDGMENTS

The authors are indebted to Mandy Evans (Rhône Merieux) for a more than capable translation from the French.

REFERENCES

1. Ministère de la Santé et de la Famille. Direction de la Pharmacie et du Médicament (1978). "Pratiques de bonne fabrication. Instruction du 30 octobre 1978 relative à la mise en oeuvre du système international de certification de la qualité des produits pharmaceutiques." Maisonneuve.
2. U.S. Department of Agriculture (1983). "Virus Serum Toxin Act," 21 USC 151-158 and CFR Parts 101-117. Biologics Licensing and Standards Veterinary Services, APHIS, USDA, Hyattsville, Maryland.
3. International Drug Good Manufacturing Practice (1979). Interpharm. Press. Prairie City, Illinois.

4. U.S. Department of Agriculture (1983). "Virus Serum Toxin Act," 21 USC 151-158, Standard Requirements 113-51 and 113-52 and 101-117. Biologics Licensing and Standards Veterinary Services. APHIS, USDA, Hyattsville, Maryland.

5. Gayot, G. (1979). Rôle et importance des contrôles dans la normalisation des produits biologiques vétérinaires. *World Vet. Congr., Proc., 21st, 1979,* Part 7, p. 30.

6. Conseil de l'Europe (1969). Pharmacopée Européenne, Vol. I, p. 80, Maisonneuve.

7. Conseil de l'Europe (1980). Pharmacopée Européenne, 2nd ed., Part I, Chapter V.6.7. Maisonneuve.

8. Conseil de l'Europe (1980). Pharmacopée Européenne, 2nd ed., Part I, Chapter V.6.3. Maisonneuve.

9. Ministère de la Santé Publique et de la Sécurité Sociale (1973). Pharmacopée Française, 9th ed., Vol. II, pp. 324–325. Ordre National des Pharmaciens, Paris.

10. Fayet, M. T., Roumiantzeff, M., and Fargeaud, D. (1969). Analyse par centrifugation zonale en gradient de saccharose des particules virales contenues dans une préparation du virus de la fièvre aphteuse. *Rev. Immunol.* **33**, 335–344.

11. Warrington, R. E., Cunliffe, M. R., and Bachrach, M. L. (1973). Derivatives of Aziridine as inactivants for foot-and-mouth disease virus vaccines. *Am. J. Vet. Res.* **34** (8), 1087–1091.

12. Conseil de l'Europe (1971). Pharmacopée Européenne, Vol. II, 36. Maisonneuve.

13. Conseil de l'Europe (1980). Pharmacopée Européenne, 2nd ed., Part I, Chapter V.2.1.1. Maisonneuve.

14. Conseil de l'Europe (1977). Pharmacopée Européenne, Suppl., Vol. III, pp. 55–58. Maisonneuve.

15. Kaplan, M. M., and Koprowski, H. (1974). La Rage. *In* "Techniques de laboratoire," OMS, 3rd ed., pp. 88–100.

16. Mitchell, J. R., Everest, R. E., and Anderson, G. R. (1971). Sensitive procedure for detecting residual viable virus in inactivated rabies vaccines. *Appl. Microbiol.* **22**, 600–603.

17. Mateka, M. D. (1959). Uber das Verhalten des Maul—und Klauenseuchevirus bei Adsorption an Aluminium hydroxyd und Nachfolgender Elution. *Zentralbl. Bakteriol., Parasitenkd., Infektionskr. Hyg., Abt. 1: Orig.* **175**, 40–48.

18. Wittman, G. (1964). Die Interferenz Zwischen inaktiviertem und aktivem Maul und Klauenseuchevirus in Gewebskuituren. Ihre Bedeutung für die Unschädlichkeitsprüfug von Formalinvakzinen. *Zentralbl. Veterinaer med. Reihe B,* **11**, 135–142.

19. Dannacher, G., Fedida, M., Coudert, M., and Perrin, M. (1977). Les méthodes de contrôle d'innocuité du vaccin anti-aphteux. *Dev. Biol. Stand.* **35**, 271–278.

20. Anderson, E. C., Capstick, P. B., Mowat, G. M., and Leech, E. B. (1970). In vitro method for safety testing of foot-and-mouth disease vaccine. *J. Hyg.* **68**, 159–172.

21. Fayet, M. T., and Petermann, H. G. (1971). Control de inocuidad de vacunas contra la fiebre aftosa que contienen formol, saponina e hydroxido de aluminio. *Gac. Vet.* **33**, 580–590.

22. Conseil de l'Europe (1977). Pharmacopée Européenne, Suppl., Vol. III, p. 53. Maisonneuve.

23. Ministry of Agriculture, Fisheries and Food (1977). British Pharmacopoeia (Veterinary). p. 150. Univ. Press, Cambridge.

24. Appel, M., and Gillespie, J. (1972). Canine distemper virus. *Virol. Monogr.* **11**, 3–96.

25. Appel, M. and Robson, D. (1973). A microneutralisation test for canine distemper virus. *Am. J. Vet. Res.* **34**, 1459–1463.

26. Koprowski, H. (1973). The mouse inoculation test. *W.H.O. Monogr. Ser.* **23**, 85–93.

27. National Academy of Sciences, National Research Council (1981). "Methods of Examination of Poultry Biologics," Publ. No. 1038, 9CFR-113, p. 69. NAS-NRC, Washington, D.C.

28. Conseil de l'Europe (1977). Pharmacopée Européenne, Vol. III, p. 102. Maisonneuve.
29. Chappuis, G., and Terre, J. (1973). Immunisation contre la maladie de Carré et la maladie de Rubarth. II. Installation précoce de l'immunité. *Recl. Med. Vet.* **149**, 177.
30. Lucam, F., and Fedida, M. (1958). Une nouvelle méthode quantitative pour l'appréciation de l'immunité anti-aphteuse. *Bull. Off. Int. Epizoot.* **49** (9–10), 596.
31. Terre, J., Stellmann, C., Brun, A., Favre, H., and Fontaine, J. (1977). Contrôle d'activité du vaccin anti-aphteux sur bovins: Méthodes quantitatives d'extinction antigénique. *Dev. Biol. Stand.* **35**, 357–367.
32. Stellmann, C., Terre, J., Favre, H., Brun, A., and Fontaine, J. (1977). Comparison of foot-and-mouth disease vaccine potency testing in cattle in terms of the nature of diluent. *Arch. Virol.* **54**, 61–74.
33. Mackowiak, C., Fontaine, J., Terre, J., Stellmann, C., Roumiantzeff, M., and Petermann, H. G. (1966). Contróle quantitatif du vaccin anti-aphteux. Etude de la loi dose-effet et corrélation entre les doses vaccinantes 50% chez les cobayes et les bovidés. *Bull. Off. Int. Epizoot.* **65** (1–2), 131–171.
34. Seligmann, E. B. (1973). The NIH test for potency. *W.H.O. Monogr. Ser.* **23**, 279–286.
35. Lombard, M. (1979). Utilisation de sérums de bovins vaccinés dans l'évaluation des souches virales aphteuses du terrain. European Commission for the Control of Foot and Mouth Disease. F.A.O. Report of the meeting of the Research group of the Standing Technical Committee, Lindholm, Denmark.
36. Atanasiu, P. (1973). Quantitative assay and potency test of antirabies serum and immunoglobulin. *W.H.O. Monogr. Ser.* **23**, 314–318.
37. Smith, J. S., Yager, P. A., and Baer, G. M. (1973). A rapid tissue culture test for determining rabies neutralizing antibody. *W.H.O. Monogr. Ser.* **23**, 354–357.
38. Guillemin, F., Tixier, G., Soulebot, J. P., and Chappuis, G. (1981). Comparaison de deux méthodes de titrages des anticorps antirabiques neutralisants. *J. Biol. Stand.* **9**, 147–156.
39. Sohier, R. (1964). Hémagglutination et hémadsorption et leur inhibition. *In* "Diagnostic des maladies à virus," pp. 248–281. Editions Médicales Flammarion, Paris.
40. Carmichael, L. E., Joubert, J. C., and Pollock, R. V. H. (1980). Hemagglutination by canine parvovirus: Serologic studies and diagnostic applications. *Am. J. Vet. Res.* **41**, 784.
41. Quinn, P. A., and Ho, T. Y. (1978). Sensitivity of scanning electron microscopy for the detection of mycoplasma contamination. *Electron. Microsc., Pap. Int. Congr. 9th, 1978,* Vol. 2.
42. Tektoff, J., Durafour, M., Fargeaud, D., Precausta, P., and Soulebot, J. P. (1982). Particularités de la morphogénèse du virus rabique et de sa morphologie vues à l'occasion de contrôles systématiques de cultures par microscopie électronique. *Comp. Immun. Microbiol. Infect. Dis.* **5** (1–3), 9–19.
43. Fayet, M. T., Vincent-Falquet, J. C., Tayot, J. L., and Triau, R. (1978). Interest of the evaluation of tetanus antibodies by radioimmunoassay. *Commun. Int. Conf. Tetanus, 5th, 1978.*
44. Voller, A., Bidwell, D. A., and Bartlett, A. (1979). "The Enzyme Linked Immunosorbent Assay (ELISA)," A Guide with Abstracts of Microplate Applications. Dynatech Eur., Borough House, Guernsey, G. B.
45. Kristensen, F., Kristensen, B., and Lazary, S. (1982). The lymphocyte stimulation test in veterinary immunology. *Vet. Immunol. Immunopathol.* **3**, 203–277.
46. Vlaovic, M. S., Buening, G. M., and Loav, R. W. (1975). Capillary tube leukocyte migration inhibition as a correlate of cell-mediated immunity in the chicken. *Cell. Immunol.* **17**, 335–341.
47. De Bono, D. P., MacIntyre, D. E., White, D. J. G., and Gordon, J. L. (1977). Endothelial adenine uptake as an assay for cell, or complement mediated cytotoxicity. *Immunology* **32**, 221–226.

48. Gaddum, J. H. (1933). Methods of biological assay depending on a quantal response. *Med. Res. Counc.* (G. B.), *Spec. Rep. Ser.* **183**, 1–46.
49. Dannacher, G., Fedida, M., and Perraud, J. (1968). L'hypothèse de la particule unique. *Rev. Med. Vet.* **144**, 711–733.
50. Lellouch, J. (1964). Loi dose-effet dans le titrage des virus, interprétation pharmacologique ou interprétation particulaire. *Rev. Fr. Etud. Clin. Biol.* **9**, 487–489.
51. Stellmann, C., and Bornarel, P. (1972). Titrations of virus on cell culture, pharmacology or single particles hypothesis. *Arch. Gesamte Virusforsch.* **36**, 205–207.
52. Allan, W. M., and Herber, C. N. (1968). The precision of virus end point determination. *Arch. Gesamte Virusforsch.* **25**, 330–336.
53. Bliss, C. (1938). The determination of the dosage–mortality curves from small numbers. *Q. J. Pharmacol.* **11**, 191–216.
54. Brown, W. F. (1964). Variance estimation in the Reed–Muench 50% end point determination. *Am. J. Hyg.* **79**, 37–46.
55. Finney, D. J. (1964). "Probit Analysis." Cambridge Univ. Press, London and New York.
56. Irwin, W. D., and Cheeseman, E. A. (1939). On an approximate method of determining the median effective dose and its error, in the case of a quantal response. *J. Hyg.* **39**, 574–580.
57. Isaacs, A. (1957). Particle counts and infectivity titrations for animal viruses. *Adv. Virus Res.* **4**, 111–158.
58. Karber, G. (1931). L'interprétation collective des expériences pharmacologiques faites en série. *Naunyn-Schmiedebergs Arch. Exp. Patho. Pharmakol.* **162**, 480–485.
59. Litchfield, J. T., and Wilcoxon, F. (1949). A simplified method of evaluating dose effect experiments. *J. Pharmacol. Exp. Ther.* **96** (2), 99–113.
60. Pizzi, M. (1950). Sampling variation of the 50% end point determination by the Reed Muench (Behrens) method. *Hum. Biol.* **22**(3), 151–190.
61. Stellmann, C., and Terre, J. (1965). Choix d'une méthode de calculs statistiques en vue des contróles d'activité des produits biologiques sur animaux. *Bull. Soc. Sci. Vet. Lyon* **67**, 273–282.
62. Stellmann, C., Terre, J., Roumiantzeff, M., and Bornarel, P. (1965). Contrôle quantitatif du vaccin anti-aphteux. III. Analyse statistique par la méthode de Litchfield. *Rev. Med. Vet.* **116**, 609.
63. Tootil, J. P., Robinson, W. D., and Adams, A. G. (1967). Precalculated ED50, PD50 et LD50 statistics. *Symp. Ser. Immunobiol. Stand.* **10**, 183–220.
64. Lorenz, R. J. (1960). Influence de l'erreur de pipetage sur la précision de mesure de la concentration des solutions virales. *Arch. Gesamte Virusforsch.* **10**, 551–559.
65. Lorenz, R. J. (1960). Etude de la relation dose-effet et de la précision dans les titrages viraux. *Arch. Gesamte Virusforsch.* **10**, 560–577.
66. Morancy, M. J. (1956). Design of experiments. *Statistician* **7**, 143.
67. Stellmann, C., Terre, J., Favre, H., Brun, A., and Fontaine, J. (1976). Contrôle d'activité du vaccin anti-aphteux sur bovins. Etude statistique et comparaison des lois dose-effet en fonction de la nature du diluant ou du volume inoculé. *Rapp. Groupe Experts, Comm. Eur. Pharmo. 1976*, No. 15V.
68. Agnese, G., and Crovarie, P. (1959). Sulla titolazione dei virus poliomielitici con il procedimento del numero piu probabile. *Boll. Ist. Sieroter. Milan.* **38**, 9–10.
69. Chang, S. L., Berg, G., Busch, K. A., Stevenson, R. E., Clarke, N. A. and Kabler, P. W. (1958). Application of the most probable number method for estimating concentration of animal viruses by tissue culture technique. *Virology* **6**, 27–42.
70. Chochran, W. G. (1950). Estimation of bacterial densities by means of the most probable number. *Biometrics* **6**, 105–116.

71. Gart, J. J. (1964). The analysis of Poisson regression with an application in virology. *Biometrika* **51**, 517–521.

72. Fisher, R. A., and Yates, F. (1970). "Statistical Tables for Biological, Agricultural and Medical Research," 6th ed., pp. 66–80. Oliver & Boyd, Edinburgh and London.

73. Stellmann, C., and Bornarel, P. (1971). Tables de calcul des titres D50 des suspensions virales et leurs précisions. *Ann. Inst. Pasteur, Paris* **121**, 825–833.

74. Berg, G., Harris, E., Chang, S. L., and Busch, K. A. (1963). Quantitation of viruses by the plaque technique. *J. Bacteriol.* **85**, 691–700.

75. Dulbecco, R., and Vogt, M. (1954). Plaque formation and isolation of pure lines with polio-viruses. *J. Exp. Med.* **99**, 167–182.

76. Galasso, G. J., Sharp, J., and Sharp, D. G. (1964). The influence of degree of aggregation and virus quality on the plaque titer of aggregated vaccinia virus. *J. Immunol.* **92**, 870–878.

13

Pyrogenicity and Carcinogenicity Tests

C. N. WIBLIN

Vaccine Research and Production Laboratory
Public Health Laboratory Service
Centre for Applied Microbiology and Research
Salisbury, Wiltshire, United Kingdom

1. PYROGENICITY TESTING

1.1. Introduction

The parenteral administration of medicinal products may result in the induction of fever, chills, histamine release, altered vascular permeability and other unwanted sequelae in the recipient (9, 78, 126). Such reactions are termed pyrogenic responses, since fever induction was the first effect recog-

Animal Cell Biotechnology, Vol. 2

nized and easily described (11). A pyrogen may be described as any component, contaminant or otherwise, of a biological or pharmacological product which induces these effects. In practice, the pyrogen that is of greatest concern to manufacturers is contaminant bacterial endotoxin from the Enterobacteriaceae (103). The endotoxin is a lipopolysaccharide component of the cell wall of this family of Gram-negative bacteria and a dose of less than 10 ng/kg of purified endotoxin can produce a fever in man (50, 126). Pyrogenic responses can, depending on the quantity of endotoxin administered and physiological state of the recipient, be a serious danger to life.

Because of the stability and ubiquity of endotoxins, pharmaceutical and vaccine manufacturing industries now go to great lengths to ensure that inadvertent contamination of their products with endotoxin is reduced to the lowest possible levels. The correct design, operation and rigorous monitoring of manufacturing facilities, adequate training and supervision of personnel, suitable treatment of materials and equipment and use of sterile air flow cabinets have all helped enormously in the prevention of contamination of the product. Nevertheless, endotoxins remain difficult to eliminate, because of resistance to normal sterilizing temperatures, and may not be removed by normal filtration systems. Hence great importance is attached to ensuring that all starting materials—production water, solutions, vaccine seed stocks etc.—are free of pyrogenic compounds. To this end, and in conjunction with the preventive measures previously described, tests have been developed to screen such components regularly as a part of in-process control during manufacture, as well as to monitor the acceptability of the final product. Two tests are used widely in the biological manufacturing industries—the rabbit pyrogen test and the *Limulus* amoebocyte lysate assay. Other tests for pyrogens have been described, but have yet to find widespread acceptance, and will not be considered here; they are briefly reviewed by Marcus and Nelson (71).

1.2. The Rabbit Pyrogen Assay

1.2.1. Relevance of the Test

The occurrence of harmful side effects following intravenous administration of drugs was the impetus for a number of workers to investigate the nature of the cause of the effects. Seibert (102) showed that high-quality water was essential for production and/or administration of drugs and that pyrogenic responses were associated with microbiological contamination of that water. Her detection system involved the use of rabbits, and was the basis of the rabbit pyrogen assay widely used today. The test involves the measurement of a total rise in body temperatures of a group of rabbits following an intravenous administration of the substance under test.

An assumption of the test is that the threshold pyrogen dose for the rabbit is the same as for man, and that the test is therefore an acceptable indicator of potential pyrogenicity or otherwise. Analysis of the literature gives few confirmed examples of materials having passed the current rabbit test for pyrogens that subsequently proved to be pyrogenic in humans. Conversely, there are no confirmed instances of materials pyrogenic in the rabbit test that were non-pyrogenic when administered to healthy humans. Dare and Mogey (30) used a pseudomonad pyrogen preparation to show that the sensitivity of rabbits was similar to or greater than that of humans. Greisman and Hornick (50) reviewed the literature, and with others have undertaken further studies to establish this correlation using bacterial filtrates or purified endotoxins (98, 126). Their results suggested that although man was highly sensitive to endotoxin, with a very much steeper dose–response curve than that of rabbits, (1) the threshold dose of endotoxins from various bacterial sources was of a similar order for rabbits and a man on the basis of weight, and (2) threshold febrile responses could be induced in rabbits with lower doses of endotoxin than in man on a total dose basis. It should be emphasised that these studies relate only to endotoxin; documented evidence on other pyrogenic chemicals or particulates is lacking.

In the absence of confirmed contraindicative data, the tacit acceptance of the rabbit bioassay system as being truly indicative of pyrogenicity of a test sample must remain. The test system, adopted by British (later European) and U.S. regulatory authorities and described in their respective Pharmacopoeia, was for over 40 years, and in many countries still remains, the one legally acceptable method to test for pyrogenic contamination of finished parenteral preparations.

1.2.2. Assay Procedure

The test involves measuring the total rise in body temperatures of a group of rabbits after administration of a test material. For the assay to give meaningful results, the test requires that the criteria laid down in the relevant pharmacopoeia (18, 39, 113) must be followed closely. In practice, the European and U.S. pharmacopoeias differ slightly in their test requirements and the differences will be highlighted where appropriate.

Requirements of the European Pharmacopoeia

GLASSWARE AND OTHER EQUIPMENT.   Sterile plastics should be used whenever possible. Otherwise, glassware and equipment must be thoroughly washed with pyrogen-free water and heated to 250°C for a minimum of 30 min.

DILUENTS OR SOLUTIONS.   Diluents or solutions should be pyrogen-free and sterile.

THERMOMETERS.   Thermometers or other suitable electrical devices should be accurate to $\pm 0.1°C$, and during the test should be inserted about 5 cm into the rabbit rectum. Insertion depth should be constant in any one test. An electrical device, if used, should be in position 90 min before the start of the test and not disturbed during the test.

RABBITS.   Rabbits should be healthy adults of either sex, weighing at least 1.5 kg, fed normal approved diets without antibiotics and not showing any evidence of weight loss. A rabbit shall not be used for a pyrogen test if:

1. It has been used in any pyrogen test within the previous 3 days.
2. It has been used in any pyrogen test in the preceding 3 weeks where the test sample failed the test.
3. It has been used in any pyrogen test where the mean responses of the rabbits in the group exceeded $1.2°C$.

RESTRAINING BOXES.   Restraining boxes should be used in conjunction with electrical temperature devices and be so designed that the animals are re-tained with loose-fitting neck stocks allowing a natural sitting posture. Ani-mals should be put in the boxes 60 min before the test, and remain in them during the test.

PRELIMINARY TEST.   Up to 3 days before testing the sample, animals se-lected as described above, but which have not been used in the previous 2 weeks, are subjected to a full pyrogen test procedure, injecting intra-venously 10 ml/kg of body weight of pyrogen-free isotonic saline solution.

The test should be carried out in a quiet room separate from the rabbit living quarters but with similar environmental conditions. The rabbits should have been housed in this room for at least 18 hr prior to the test. Food should be withheld overnight and throughout the test; water should be withheld during the test.

Animal temperatures should be recorded throughout the test, starting 90 min before injection, and continuing for 3 hr afterwards. Any animals show-ing a temperature variation greater than $0.6°C$ must not be used in the main test.

MAIN TEST.   A group of three rabbits are used. The test sample (dissolved or diluted in pyrogen-free isotonic saline if necessary) is warmed to $38.5°C$ prior to injection into the marginal ear vein of each rabbit over a time interval of not more than 4 min. The volume of injection must be not less than 0.5 ml/kg, and not more than 10 ml/kg. (The time interval for injection and volume of test sample injected are subject to the specifications attached to that test sample and so may vary outside the stated limits.)

INITIAL AND MAXIMUM TEMPERATURES. The initial temperature of each rabbit is the mean of two temperature readings taken at an interval of 30 min in the 40 min preceding the test sample injection. The maximum temperature of each rabbit is the highest temperature reading taken in the following 3 hr, the readings being taken at intervals not greater than 30 min. The difference between the initial and maximum temperatures of each rabbit is the response. Where the response is negative, the figure is taken as zero.

Rabbits with a temperature variation greater than 0.2°C between successive readings in initial temperature determination are excluded from the test. In any one test the initial temperatures of the rabbits should not differ by more than 1°C; any outside that limit are excluded from the test. All rabbits with initial temperatures greater than 39.8°C and less than 38°C are excluded from the test.

Interpretation of Test Results. The test material passes the pyrogen test if the summed response of the group of three rabbits does not exceed 1.15°C. The material fails if the summed response exceeds 2.65°C. Within that range, the test must be repeated using another group of three rabbits. The test may be repeated up to three times in all, depending on the results obtained (Table 1). If after the third repeat test the cumulative summed responses exceed 6.6°C, the material fails.

Requirements of the U.S. Pharmacopoeia. Differences with the European Pharmacopoeia occur in the following requirements:

THERMOMETERS. The requirements are the same as those for European thermometers, except that the device must reach a maximum reading in 5 min, and should be inserted to a depth of not less than 7.5 cm. There is no specified limit on the length of time a thermistor probe should be in position prior to commencement of the test.

RABBITS. The requirements are the same as for European rabbits, except that a rabbit shall not be used for a pyrogen test if:

1. It has been used in any pyrogen test within the previous 2 days.
2. It has been used in any pyrogen test in the preceding 2 weeks where the test sample failed the test.

TABLE I   Summed Response Limits in Initial and Repeat Pyrogen Tests

Total no. of rabbits	Test material passes if summed response does not exceed	Test material fails if summed response exceeds
3	1.15°C	2.65°C
6	2.80°C	4.30°C
9	4.45°C	5.95°C
12	6.60°C	6.60°C

3. It has been used in any pyrogen test in the preceding 2 weeks where its temperature rise was 0.6°C or more.

PRELIMINARY TEST.   Before using a rabbit for the first time in a pyrogen test, and not more than 7 days before the test, the animal should be subjected to a full pyrogen test excepting the injection step.

The test should be carried out in a quiet room separate from the rabbit living quarters, but with similar environmental conditions. Food should be withheld during the test; water may be restricted during the test. Any animals showing a temperature variation of 0.6°C or more should not be used in the main test.

MAIN TEST.   A group of three rabbits must be used. The test sample (dissolved or diluted as directed) is warmed to 37° ± 2°C prior to injection into the marginal ear vein of each rabbit over a time interval of not more than 10 min. The volume of injection should be 10 ml/kg, unless otherwise specified for the particular test sample.

INITIAL AND MAXIMUM TEMPERATURES.   The initial (control) temperature is the temperature reading taken within 30 min prior to test sample injection. No minimum number of readings is stated, nor the time interval between them. The maximum temperature of each rabbit is the highest temperature reading taken in the following 3 hr, the readings being taken at 1, 2, and 3 hr subsequent to injection. The difference between initial and maximum temperatures is the response. Where the response is negative, the figure is taken as zero.

Rabbits in any one test group should not have initial temperatures that vary by more than 1°C from each other; any outside that limit must be excluded from the test. All rabbits with initial temperatures exceeding 39.8°C should be excluded from the test. No lower limit is stated, and allowable individual variation between successive readings in initial temperature determination is not stated.

Interpretation of Test Results. The test material passes the pyrogen test if the summed response of the group of three rabbits does not exceed 1.4°C, and if no rabbit shows an individual response greater than 0.6°C. If one or both of these criteria are not met, the test is repeated using five other rabbits. If not more than three of the eight rabbits show individual responses of 0.6°C or higher, and if the summed response of the eight animals is not greater than 3.7°C, the material under test is considered to have met the requirements for the absence of pyrogens.

Comparison between Pyrogen Test Specifications of the European and U.S. Pharmacopoeia. In general, the differences between the two systems are small. However, the European method is to be preferred for a number of reasons:

1. The absence in the U.S. Pharmacopoeia of a preliminary test involving intravenous administration of material may fail to exclude animals that would react adversely to that type of handling and experimentation.

2. The recording of temperatures at hourly rather than half-hourly intervals post-injection can result in test material passing the U.S. pyrogen test, but failing the European test if the maximum temperature recorded falls on a half-hour reading. This can be overcome by the use of thermistor probes and continuous readout. Their use is desirable in any event, since maximum temperature readings may occur at times other than at half-hour or hourly intervals post-injection. Also, their use means the animals are not subjected to additional handling stresses.

3. The U.S. pyrogen test involves a presumptive test with three animals which, if positive, means using a further five animals to complete the test. As explained by Marcus and Nelson *(71)*, the test assumes that if the presumptive test passes, then a full test with eight animals would not have failed either. With certain materials that might cause false positive results, this has led to the testing of more animals than necessary and picking the 'best' three results. The European pyrogen test, with an 'intermediate' temperature range and serial increases in rabbit numbers, is arguably a more realistic system for interpretation of the results.

1.2.3. Sensitivity of the Rabbit Pyrogen Test

The sensitivity of the rabbit pyrogen test, as expressed in the various pharmacopoeias, is based on the clinical observation that infusions of up to 1 litre/day do not produce any pyrogenic response in humans when the same preparations do not increase the temperature of a rabbit by 0.6°C if administered to the animal at a dose of 10 ml/kg.

The rabbit model for pyrogenicity has an advantage over *in vitro* systems in that in theory it should detect any pyrogenic activity in a preparation, and not that due solely to the presence of endotoxin; there are reports of positive rabbit pyrogen tests on samples with no detectable microbial or endotoxin contamination (117). However, published data on the ability of the rabbit to detect such pyrogens and the threshold sensitivity to them are lacking. Only for endotoxin have numerous studies been done. Although there are variations between laboratories in the findings reported—indicative of differences in animals, operator techniques, endotoxin source, preparation and purity—in general it is agreed that the rabbit is sensitive to endotoxin threshold doses of between 0.5 and 10 ng/kg. Such levels are defined as the minimum concentrations of endotoxin that cause a temperature rise of 0.6°C within 3 hr of injection, provided the volume does not exceed 10 ml/kg (21, 26, 30, 50, 98, 116).

The frequency with which false negative test results occur is difficult to establish without careful and prolonged retrospective analyses, which have been carried out rarely. In one reported study, the theoretical frequency was less than 0.3% (117). Given that the test rabbits employed do respond to pyrogenic materials—an estimated 2–5% may not do so (118), species variation is known to exist (115), and some rabbit colonies can become refractory to low levels of pyrogen (77)—confirmed reports of materials causing febrile reactions in humans after a negative rabbit test are extremely rare indeed, so the actual frequency of false negatives is probably much lower.

1.2.4. Factors Affecting the Rabbit Pyrogen Test

False reactions can result from a number of variables in the rabbit bioassay system. Variation within and between strains leading to false negative reactions has been mentioned above. The rabbit is very sensitive to various conditions that govern hyperthermia, such as stress factors and physiological state. It is essential that all animals are in good health, are housed and tested under uniform temperature conditions and are allowed ample time to settle into living quarters before preliminary testing. At no time during any preliminary or main testing procedures should the animals be disturbed by undue noise, unnecessary handling or the presence of strangers. Ideally, the same operator(s) should carry out the tests on a particular batch of rabbits at all times. Preliminary screening must be carried out rigorously to exclude unreliable animals.

False positive reactions can result from the administration of a thermogenic agent. Body temperature is controlled by the hypothalamus, and as such is sensitive to sodium–calcium levels. Any alteration in that balance, e.g. calcium solutions or chelating agents under test, may cause temperature variations (15). The same author points to the use of antilymphocyte serum, which if used in rabbits would react with the rabbit lymphocytes and activate macrophages to release pyrogen. A number of drugs, for example sulphur in oil, are known to be thermogenic [Whittet (*122*), quoted by Marcus and Nelson (*71*)]. Under these circumstances, the rabbit may not prove to be a suitable model for testing such agents, and another more suitable system should be chosen if available.

1.2.5. Quantitative Assessment of Pyrogen Content

The rabbit pyrogen test as applied is qualitative in nature. Quantitative assessments of the actual pyrogen content in a test sample—positive or negative—would require the use of dilutions of a reference standard pyrogen (endotoxin), and assay of serial dilutions of the test sample. Clearly the necessity to use an even greater number of animals would make such assessments cumbersome and expensive. Nevertheless, an attempt at quantitation is desirable, and the acceptance of a reference endotoxin source and its desired level of sensitivity in rabbits is currently the subject of much debate in the United Kingdom. Inclusion of this positive control, which need not necessarily be carried out in every test, would enable an upper limit to be set on endotoxin concentration, which no negative test sample would exceed. A secondary advantage of the inclusion of this positive control is that it would guard against rabbit species and colony variation and detect the onset of tolerance within a rabbit colony, hence avoiding the occurrence of false negative test results for those reasons.

Such an attempt has very recently been proposed by the U.S. Pharmacopoeial Forum (*87*). Significant changes in the rabbit pyrogen test are detailed in an attempt to classify the rabbit test colony into 'adequate', 'less sensitive' and 'extra sensitive' categories. Specifically, it is proposed that 10 rabbits or 10% of the colony (whichever is greater) be divided into two groups, one injected with 10 endotoxin units/kg (EU/kg; for a definition of EU see Section 1.3.5 below), the other with 2.5 EU/kg. All injections are made in a volume of 10 ml/kg rabbit, and the number of positive responses is scored. Colony sensitivity is defined in Table II:

1. If the response is 'adequate' then the pyrogen test remains as specified in the U.S. Pharmacopoeia.

2. If the colony is found to be 'less sensitive', then the test is modified such

TABLE II   Rabbit Colony Sensitivity to Endotoxin

Colony sensitivity	Percentage of positive responders in 'high' (10 EU/kg) group	Percentage of positive responders in 'low' (2.5 EU/kg) group
Adequate	40–60	20
Less sensitive	<40	<20
Extra sensitive	>60	>20

that it fails if any of the first three rabbits have temperature rises greater than 0.5°C, or the combined temperature rises exceed 1.2°C. For an additional five rabbits, the test fails if more than three of the eight rabbits exceed a 0.5°C rise, or if the summed response exceeds 3.5°C.

3. If the colony is 'extra sensitive' then the corresponding figures for the first three rabbits are 0.7° and 1.5°C respectively, and for the additional five rabbits these become 0.7° and 3.8°C respectively.

4. It should be emphasised that at the time of writing these revisions are proposals only and comments are invited from all interested parties. Nevertheless, it is likely that some such revisions will be incorporated into rabbit pyrogen test protocols in future.

1.3. The Limulus Amoebocyte Lysate Assay

1.3.1. Background

Intravenous coagulation in the horseshoe crab *Limulus polyphemus* was first reported by Loeb in 1902 (68), but it was not until 1964 that Levin and Bang (66) showed that the aetiology of this disease was endotoxin-mediated. Later, the same workers showed that a lysate of circulating amoebocyte blood cells would react with endotoxin to form gel under suitable conditions (67). The reaction was enzymatic, requiring endotoxin, divalent cations (Ca^{2+}), a 'proclotting' enzyme and clottable proteins (48, 105, 106, 110). It is thought that the calcium ions and endotoxin together activate the proclotting enzyme, which then cleaves each clottable protein into at least two sub-unit chains, one soluble, the other insoluble. Oxidation of sulphydryl groups on the insoluble sub-unit chains results in the formation of disulphide bridges, thus forming the gel (27, 82, 83).

Although first used to detect endotoxaemia in patients, the potential value of the system in biological manufacturing areas was recognized quickly, and *Limulus* amoebocyte lysate (LAL) was soon to be produced on a large scale. Although the basis of the production system was very simple—collection of amoebocytes by heart puncture, washing, concentration and lysis of the cells, removal of cellular debris and retention of the supernatant lysate fraction for freeze-drying—the processes were subject to a range of factors that adversely affected the efficacy and sensitivity of LAL from batch to batch. This was of great concern to all potential users of the method for pyrogen (endotoxin) screening, and one reason why general acceptance of the test was slow in coming until improvements in, and standardization of, production methods had been developed.

1.3.2. Assay Procedure

The basis of the LAL pyrogen test methodology is very simple. A positive test is defined as the formation of a firm gel capable of maintaining its integrity when the test tube is inverted through 180° after 60 min incubation of the reaction mixture at 37°C. This is indicative of the presence of endotoxin in quantities sufficient to cause a pyrogenic response if the test material was administered parenterally to man.

Requirements of the Test. The successful performance of a LAL test requires very strict adherence to the protocol if accurate and reliable results are to be obtained:

GLASSWARE AND OTHER EQUIPMENT. Glassware and other equipment must be sterile and pyrogen-free (heated to 250°C for a minimum of 30 min). Sterile plasticware should be used whenever possible.

DILUENTS OR SOLUTIONS. Diluents or solutions used to reconstitute materials or dilute samples must be sterile and pyrogen-free.

pH. The pH of the test samples must lie, or be brought within, the range 6.0–8.0, using sterile pyrogen-free acid or alkali solutions.

TEST SAMPLES. Test samples must be collected aseptically into pyrogen-free containers and stored at 4°C until use. For quantitative assessment of endotoxin concentration in a test sample, the sample must be serially diluted. American authorities require serial dilutions in quadruplicate (113).

REFERENCE STANDARD ENDOTOXIN. The reference standard endotoxin should be used to confirm the labelled potency of the LAL preparation, to guard against non-specific interference with the test by the sample under examination (positive control) and whenever quantitative assessments of endotoxin concentration are desired. Serial dilutions within the appropriate concentration ranges must be made (in quadruplicate to meet U.S. requirements). Thorough mixing is essential to ensure homogeneity of the preparations.

NEGATIVE CONTROLS. Negative controls should always be included to guard against any non-specific interference with the gelation process.

THE LAL PREPARATION. The LAL preparation is provided as a sterile freeze-dried preparation stable at temperatures up to 25°C. Reconstitution for use must be carried out with minimal agitation, and once reconstituted the preparation should be dispensed in aliquots and stored at −20°C until use.

REACTION MIXTURES. Reaction mixtures should be handled with care. Improper handling can negate the gelation reaction irreversibly.

TEST INCUBATION TIME.   The test incubation time should not exceed the recommended limits—usually 60 min, though this may be shorter if the test is performed on a microscale (R. Tattersall, personal communication). Otherwise, a firm gel may form at endotoxin levels below those held to be pyrogenic in man, giving a false positive result. Equally, reading the test before the expiration of the time limit may give false negative results.

Characterization of the Test.   Aliquots of the appropriately reconstituted LAL preparations are added to either test-tubes or other suitable containers if being performed on a microscale. Similar volumes of test sample, endotoxin reference standard, both diluted if and as required, and negative control material are then added to the LAL. The mixtures, after gentle mixing, are incubated undisturbed at 37°C for the specified time period. A typical set of reaction mixtures is listed in Table III.

A positive reading is characterized by the formation of a firm gel that remains intact when the test-tube is inverted through 180°. In a microscale application, where the reaction mixtures are discrete blobs on the bottom of a petri dish, gelation is tested by adding a suitable dye solution to the edge of the blob. If the dye diffuses throughout the blob the test is negative; if the dye forms a coloured collar around a clear blob, the test is positive (R. Tattersall, personal communication). The right-hand column of Table III indicates a typical set of results.

Interpretation.   The lysate sensitivity is taken as the lowest concentration of reference standard that forms a firm gel with the lysate. The sensitivity should reflect the specification of the LAL preparation used. In the example in Table III, the lysate sensitivity is 0.125 ng/ml.

TABLE III   Typical Series of Reaction Mixtures and Results after Incubation

	Reaction mixture	Results
1. Negative control	LAL + pyrogen-free water	−
2. Positive control	LAL + reference standard endotoxin (2 ng/ml) +undiluted test sample	+
3. Reference standard	LAL + reference standard endotoxin 0.25 ng/ml	+
4. Reference standard	LAL + reference standard endotoxin 0.125 ng/ml	+
5. Reference standard	LAL + reference standard endotoxin 0.06 ng/ml	−
6. Reference standard	LAL + reference standard endotoxin 0.03 ng/ml	−
7. Test sample	LAL + undiluted sample	+
8. Test sample	LAL + ×2 dilution	+
9. Test sample	LAL + ×4 dilution	+
10. Test sample	LAL + ×8 dilution	−

If the negative control gels, then either water, glassware, plasticware or the LAL is contaminated. All possibilities must be checked.

If the positive control does not gel, either the lysate has deteriorated—in which case the reference standards would not gel either—or the test sample inhibits gel formation.

Originally, the LAL test was qualitative in nature. However, by use of a reference endotoxin standard to obtain a figure of lysate sensitivity and by assaying serial dilutions of test sample, a quantitative assessment of endotoxin content is possible. In the above example, the concentration is the product of the lysate sensitivity and the lowest dilution of test sample where gelation is seen, i.e. $0.125 \times 4 = 0.5$ ng/ml.

1.3.3. Specificity of LAL for Endotoxin and Other Pyrogens

LAL will detect endotoxins from *Escherichia*, *Klebsiella*, *Serratia*, *Salmonella*, *Shigella* and other members of the Enterobacteriaceae. Although there have been reports and subsequent controversy surrounding the ability of compounds other than endotoxins from this family of bacteria to initiate gel formation, confirmation of this is lacking, with the possible exception of trypsin. Since the 'proclotting' enzyme in LAL preparations may not be dissimilar to trypsin in its action, reported initiation by this enzyme may not be unexpected (*106, 110*). Peptidoglycan has been reported to cause gelation, but this has not been observed when using a range of whole gram-positive bacteria (*60, 124*). Synthetic poly 1: poly C, thrombins and protease have also been implicated (*40, 126*), but the difficulty of proving that the various preparations were endotoxin-free made the observations the subject of some doubt (*128*).

All evidence to date supports the conclusion that a positive result in a *Limulus* assay is indicative of endotoxin contamination. It cannot be emphasised strongly enough, however, that the LAL test will not detect pyrogenic contamination that may be due to fungi, yeasts, Gram-positive bacteria, chemicals or particulate materials. The rabbit pyrogen test may be needed to exclude their presence if suspected in a preparation.

1.3.4. Inhibitory Agents

The test may be negated by substances that mask the gelation reaction. A classification of these is necessary to outline the limits of applicability of the system, and hence its value in pyrogen testing.

Viral vaccines have been reported to give negative results in LAL tests, yet are positive in the rabbit test (*95*). False negative tests can result in the presence of glutathion, lipids, 20% chloramphenicol monosuccinate (*64*), metal chelators (*27*), calcium ions in high concentration and tetracyclines

(116). Since the gelation reaction is enzymic in nature, any material that can inhibit enzymic actions, denature protein or detoxify endotoxin will modify the reaction. Certain vaccines or other biological products may contain preservatives such as phenol, thiomersal or sodium bisulphite, all of which can inhibit gelation if present above certain concentrations (28, 77), and this should be borne in mind when such products are put forward for evaluation. In some instances, the inhibition may be overcome by dilution and/or pH adjustment, e.g. human serum albumin, but this is not always possible, and the presence of such a range of potential inhibitors makes the inclusion of a positive control essential in an LAL assay. If inhibition persists, the rabbit bioassay should be used. An additional important point is that inhibition by a pharmaceutical may be due to the vehicle of administration; tests should not be restricted to the active compound, but on the pharmaceutical preparation, e.g. hydrocortisone (117).

1.3.5. Sensitivity of the LAL System

A major concern of all potential users of the LAL assay was to demonstrate that the sensitivity of the lysate was equivalent to, or greater than, that of the rabbit test. For any comparison to be valid, it must be remembered that the rabbit pyrogen test is dose-dependent (the amount of pyrogen injected per kilogram of body weight), whereas the LAL test is concentration-dependent (gelation time varies with endotoxin concentration). Therefore the sensitivity of the former system is dependent on a constant test volume, the latter on a constant incubation period. For example, a rabbit endotoxin threshold dose could be a positive febrile response to a test dose of 10 ml/kg, and the LAL endotoxin threshold dose could be gelation within 60 min at 37°C.

Whilst early lots of LAL were of somewhat variable quality, production methodology is now much improved and standardized, and in a wide range of reports on the comparative sensitivities of the LAL and rabbit systems, the findings are that the endotoxin threshold dose is at least 10 times lower for the LAL system (25, 26, 36, 54, 64, 65, 72, 94, 116). Furthermore, there were high levels of agreement between the two test systems, so long as the LAL test conditions in particular were adhered to strictly. Where differences arose, it was usually the LAL test system giving a positive result when a rabbit result was negative. This is to a great extent a reflection of the increased sensitivity of the former, but it is known also that minimum detectable concentrations in *Limulus* and rabbit tests differ amongst endotoxin preparations (121).

In 1973, the Bureau of Biologics of the Food and Drug Administration (FDA) issued notices (41, 42) authorizing the use of the method for in-line testing of parenteral and biological products, but not as a final test for approval of the product; the USP rabbit pyrogen test was still mandatory. As a result, the LAL test found extensive use for the monitoring of production water and other solutions, in-process controls, and assessment of the quality of vaccines, radiopharmaceuticals and other drugs (25, 29, 63, 91).

In 1976, the FDA issued specifications for LAL in freeze-dried form calling for a sensitivity of 0.3 ng/ml against a reference endotoxin preparation of purified *K. pneumoniae* lipopolysaccharide lot 1B; current LAL preparations have sensitivities to a number of specific lots of endotoxin in the range 0.01–0.5 ng/ml, and may detect picogram quantities of potent endotoxin preparations of *E. coli*. With improved production methods and the greatly enhanced sensitivity, guidelines were issued to establish the sensitivity of each lot of LAL by testing in parallel with a reference LAL obtained from the FDA (43) or tested against a reference standard endotoxin EC-2, derived from *E. coli*. Later in the same year, the FDA licensed the first commercial LAL preparation for use as an alternative to the rabbit pyrogen test for endotoxin detection in finished biological products and medical devices (44) and additional standards were published 3 years later (45). Since that date the test has assumed ever-increasing importance in the United States, culminating in the publication by the U.S. Pharmacopoeia of revised official monographs in 1983 requiring that from November of that year the LAL test be mandatory for pyrogen testing of water for injection and a range of radiopharmaceuticals.

The EC-2 standard had a defined potency of 5 endotoxin units per nanogram, 1 EU being defined as 0.2 ng of EC-2. LAL preparations originally had labelled sensitivities in nanograms per millilitre, but from June 1982 these sensitivity figures were replaced by endotoxin units per millilitre. Conversion from ng/ml to EU/ml involves multiplication by 5; a sensitivity of 0.02 ng/ml (EC-2) becomes 0.1 EU/ml. By the end of 1982, supplies of the EC-2 standard were becoming exhausted, and during 1983 it has been replaced by a new reference standard endotoxin EC-5 with a potency of 10 EU/ng.

1.4. Comparison of the Rabbit and LAL Pyrogen Test Systems

The rabbit pyrogen test has been, until recently, the only legally acceptable method of testing for pyrogen contamination of parenterals for some 40 years. Nevertheless, it has always been recognized as having many drawbacks:

1. The test involves the use of large numbers of rabbits each year. If an acceptable *in vitro* test exists, such large-scale usage of animals is not morally justifiable.

2. The test is very expensive. Large numbers of animals must be main-

tained, and their upkeep makes large demands on facilities, staff and equipment.

3. Each test may involve anything up to 12 animals, excluding preliminary testing. This is laborious and time-consuming, and only a small number of tests are possible each day.

4. The system is of little value as an in-process control, since in the event of a positive result, this is often received too late for preventive action.

5. The test is impossible to perform in a truly quantitative way.

6. Rabbits are subject to species variation, stress conditions and acquisition of tolerance, all of which can affect sensitivity to pyrogens.

7. Relatively large quantities of test material are required.

There is universal agreement that the *in vitro* LAL test system is a very simple, rapid, inexpensive and sensitive method that can be carried out on microscale for the detection of endotoxin, and thus is not subject to the drawbacks listed above. That it has been adopted for the in-line testing of parenteral products and their components is sensible and of great advantage to manufacturers. Its acceptance as a test in place of the rabbit bioassay by the FDA has yet to be mirrored by British and European counterparts, however. In part this is because the latter have yet to agree on an acceptable reference standard endotoxin and desired lysate sensitivity; results and recommendations from study groups set up as long ago as 1981 have yet to be announced.

Since all the evidence points to the scientifically and legally acceptable rabbit test as having successfully screened parenteral preparations for over four decades at threshold pyrogen dose levels of 0.5–10 ng/kg, then the LAL endotoxin threshold sensitivity should reflect those levels. To demand considerably lower threshold sensitivities may be an advantage in that a genuine negative LAL test makes it very unlikely that the rabbit test on the same sample would be positive. On the other hand, unless the conditions are established correctly, false positives could occur, resulting in the unnecessary rebuttal of material. The natural tendency of regulatory authorities to recommend the lowest acceptable levels of endotoxin in the interests of enhanced safety must be tempered by the knowledge of the endotoxin levels that humans can tolerate safely.

More specifically regarding the use of LAL, the test detects only endotoxins. Acceptance in place of a rabbit test implies tacit acceptance that the only pyrogens in parenteral products are endotoxins from Enterobacteriaceae. This may not necessarily be the case; Gram-positive micro-organisms, fungi, yeasts, chemicals and particulates that may be pyrogenic will not be detected. Additonally, the LAL test conditions are particularly restricted, and it must be borne in mind that some materials cannot be tested without invalidating that system, i.e. vaccine preservatives and inactivants,

tetracycline hydrochloride which precipitates at pH 6–8 (*116*). If it is suspected that such factors may be present, then the rabbit test must remain
the method of choice. In certain instances, for example bacterial vaccine
preparations, it may not be possible to perform a valid pyrogen test by either
method, and indeed the United Kingdom Medicines Amendment Regulations (*75*) specifically allow for exclusion of pyrogen tests in the case of
vaccines and insulin preparations. This blanket exclusion for vaccines is
unnecessarily harsh, since many can be tested quite satisfactorily by either
test method. Given these considerations it is important to select a test
system with reference to the sample to be tested; the LAL test is excellent
for in-process and end-product testing of a wide range of sterile solutions,
viral vaccine preparations, drugs etc., but it does not have the wide-ranging
detection capability of the rabbit test despite the inferior sensitivity of the
latter.

2. CARCINOGENICITY TESTING

2.1. Introduction

Developments in a range of scientific fields over the last 25 years have
made it increasingly clear that many human and animal tumours are likely to
be caused by natural or man-made chemicals, viruses or other biological
products. It is also likely that the same agents may cause other hereditary
defects in individuals exposed to them, resulting in the accumulation of
potentially serious genetic alterations in the human gene pool. These genetic
changes may manifest themselves later in the life of the individual or perhaps only in subsequent generations.

The awareness of these problems has led to the development of many
methods to minimize exposure of humans to known hazardous chemicals (*4*),
and to the increasingly comprehensive testing of many others for potential
carcinogenic activity. Nevertheless, for many chemicals little is known of
adverse long-term effects; animal carcinogenicity tests are costly and very
laborious, and because only relatively few animals can be used, of limited
sensitivity. Over the last 10 years, a wide range of shorter term tests have
been developed as screens for carcinogenic and mutagenic activity, some of
which are now required as a part of the normal testing procedures for chemicals by many regulatory authorities. A number of these tests, using both *in
vivo* and *in vitro* systems, have been reviewed and discussed in great detail
by Hollstein and McCann (*56*).

2.1.1. Cell Culture Products

In the case of products obtained from animal cell cultures, progress has
not mirrored the intense activity of the chemical industry. Few products are

subjected to the type of test required of chemicals, in the main because evidence for the potential carcinogenic or mutagenic activity of the vast majority of biological products is lacking, but partly because the products do not lend themsleves to the commonly applied tests for chemicals—a few fungal toxins are reported exceptions, proving positive in some tests, and these will be described below. Although the oncogenic potential of a wide spectrum of DNA and RNA viruses has been well documented in experimental and natural hosts (16, 20, 35, 76, 96, 104, 108, 112), the realization that viruses may be involved in human carcinogenesis was slow to materialize. Intensive investigations now suggest that three groups in particular may be closely associated with human neoplasms—retroviruses, herpesviruses and hepatitis virus B (33, 38, 73, 81, 92, 97). Since the two last named are viruses for which vaccines are highly desirable, their association with oncogenicity has made it important in the eyes of current regulatory authorities to ensure that any potential vaccines for human use do not contain those viral elements responsible for inducing the oncogenic state in infected cells.

2.1.2. Production Cells

Although the products of animal cell cultures have seldom been screened, the cells used in the production processes have attracted far more attention. Although some early adenovirus vaccines were produced in human tumour cells (53) with no discernable unwanted sequelae in over 20 years, it was decided early in the history of tissue culture cell vaccine production to use only cells derived from normal tissues and that cells from a neoplastic source would be unacceptable (53). Only non-human primary cells or (later) human diploid cells became acceptable substrates. The rationale was that the cancer cell might carry a derepressed oncogene or an oncogenic agent associated with the nucleic acid of the neoplastic cell; in either instance, the gene(s) could contaminate the vaccine virus preparation or even become integrated into the vaccine virus nucleic acid. By the same token, use of normal cells for vaccine virus production had to be monitored carefully to ensure that the cells did not become transformed at any stage in the production process. Transformation, be it induced spontaneously, by viruses or by any other agent, is a process associated with a number of changes in the properties of animal cells, one of which may be the acquisition of tumorigenicity (69). Hence, again there is the possibility that the gene(s) responsible for the transforming event(s) could contaminate or become incorporated into the vaccine product.

It must be emphasized, however, that if normality is to be related to the condition of cells *in vivo*, then any cell growing *in vitro* is in an abnormal situation, with none of the usual homeostatic controls upon it. Nevertheless, such cells do exhibit certain characteristics, and a number of parameters of acceptability (normality) have been applied to them wherever possible.

2.2. Testing of Production Cells

2.2.1. Karyology

Primary Cells. Cells from the normal animal rarely show deviations from the diploid karyotype (57) and this remains true of primary and low passage cell cultures (51, 111). With increasing age, the frequency of abnormalities does increase (120) and transformed cells invariably have altered karyotypes (69). The screening of karyotypes of primary and low passage cell cultures—despite the theoretical advantage of indicating the onset of an abnormal state—is not routinely required or recommended. Historically, the seriousness of the diseases far outweighed any theoretical risks from the substrate and proper karyological screening was not carried out. Furthermore, since primary cell cultures for vaccine virus production should not involve the use of cells of high passage doubling level, prospective karyological screening would be impractical. Primary cultures of chicken, duck, monkey and rabbit origins have been used to produce millions of doses of live and inactivated virus vaccines without any regard to the karyological status of those cells. At the same time, there has not been the slightest evidence to suggest that any untoward effects have resulted from the use of the cells, despite the fact that they may not be strictly diploid and that abnormalities may occur (120) during their production life.

Human Diploid Cells. In the case of human cell strains, Hayflick and Moorhead (51) showed that the cells maintained a normal diploid karyology until senescence at about 40 ± 10 generations. During the phase of decline, aneuploidy and increasing chromosomal abnormalities occur (100). Since such cell strains (WI-38, MRC-5) offered great advantages in terms of standardization and screening potential, their use for vaccine manufacture became accepted on a wide scale. Nevertheless, the theoretical possibilities of oncogenes remained, and karyological monitoring of every lot of cells used in vaccine production was a requisite of the standards of acceptability (80).

The accumulation of vast amounts of cytogenetic data on these two cell strains, all of which showed that karyological changes were minimal throughout the useful vaccine production life of the cells (30 passage doublings), together with the fact that normal human cell strains do not appear to spontaneously transform, nor transform with chemical carcinogens [but see Milo and diPaolo (79)], did encourage a reassessment of the need to monitor every cell lot extensively. Current guidelines reflect this thinking, and whilst the recommendations still require a wide-ranging karyological analysis of a passaged human diploid cell population before acceptance as a substrate for vaccine production, and of the secondary cell seed stocks laid down for production, the monitoring of cells used for each production lot is considerably reduced to a confirmation of identity of the cell strain (58). Given that the human diploid cell system can be well characterized prior to use in vaccine production, the question of cell identity from lot to lot becomes the major requirement for consistency.

Value of Karyological Screening.   Two general points should be made about the usefulness of karyological screening. Firstly, it has been suggested that chromosomal abnormalities may also give an indication of the effects of exposure to irradiation, chemicals, viruses and mycoplasmas. That such agents can induce changes, often leading to transformation and tumorigenicity, is well documented (17, 46, 69, 99), but whether karyological analyses are a meaningful way of detecting such exposure is doubtful. There is no evidence that consistent alterations occur, and minor karyological changes or gene derepression would not be detected by normal methods. In any event, exposure to all four agents would be guarded against during the normal in-line and end-product quality control test procedures available and required by licensing authorities.

Secondly, although karyology may give an indication of normality, it must be borne in mind that changes that do occur need not necessarily imply transformation and acquisition of tumorigenicity—as will be discussed in greater detail in a subsequent section, the two features are separate—and conversely, since only gross alterations are detectable, the possibility of derepression of an oncogene may occur without any obvious sign. Despite the wealth of data available—particularly for human diploid cell strains—karyological analyses cannot necessarily give a true indication of suitability or otherwise of the cells under scrutiny.

2.2.2. Growth in Agar Suspension

Normal cells will not grow and form colonies when plated in media containing 0.3–0.5% agar (70). A wide range of cell types transformed by a wide range of DNA and RNA oncogenic viruses acquire the ability to grow progressively in this system, as also do spontaneously transformed cells, or cells originally derived from neoplastic tissue (69). In the case of virally transformed cells at least, this characteristic is acquired very shortly after infection. The test can be used as an additional method for monitoring the state of cells destined for use as a substrate for vaccine production, although certain provisos should be borne in mind:

(1) Some cell isolates and types may not exhibit this characteristic even after transformation has occurred.

(2) The test is, in itself, a selective one. Although it can be shown that it is not due to selective toxicity of the agar, since normal cells attached to glass microfibres in agar will form colonies (70), the plating efficiencies of virally transformed cloned cell populations in agar suspension may not be particularly high. This indicates that a proportion of the transformed cell population may not exhibit this characteristic.

(3) Growth in agar, whilst indicative of transformation, may not be indicative of acquisition of tumorigenicity by the cells. Colonies of virally transformed hamster or mouse cells picked from agar may not produce tumours in syngeneic hosts immediately, or at all, depending on cultural conditions (123; I. A. Macpherson, unpublished results).

2.2.3. Tumorigenicity

Animal Test Systems. The best test for the absence of tumorigenic potential of vaccine production cells is to demonstrate their inability to produce tumours when inoculated in large numbers into suitable hosts. Ideally, the hosts should be isogeneic, though this is only possible when the cells have been derived from inbred animals, or are returned to the autologous host. In such cases, it can be shown quite clearly that the tumorigenic potential of normal cells and their transformed derivatives are very different, even though in some cases the tumorigenic potential of the latter was not apparent immediately (12, 13, 31, 109, 119, 125). Obviously, for most of the cell sources used for human vaccine production the criterion cannot yet be fulfilled, but the finding that heterografts of neoplastic cells would produce tumours in the hamster cheek pouch provided a means for testing such cells (47). Because of the relative absence of access to the area by lymphoid cells, the cheek pouch protects the heterograft from the animals' graft rejection mechanisms (14). At least 10^6 normal cells inoculated intradermally into that area should not produce tumours; tumorigenic cells from various sources will do so. Similar results are found by inoculating suckling guinea pigs subcutaneously.

Alternative methods of testing cells in heterologous hosts are possible, by treating animals with whole body irradiation (62), treatment with anti-lymphocyte serum (88) or the use of thymectomized animals (114). These tests may have an advantage over the guinea pig in that the period of postnatal susceptibility is lengthened and the latency period considerably reduced—suckling hamsters may have to be observed for several months (49). During that time spontaneous tumours are likely to appear in the hamsters, thus obscuring the test results. By using animals treated as described, the holding period post-inoculation need be no more than 4–6 weeks. Positive controls should always be included.

Vaccine Cell Testing Requirements. Human diploid cells are tested for tumorigenicity as a part of assessing acceptability as a substrate for vaccine production. Subcutaneous inoculation of immunologically deficient mice with 5×10^7 cells is recommended, using HeLa cell-inoculated animals as positive controls and observing all animals for 3 weeks (58). Production cells need not be so tested. For most regulatory authorities, tumorigenicity tests on primary cells used for vaccine production are not required, and when they are, often have involved the use of immunologically competent heterologous animals; hence negative results are meaningless. The double standard for primary and human diploid cells originates partly in the fact that prospective testing of the former is often impractical, and also in the knowledge that they have been used for several decades with no untoward effects, even though it was subsequently discovered that early batches of vaccine cells were contaminated with oncogenic or other adventitious viruses, and that any primary cell may contain retrovirus genomes. The former problem has to a large extent been overcome by adequate screening measures and use of closed colonies—even monkey cells may soon be produced in such a way; the latter remains more refractory to elimination and is equally applicable to diploid cell strains.

Despite the right and proper concern of manufacturers and regulatory authorities about administering vaccines containing known or unknown, extrinsic or intrinsic oncogenic viruses derived from contaminated cells, it is the finding of numerous investigators that only cells that are derived from neoplastic tissue, or are previously transformed by an oncogenic agent, will produce tumours in the relevant tests. The likelihood of normal primary, or human diploid cells being positive in a tumorigenicity test is remote, and there is no documented evidence that harm has come to any of the millions of vaccinees receiving material derived from either source in almost 30 years that could be due to any contaminant agent in the cells concerned. Whilst reactivation of an endogenous retrovirus might occur, the event is extremely infrequent, would almost never be picked up by any current tumour (or karyotype) test, and the clinical significance would be almost impossible to establish (85). Nevertheless, prospective testing of human diploid cells is justifiable during characterization for vaccine production and takes little time or animals. Subsequent testing of production cells on a lot-to-lot basis, or of primary cell vaccine cultures, cannot be done prospectively and would represent significant expenditures in terms of money, animals and effort for no useful purpose.

2.3. Testing of End Product

2.3.1. Ames Test

Biological products from animal cell cultures are seldom tested for potential mutagenicity or carcinogenicity. A few—mitomycin C and aflatoxin B_1 (both potent carcinogens)—have been subjected to the Ames test, which is a versatile and sensitive method that detects almost all organic chemicals known to be human carcinogens (56). The widespread and systematic use of the test, and a comparison with the same compounds tested *in vivo*, shows a near-absolute relationship between the Ames test for mutagenesis and tests for carcinogenic potential *in vivo* (74).

The test, its usefulness and its limitations have been described in numerous publications (5–8). The system makes use of a number of strains of *Salmonella typhimurium* which contain mutations in the genes controlling histidine synthesis and are unable to grow in the absence of that amino acid. For routine screening, different doses of the chemicals under test are mixed with a bacterial test strain (usually four are employed). To this mixture is added rodent or human tissue homogenates that will metabolize chemicals to their active mutagenic forms, if necessary. The full reaction mixture is plated out on medium with a small concentration of histidine, sufficient to allow growth for expression of mutation, but not full colony formation. After 2–3 days, the numbers of revertant colonies are counted. The test can detect nanogram quantities of mutagen in some cases. Spontaneous reversions can occur, but their frequency in control plates is significantly lower than in test plates for a positive sample.

The Ames test is of undoubted value for the testing of chemical compounds (biological products or otherwise). As a prokaryotic system, it is unlikely to be of value in any screening of viruses or viral products that may conceivably be mutagenic or carcinogenic. Even if the products infected the bacterial test strains, there would be no guarantee of correct, or indeed any, expression of viral genes. It is possible that a test similar to the Ames system, but using eukaryotic cells, could be designed. Certainly, there are a large number of short-term tests using eukaryotic organisms and mammalian cells in culture which have been applied to the testing of chemical compounds [for a full review, see Hollstein and McCann (56)] including growth in soft agar suspension. It remains possible that such a test could be adapted to screen for the presence of oncogenes in sub-unit or inactivated preparations of vaccines against viruses of likely oncogenic potential, e.g. herpesviruses, but any such development would presuppose a high efficiency of infection with the virus or sub-unit material—a situation difficult to achieve in the latter instance—and that the cell system chosen was permissive for any potentially active viral genes.

2.3.2. In Vitro *Detection of Contaminant Nucleic Acid*

Unwanted nucleic acid contamination of biological material can arise from the cells from which the material is produced, or in the case of sub-unit viral vaccines, from the virus itself. Since it is current dogma that primary cells and human diploid cells are non-oncogenic, cellular contamination of the product may be of concern only if the material is produced in transformed or frankly neoplastic cells. Whilst use of such substrates is not accepted generally, interferon is produced in human lymphoblastoid cells (*10*) and it may be that transformed heteroploid cell lines may find favour for vaccine production if the level of contamination is quantifiable and an assessment of the likely risk is possible (see Section 2.4 below).

Similar arguments pertain to certain vaccines against viruses that are known to have latent and/or oncogenic properties, e.g. herpes simplex virus, cytomegalovirus. Effective vaccines against both viruses are a matter of increasing urgency, and indeed experimental whole killed vaccines of both types have been used, with no detectable side effects (*22, 34, 37, 61, 90*). Whilst it is difficult to justify the exclusion of such vaccines for the treatment of persons already infected with these viruses, inactivation conceivably may not imply loss of transforming capability, and it remains unlikely that authorities will license anything other than sub-unit viral nucleic acid-free herpesvirus vaccines for general use. It is therefore important to quantify the level of contamination with viral nucleic acid.

Production of any biological material invariably involves a series of purification and (in the case of micro-organisms) inactivation processes. Such sequential processes can be designed to effect an efficient, successive reduction of actual or potential oncogenic moieties; hence the cumulative effects of each preparative stage should give a high degree of probability that the final product will be safe in the field. That this is the case has been clearly shown by such studies as following radiolabelled virus or cells through the purification and inactivation processes, or by the deliberate introduction of contaminant viruses to, and their subsequent removal from, crude starting material (*10, 52*).

The most sensitive technique for detection of nucleic acid contamination of end-product material involves the use of nick-translated ^{32}P-labelled DNA probes of high specific activity (up to 10^8 cpm/μg) in hybridization experiments with test material (*3, 93, 107*). Whilst the greatest levels of sensitivity would be best achieved using gene-specific probes of high purity (corresponding to that part of the viral nucleic acid where the potential oncogene resides, if known), there is no reason to doubt that less specific probes cannot be utilised to screen potential sub-unit vaccines or other biological products. The system has been used to screen for the presence of endogenous viral sequences in cells, detecting as little as one genome equivalent per 40 cells (55) and to detect less than 10 picograms (10^{-11}g) of DNA in sub-unit vaccine preparations (A. Buchan, personal communication).

However, it must be recognized that even such low levels still represent a considerable quantity of contaminant nucleic acid sequences. On the basis that a gene comprises approximately 1000–1500 base pairs, and therefore has a molecular weight of about 10^6, it can be deduced that if the final product contains less than 10^{-14} g/ml nucleic acid, no complete (onco)gene is likely to be present. It is not yet feasible to detect such levels, and this should be borne in mind when assessing the suitability of relevant potential sub-unit vaccines screened by hybridization techniques.

2.4. Use of Continuous Cell Lines in Human Vaccine Manufacture

The use of continuous cell lines in vaccine manufacture was condemned in the early 1960s, since a characteristic of such cultures was an unlimited life *in vitro*. Since this was a transformed characteristic, it became equated with the acquisition of tumorigenicity and hence the possibility of incorporation of the oncogene(s) into the final product. Thus vaccine manufacture involved the use of non-human primary cells or of human diploid cells only. Nevertheless, this condemnation of continuous cells was the result of hypothetical speculation and not proper investigation, and the accumulated evidence to date would not support the validity of that condemnation.

2.4.1. Vaccines Produced in Continuous Cell Lines

A number of human vaccines were produced in cells of distinct neoplastic origin in the early days of tissue culture vaccine manufacture. Adenovirus vaccine (53) and poliovirus vaccines (23) were produced in HeLa cells. There has been no evidence to suggest that the populations receiving such vaccines have suffered any adverse reactions in over 20 years. In the latter instance, the tumorigenic potential of the cell was limited to a non-metastasizing growth at the inoculation site in immunologically impaired animals. Viable cells administered intravenously in large numbers, ruptured cells or the vaccines themselves had no adverse effects under any circumstances (101).

Continuous cell lines have been used extensively in the manufacture of veterinary vaccines, e.g. foot-and-mouth disease (FMD) and Newcastle disease. The continuous cell line used (baby Syrian hamster kidney BHK 21) has been extensively tested in a wide range of mammals and birds with no adverse effects, and millions of doses of vaccine given to animals have caused no discernable unwanted sequelae. It has to be admitted that many of these animals are killed for food purposes, but others have been allowed to exist for their full life-span (10).

2.4.2. Interferon Produced in Continuous Cell Lines

Interferon is produced commercially in human lymphoblastoid cells in suspension culture (10). Of heteroploid karyotype, the cells are derived from a Burkitt lymphoma, and do contain about 50% of the Epstein–Barr virus genome, but they do not liberate any infectious virus. Under normally applied procedures, the cells do not seem to be tumorigenic in immunologically deficient mice, although Nilsson et al. (84) have reported otherwise. The purification processes are such that the final product contains little or no detectable cell proteins or nucleic acid (and hence Epstein–Barr virus DNA), and the interferon appears safe and effective in treated persons.

Detection of residual cellular DNA by hybridization techniques would be the most sensitive method, using labelled whole cell DNA as the test probe. As with the detection of contaminant viral nucleic acid sequences, the greatest sensitivity would be achieved by using gene-specific probes, but it must be recognized that the sites of potential oncogenes or retrovirus genomes on mammalian or avian chromosomes are not yet mapped precisely.

2.4.3. Transformation and Tumorigenicity

When normal cells in vitro undergo transformation, there are changes in some or all of the characteristics associated with normality. One of these is the acquisition of an unlimited life in vitro, and another may be the acquisition of transplantability. However, it is important to realize that transformation does not necessarily equate with tumorigenicity. This has been carefully analysed by Aaronson and Todaro (1, 2) using Balb/c mouse embryo cells passaged under strict transfer schedules. Cells were passaged at 3-day intervals and seeding inocula were 3×10^5 or 1.2×10^6 cells per 50-mm Petri dish. After an initial drop in growth rate (cell doubling time), this rate began to increase again and the cells became an established continuous line. However, by maintaining the cells on the strict transfer schedules, it was possible to see a variation in the growth parameters of the cells maintained at low cell densities (3-day transfer of 3×10^5 cells—3T3) and at high cell densities (3-day transfer of 1.2×10^6—3T12). The former only reached saturation densities of 5×10^4 cells/cm^2, the latter 2×10^5 cells/cm^2. After intervals of 30, 100 and 200 generations in culture, the contact-sensitive 3T3 and contact-insensitive 3T12 were inoculated into newborn and immunologically compromised Balb/c mice. Only cells on a 3T12 transfer schedule proved tumorigenic. Hence acquisition of tumorigenicity is purely a function of the cultural conditions employed and not of the establishment of a continuous cell line. Passage at high cell transfer levels encourages the selection of less contacted-inhibited cells, and there is a good correlation between contact insensitivity and tumorigenicity (89).

The inference of this study is that any newly derived continuous cell line, if maintained under the correct cultural conditions, should not acquire tumorigenic potential. Affirmation of this for cells of other species is still lacking, although there is evidence that the observation holds true for clones of the Syrian hamster line BHK 21/13 (59, 127) and rat embryo cells (C. N. Wiblin, unpublished observations).

2.4.4. Use of Primary Cells with the Capacity for Infinite Life

With the exception of human diploid fibroblasts, normal cells from most species may transform into cell lines. Human vaccines have been, or are still, produced in primary cells of duck, rabbit, dog and monkey origins. All of these are capable of transforming into continuous cell lines, for example RK13 (rabbit) and BSC-1 (monkey). Thus for several decades, cells have been used in human vaccine manufacture with complete safety that have the potential for an infinite life *in vitro*, so it is arguable that the principle of using continuous cell lines has been established (86).

2.4.5. Advantages of Continuous Cell Lines in Vaccine Manufacture

An immediate advantage of continuous cell lines is that, unlike primary cells but like human diploid cell strains, it is possible to screen them thoroughly for adventitious agents, karyology, tumorigenicity and any other characteristics deemed relevant for excluding oncogenic potential. Once satisfactorily screened, a cell seed stock can be stored down, and a cell seed lot system utilized in subsequent production procedures.

It is often difficult to obtain biological products in sufficient quantities or of suitable quality at reasonable cost with existing techniques. Interferon, as described above, may be prepared in the most cost-effective way from a continuous cell line rather than from human fibroblasts or buffy coat cells. Rabies vaccine, currently produced in human diploid cells, is extremely expensive. Yields of rabies virus from BHK 21/13 cells, particularly when grown in deep suspension culture, are very much higher (19). Poliovirus vaccine is produced in primary monkey cells, and quite apart from the moral implications of decimating the world's monkey stocks, the maintenance of the animals—particularly in closed colonies—represents a significant expense. Whilst attenuated poliovirus vaccines, with their low antigenic loading per vaccine dose, continue to be used on a wide scale this may not matter, but there are strong arguments for the expansion of use of the killed polio vaccine, particularly in tropical areas (24, 32). Since this entails the use of a much higher antigenic loading per vaccine dose, an economically viable

method of producing killed vaccine may well depend upon using a transformed line capable of growth in suspension culture, for example HeLa cells,
in which yields of poliovirus antigen are far higher than in human diploid or
primary cells (10).

Mention has been made of deep suspension culture. Many continous cell
lines can be grown in this manner, whereas human diploid cells and many
primary cells are anchorage-dependent. This makes the continuous cell line
the most promising for large-scale production and technological simplicity.
In the case of interferon, vessels of up to 1000 litres are utilized, but even
larger volumes have been used for manufacture of rabies vaccine in BHK
21/13 cells for veterinary use.

Despite the potential advantages of continuous cell lines, and the lack of
any evidence to support the hypothetical properties of transfer of oncogenicity to the host, current thinking is still against the use of such cells for
human vaccine manufacture. Nevertheless, scientific, moral and financial
pressures are gradually forcing a more realistic approach to be taken to the
re-use of certain continuous cell lines for particular biological products. The
only relevant criterion for exclusion would be on the grounds of tumorigenicity. If this is not apparent, then the manufacturer of biological
products is being denied a wide range of excellent substrates for the preparation of human vaccines in the most cost-effective and economic manner.
Even if the cells are tumorigenic, or at least derived from a neoplastic
source, the oncogenicity of the product can still be tested following the
various highly selective purification and/or inactivation procedures. With
the range of sensitive techniques now available for detecting known foreign
nucleic acid sequences in viral or sub-unit material, it is now possible to
obtain an indication of the level of risk that such material might be contaminated with cellular DNA. The production of interferon has now been accepted following such studies; it is to be hoped that other inactivated or sub-unit
virus vaccines produced in continuous cell culture will also be accepted in
the near future, subject to stringent control of the cells and their cultural
regimens. The practical and moral advantages of including such cell lines in
the battery available for human vaccine manufacture makes it desirable that
sound and impartial investigative studies should be undertaken to confirm or
deny that acceptance.

ACKNOWLEDGMENTS

I should like to thank Drs. J. Roff and R. Tattersall for their constructive comments on this
subject, and to acknowledge with gratitude the patience and good humour of Mrs. L. Hart, who
typed the manuscript.

REFERENCES

1. Aaronson, S. A., and Todaro, G. J. (1968). Basis for the acquisition of malignant potential by mouse cells cultivated *in vitro*. *Science* **162**, 1024–1026.
2. Aaronson, S. A., and Todaro, G. J. (1968). Development of 3T3-like lines from Balb/c mouse embryo cultures: Transformation susceptibility to SV40. *J. Cell. Physiol.* **72**, 141–148.
3. Alwine, J. C., Kemp, D. J., Parker, B. A., Reiser, J., Renart, J., Stark, G. R., and Wahl, G. M. (1979). Detection of specific RNA's or specific fragments of DNA by fractionation of gels and transfer to diazobenzyloxymethyl paper. *In* "Methods in Enzymology" (R. Wu, ed.), Vol. 68, pp. 220–242. Academic Press, New York.
4. Ames, B. N. (1979). Identifying environmental chemicals causing mutations and cancer. *Science* **204**, 587–593.
5. Ames, B. N., Durston, W., Yamasaki, E., and Lee, F. (1973). Carcinogens are mutagens; a simple test system combining liver homogenates for activation and bacteria for detection. *Proc. Natl. Acad. Sci. U.S.A.* **70**, 2281–2285.
6. Ames, B. N., Lee, F., and Durston, W. (1973). An improved bacterial test system for the detection and classification of mutagens and carcinogens. *Proc. Natl. Acad. Sci. U.S.A.* **70**(3), 782–786.
7. Ames, B. N., and McCann, J. (1976). *In* "Screening Tests in Chemical Carcinogenesis" (R. Montesano, H. Bartsen, and L. Tomatis, eds.), Vol. 12, pp. 493–504. Int. Agency Res. Cancer, Lyon.
8. Ames, B. N., McCann, J., and Yamasaki, E. (1975). Methods for detecting carcinogens and mutagens with the *Salmonella*/mammalian microsome mutagenicity test. *Mutat. Res.* **31**, 347–364.
9. Atkins, E., and Bodel, P. (1972). Fever. *N. Engl. J. Med.* **286**, 27–34.
10. Beale, A. J. (1979). Choice of cell substrate for biological products. *Adv. Exp. Med. Biol.* **118**, 83–97.
11. Bennett, I. L., and Cluff, L. E. (1957). Bacterial pyrogens. *Pharmacol. Rev.* **9**, 427–475.
12. Berwald, Y., and Sachs, L. (1963). *In vitro* cell transformation with chemical carcinogens. *Nature (London)* **200**, 1182–1184.
13. Berwald, Y., and Sachs, L. (1965). *In vitro* transformation of normal cells to tumour cells by carcinogenic hydrocarbons. *J. Natl. Cancer Inst. (U.S.)* **35**, 641–661.
14. Billingham, R. E., Ferrigan, L. W., and Silvers, W. K. (1960). Cheek pouch of the Syrian hamster and tissue transplantation immunity. *Science* **132**, 1488.
15. Binder, B. (1977). Joint discussion—Pyrogenicity test systems. *Dev. Biol. Stand.* **34**, 70–71.
16. Bittner, J. J. (1936). Some possible effects of nursing on the mammary gland tumour incidence in mice. *Science* **84**, 162.
17. Borenfreund, E., Krim, M., Sanders, F. K., Sternberg, S. S., and Bendich, A. (1966). Malignant conversion of cells *in vitro* by carcinogens and viruses. *Proc. Natl. Acad. Sci. U.S.A.* **56**, 672–679.
18. British Pharmacopoeia (1968). "Test for Pyrogens," pp. 1348–1349. Pharmaceutical Press, London.
19. Capstick, P. B., Garland, A. J., Chapman, W. G., and Masters, R. C. (1965). Production of foot and mouth disease virus antigen taken from BHK 21 clone 13 cells grown and infected in deep suspension cultures. *Nature (London)* **205**, 1135–1136.
20. Casto, B. C. (1973). Biologic parameters of adenovirus transformation. *Prog. Exp. Tumor. Res.* **18**, 166–198.
21. Castor, G. B., Kantor, N., Knoll, E., Blakely, J., Neilson, J. K., Randolph, J. O., and

Kirschbaum, A. (1971). Characteristics of a highly purified pyrogenic lipopolysaccharide. *J. Pharm. Sci.* **60**, 1578–1580.

22. Chapin, H. B., Wong, S. C., and Reapsome, J. (1962). The value of tissue culture vaccine in the prophylaxis of recurrent attacks of herpetic keratitis. *Am. J. Ophthalmol.* **54**, 255–265.

23. Clausen, J. J., and Syverton, J. T. (1962). Comparative chromosomal study of 31 cultured mammalian cell lines. *J. Natl. Cancer Inst. (U.S.)* **28**, 117–146.

24. Cockburn, W. C., and Drozdov, S. G. (1970). Poliomyelitis in the world. *Bull. W.H.O.* **42**, 405–417.

25. Cooper, J. F., Hochstein, H. D., and Seligmann, E. D. (1972). The *Limulus* test for endotoxin (pyrogen) in radiopharmaceuticals and biologicals. *Bull. Parenter. Drug Assoc.* **26**, 153–162.

26. Cooper, J. F., Levin, J., and Wagner, H. N. (1970). Quantitative comparison of *in vitro* and *in vivo* methods for detection of endotoxin. *J. Lab. Clin. Med.* **78**, 138–148.

27. Cooper, J. F., and Neely, M. E. (1980). Validation of the LAL test for endproduct evaluation. *Pharm. Technol.* **4**, 72–79.

28. Cooper, J. F., and Pearson, S. M. (1977). Detection of endotoxin in biological products by the *Limulus* test. *Dev. Biol. Stand.* **34**, 7–13.

29. Daoust, D. R., Orlowski, S. J., McMahon, G., Weber, C. J., Shaw, W. C., Fisher, T. C., Bennett, C. R., and Gray, A. (1976). *Limulus* amoebocyte lysate test as a method for detection of endotoxins and endotoxin-like materials. *Bull. Parenter. Drug Assoc.* **30**, 13–20.

30. Dare, J. G., and Mogey, G. A. (1954). Rabbit responses to human threshold doses of a bacterial pyrogen. *J. Pharm. Pharmacol.* **6**, 325–332.

31. diPaolo, J. A., and Donovan, P. J. (1967). Properties of Syrian hamster cells transformed in the presence of carcinogenic hydrocarbons. *Exp. Cell Res.* **48**, 361–377.

32. Dömok, I., Balayan, M. S., Fayinka, O. A., Skrtic, N., Soneji, A. D., and Harland, P. S. (1974). Factors affecting the efficacy of live poliovirus vaccine in warm climates. *Bull. W.H.O.* **51**, 333–347.

33. Duff, R., and Rapp, F. (1973). Induction of oncogenic potential by herpes simplex. *Perpsect Virol.* **8**, 189–210.

34. Dundarov, S., Andonov, P., Bakalov, B., Nechev, K., and Tomov, C. (1982). Immunotherapy with inactivated polyvalent herpes vaccines. *Dev. Biol. Stand.* **52**, 351–358.

35. Eddy, B. E. (1962). Tumours produced in hamsters by SV40. *Fed. Proc., Fed. Am. Soc. Exp. Biol.* **21**, 930–935.

36. Eibert, J. (1972). Pyrogen testing: Horseshoe crabs versus rabbits. *Bull. Parenter. Drug Assoc.* **26**, 253–260.

37. Elek, S. D., and Stern, H. (1974). Development of a vaccine against mental retardation caused by cytomegalovirus infection *in vitro*. *Lancet* **1**, 1–5.

38. Epstein, M. A., Achong, B. G., and Barr, Y. M. (1964). Virus particles in cultured lymphoblasts from Burkitt's lymphoma. *Lancet* **1**, 702–703.

39. European Pharmacopoeia (1971). "Test for Pyrogens," Vol. II, pp. 58–60. Maisonneuve S.A., France.

40. Favero, M. S., Carson, L. A., Bond, W. W., and Petersen, N. J. (1971). *Pseudomonas aeruginosa*; growth in distilled water from hospitals. *Science* **173**, 836–838.

41. Federal Register (1973). Status of biological substances used for detecting bacterial endotoxins. *Fed. Regist.* **38**, FR 1404.

42. Federal Register (1973). Limulus amoebocyte lysate. Additional standards. *Fed. Regist.* **180**, FR 26130–26132.

43. Federal Register (1977). Availability of a proposed guideline to establish the sensitivity of each lot of *Limulus* amoebocyte lysate. *Fed. Regist.* **42**, FR 23167.

44. Federal Register (1977). Licensing of *Limulus* amoebocyte lysate: Use as an alternative for the rabbit pyrogen system. *Fed. Regist.* **42**, FR 57749–57750.

45. Federal Register (1980). Additional standards for *Limulus* amoebocyte lysate. *Fed. Regist.* **45**, FR 32296–32300.

46. Fogh, J., and Fogh, H. (1973). Chromosome changes in cell culture induced by mycoplasma infection. *Ann. N.Y. Acad. Sci.* **225**, 311–329.

47. Foley, G. E., Handler, A. H., Adams, R. A., and Craig, J. M. (1962). Assessment of potential malignancy of cultured cells; further observations on the differentiation of "normal" and "neoplastic" cells maintained *in vivo* by heterotransplantation in Syrian hamsters. *Natl. Cancer Inst. Monogr.* **7**, 173–204.

48. Gaffin, S. L. (1976). The clotting of the lysed white cells of *Limulus* induced by toxin. Preparation and characterization of clot-forming proteins. *Biorheology* **13**, 273–280.

49. Glathe, H., and Abel, C. (1977). Evidence of tumorigenic activity of candidate cell substrates in vaccine production by the use of antilymphocyte serum. *Dev. Biol. Stand.* **34**, 145–148.

50. Greisman, S. E., and Hornick, R. B. (1969). Comparative pyrogenic reactivity of rabbit and man to bacterial endotoxin. *Proc. Soc. Exp. Biol. Med.* **131**, 1154–1158.

51. Hayflick, L., and Moorhead, P. S. (1961). The serial cultivation of human diploid cell strains. *Exp. Cell Res.* **25**, 585–621.

52. Hilfenhaus, J., Moser, H., Hermann, A., and Mauler, R. (1982). Herpes simplex virus subunit vaccine: Characterization of the virus strain used, and testing of the vaccine. *Dev. Biol. Stand.* **52**, 321–331.

53. Hilleman, M. R. (1968). Cells, vaccines, and the pursuit of precedent. *Natl. Cancer Inst. Monogr.* **29**, 463–470.

54. Hochstein, H. D. (1981). The LAL test versus the rabbit pyrogen test for endotoxin detection. *Pharm. Technol.* **5**, 37–42.

55. Holland, J. J., Villareal, L. P., Welsh, R. M., Oldstone, M. B. A., Koline, D., Lazzarini, R., and Scholnick, E. (1976). Long-term persistent vesicular stomatitis virus and rabies infection in cells *in vitro*. *J. Gen. Virol.* **33**, 193–211.

56. Hollstein, M., and McCann, J. (1979). Short term tests for carcinogens and mutagens. *Mutat. Res.* **65**, 133–226.

57. Hsu, T. C. (1961). Chromosomal evolution in cell populations. *Int. Rev. Cytol.* **12**, 69–161.

58. Jacobs, J. P., Magrath, D. I., Garrett, A. J., and Schild, G. C. (1981). Guidelines for the acceptability, management and testing of serially propagated human diploid cells for the production of live virus vaccines for use in man. *J. Biol. Stand.* **9**, 331–342.

59. Jarrett, O., and Macpherson, I. A. (1968). The basis of the tumorigenicity of BHK 21 cells. *Int. J. Cancer* **3**, 654–662.

60. Jorgenson, J. H., and Smith, R. F. (1974). Measurement of bound and free endotoxin by the *Limulus* assay. *Proc. Soc. Exp. Biol. Med.* **146**, 1024–1031.

61. Kern, A. B., and Schiff, B. L. (1964). Vaccine therapy in recurrent herpes simplex. *Arch. Dermatol.* **89**, 844–845.

62. Klein, G., and Klein, E. (1962). Antigenic properties of other experimental tumours. *Cold Spring Harbor Symp. Quant. Biol.* **27**, 463–470.

63. Kreeftenberg, J. G., Loggen, H. G., van Ramshorst, J. D., and Beuvery, E. C. (1977). The *Limulus* amoebocyte lysate test micromethod and application in the control of sera and vaccines. *Dev. Biol. Stand.* **34**, 15–20.

64. Kruger, D., Hebold, G., and Zimmerman, G. (1975). Ein Beitrag zur *in vitro* Pyrogenprufung mittels des Limulus Tests. *Arzneim.-Forsch.* **25**, 160–162.
65. Levin, J. (1982). The Limulus test and bacterial endotoxins: Some perspectives. *In* "Endotoxins and their Detection with the *Limulus* Amoebocyte Lysate Test" (S. W. Watson, J. Levin, and T. Novitsky, eds.), pp. 7–24. Alan, R. Liss, Inc., New York.
66. Levin, J., and Bang, F. B. (1964). The role of endotoxin in the extracellular coagulation of *Limulus* blood. *Bull. John Hopkins Hosp.* **115**, 265–274.
67. Levin, J., and Bang, F. B. (1968). Clottable protein in *Limulus*; its localization and kinetics of its coagulation by endotoxin. *Thromb. Diath. Haemorrh.* **19**, 186–197.
68. Loeb, L. (1902). On the blood lymph cells and inflammatory processes of *Limulus*. *J. Med. Res.* **7**, 145–158.
69. Macpherson, I. A. (1970). The characteristics of animal cells transformed *in vitro*. *Adv. Cancer Res.* **13**, 169–215.
70. Macpherson, I. A., and Montagnier, L. (1964). Agar suspension culture for the selective assay of cells transformed by polyoma virus. *Virology* **23**, 291–294.
71. Marcus, S., and Nelson, J. R. (1977). Detection of endotoxin in biological products by the *Limulus* test. *Dev. Biol. Stand.* **34**, 7–13.
72. Mascoli, C. C., and Weary, M. E. (1979). *Limulus* amoebocyte lysate (LAL) test for detecting pyrogens in parental injectable products and medical devices: Advantages to manufacturers and regulatory authorities. *J. Parenter. Drug. Assoc.* **33**, 81–95.
73. Maupas, P., and Melnick, J. L. (1981). Hepatitis B infection and primary liver cancer. *Prog. Med. Virol.* **27**, 1–5.
74. McCann, P., and Ames, B. N. (1977). *In* "Origins of Human Cancer" (H. Hiatt, J. D. Watson, and J. A. Wisten, eds.), pp. 1431–1450. Cold Spring Harbor Lab., Cold Spring Harbor, New York.
75. Medicines (1977). "Standard Provisions for Licences and Certificates," Amendment Regul. No. 675, p. 9. H. M. Stationery Office, London.
76. Melendez, L. V., Hunt, R. D., King, N. W., Daniel, M. D., Garcia, F. G., and Frazer, C. A. O. (1969). *Herpesvirus saimiri*. II. Experimentally induced malignant lymphoma in primates. *Lab. Anim. Care* **19**, 378–386.
77. Mills, D. F. (1978). "Pyrogent (Limulus Amoebocyte Lysate) for Detection of Endotoxins." Mallinckrodt Inc., Missouri.
78. Milner, K. C., Rudbach, J. A., and Ribi, E. (1971). General characteristics. *In* "Microbial Toxins" (G. Weinbaum, S. Kadis, and S. J. Ajl, eds.), Vol. **4**, pp. 1–65. Academic Press, London.
79. Milo, G. E., and diPaolo, J. A. (1978). Neoplastic transformation of human diploid cells *in vitro* after chemical carcinogen treatment. *Nature (London)* **275**, 130–132.
80. Moorhead, P. S., Nicholls, W. W., Perkins, F. T., and Hayflick, L. (1974). Standards of karyology for human diploid cells. *J. Biol. Stand.* **2**, 95–101.
81. Naib, Z. M., Nahmias, A. J., and Josey, W. E. (1966). Cytology and histopathology of cervical herpes simplex infection. *Cancer* **19**, 1026–1031.
82. Nakamura, S., Takagi, T., Iwanaga, S., Niwa, M., and Takahashi, K. (1976). Amino acid sequence studies on the fragments produced from horseshoe crab coagulogen during gel formation; homologies with primate fibrinopeptide B. *Biochem. Biophys. Res. Commun.* **72**(3), 902–908.
83. Nakamura, S., Takagi, T., Iwanaga, S., Niwa, M., and Takahashi, K. (1976b). A clottable protein (coagulogen) of horseshoe crab haematocytes. Structural change of its polypeptide chain during gel formation. *J. Biochem. (Tokyo)* **80**, 649–652.
84. Nillson, K., Giovanella, B. C., Stehlin, T. S., and Klein, J. (1977). Tumorigenicity of human haematopoietic cell lines in athymic nude mice. *Int. J. Cancer* **19**, 337–344.

85. Parks, W. P., and Hubbell, E. S. (1979). Quantitation in the evaluation of cell substrates for viral vaccine production. *Adv. Exp. Med. Biol.* **118**, 23–33.
86. Petricciani, J. C. (1979). Cell substrates for biologics production: Factors affecting acceptability. *Adv. Exp. Med. Biol.* **118**, 9–21.
87. *Pharmacopoeial Forum* (1983). Characterization of rabbit colonies for the pyrogen test. Sept.–Oct., pp. 3559–3561.
88. Phillips, B., and Gazet, J.-C. (1967). Growth of two human tumour cell lines in mice treated with antilymphocyte serum. *Nature (London)* **215**, 548–549.
89. Pollack, R. E., Green, H., and Todaro, G. J. (1968). Growth control in cultured cells: Selection of sublines with increased sensitivity to contact inhibition and decreased tumour-producing ability. *Proc. Natl. Acad. Sci. U.S.A.* **60**, 126–133.
90. Plotkin, S. A., Farquhar, J., and Hornberger, E. (1976). Clinical trials of immunization with the Towne 125 strain of human cytomegalovirus. *J. Infect. Dis.* **134**, 470–475.
91. Rastogi, S. C., Hochstein, H. D., and Seligmann, E. G. (1977). Statistical determination of endotoxin content in influenza virus vaccine by the *Limulus* amoebocyte lysate test. *J. Clin. Microbiol.* **6**, 144–148.
92. Reitz, M. S., Kalyanaraman, V. S., Robert-Guroff, M., Popovic, M., Sarngedharan, M. G., Sarin, P. S., and Gallo, R. C. (1983). Human T-cell leukemia/lymphoma virus: The retrovirus of adult T-cell leukemia/lymphoma. *J. Infect. Dis.* **147**(3), 399–405.
93. Rigby, P. W., Diechmann, M., Rhodes, C., and Berg, P. (1977). Labelling deoxyribonucleic acid to high specific activity *in vitro* by nick-translation with DNA polymerase I. *J. Mol. Biol.* **113**, 237–251.
94. Ronneberger, H. J. (1977). Comparison of the pyrogen tests in rabbits and with *Limulus* lysate. *Dev. Biol. Stand.* **34**, 27–32.
95. Ronneberger, H. J., and Stark, J. (1974). Der Limulustest in Vergleich mit dem Pyrogentest am Kaninchen. *Arzeim. Forsch.* **24**, 933–934.
96. Rous, P. (1911). A sarcoma of the fowl transmissible by an agent separable from the tumour cells. *J. Exp. Med.* **13**, 397–411.
97. Royston, I., and Aurelian, L. (1970). Immunofluorescent detection of herpesvirus antigens in exfoliated cells from human cervical carcinoma. *Proc. Natl. Acad. Sci. U.S.A.* **67**(1), 204–212.
98. Rudbach, J. A., Akiya, F. I., Elin, R. J., Hochstein, H. D., Louma, M. K., Milner, E. C. B., Milner, K. B., and Thomas, K. R. (1976). Preparation and properties of a national reference endotoxin. *J. Clin. Microbiol.* **3**, 21–25.
99. Ruddle, F. H. (1961). Chromosome variation in cell populations derived from pig kidney. *Cancer Res.* **21**, 885–894.
100. Saksela, E., and Moorhead, P. S. (1963). Aneuploidy in the degeneration phase of serial cultivation of human cell strains. *Proc. Natl. Acad. Sci. U.S.A.* **50**, 390–395.
101. Salk, J. (1979). The spectre of malignancy and criteria for cell lines as substrates for vaccines. *Adv. Exp. Med. Biol.* **118**, 107–113.
102. Seibert, F. B. (1923). Fever-producing substance found in some distilled waters. *Am. J. Physiol.* **67**, 90–104.
103. Seibert, F. B. (1963). Pyrogens from an historical viewpoint. *Transfusion* **3**, 245–249.
104. Shope, R. E. (1933). Infectious papillomatosis of rabbits. *J. Exp. Med.* **58**, 607–624.
105. Solum, N. O. (1970). Some characteristics of the clottable protein of *Limulus polyphemus* blood cells. *Thromb. Diath. Haemorrh.* **23**, 170–181.
106. Solum, N. O. (1973). The coagulogen of *Limulus polyphemus* haemocytes. A comparison of the clotted and non-clotted forms of the molecule. *Thromb. Res.* **2**, 55–70.
107. Southern, E. M. (1975). Detection of specific sequences among DNA fragments separated by gel electrophoresis. *J. Mol. Biol.* **98**, 503–517.

108. Stewart, S. E., Eddy, B. E., and Borgese, N. (1958). Neoplasms in mice inoculated with a tumour agent carried in tissue culture. *J. Natl. Cancer Inst. (U.S.)* **20**, 1223–1243.

109. Stoker, M. P. G., and Macpherson, I. A. (1961). Studies on transformation of hamster cells by polyoma virus *in vitro*. *Virology* **14**, 359–370.

110. Sullivan, J. D., and Watson, S. W. (1975). Purification and properties of the clotting enzyme from *Limulus* lysates. *Biochem. Biophys. Res. Commun.* **66**(2), 848–855.

111. Tjio, J. H., and Puck, T. T. (1958). Genetics of somatic mammalian cells. II. Chromosomal constitution of cells in tissue culture. *J. Exp. Med.* **108**, 259–268.

112. Todaro, G. J., Sherr, C. J., Sen, A., King, N., Daniel, M. D., and Fleckenstein, B. (1978). Endogenous new world primate type C viruses isolated from an owl monkey (*Aotus trivirgatus*). *Proc. Natl. Acad. Sci. U.S.A.* **75**(2), 1004–1008.

113. *U.S. Pharmacopoeia* (1980). 20th revision, pp. 888–889 and 902–903. U.S. Pharmacopoeial Convention Inc., Maryland. (Subsequent revisions to the official monographs for Water for Injection and certain radiopharmaceuticals appear in the 4th Supplement Addendum A, USP-NF, 1983.)

114. Vandeputte, M., Denys, P., Leyten, R., and de Somer, P. (1963). The oncogenic activity of the polyoma virus in thymectomized rats. *Life Sci.* **7**, 475–478.

115. van Dijck, P., and van der Voorde, H. (1977). Factors affecting pyrogen testing in rabbits. *Dev. Biol. Stand.* **34**, 57–63.

116. van Noordwijk, J., and de Jong, Y. (1976). Comparison of the *Limulus* test for endotoxin with the rabbit test for pyrogens of the European Pharmacopoeia. *J. Biol. Stand.* **4**, 131–139.

117. van Noordwijk, J., and de Jong, Y. (1977). Comparison of the *Limulus* amoebocyte lysate (LAL) test with the rabbit test; false positives and false negatives. *Dev. Biol. Stand.* **34**, 39–43.

118. van Ramshorst, J. D. (1977). Joint discussion—Pyrogenicity test systems. *Dev. Biol. Stand.* **34**, 70.

119. Vogt, M., and Dulbecco, R. (1963). Steps in the neoplastic transformation of hamster embryo cells by polyoma virus. *Proc. Natl. Acad. Sci. U.S.A.* **49**, 171–179.

120. Wallace, R. E., Vasington, P. J., Petricciani, J. C., Hopps, H. E., Lorenz, D. E., and Kadanka, Z. (1973). Development and characterization of cell lines from sub-human primates. *In Vitro* **8**, 333–341.

121. Weary, M., Pearson, F. C., Bohon, J., and Donohue, G. (1982). The activity of various endotoxins in the USP rabbit test in three different LAL tests. *In* "Endotoxins and Their Detection with the Limulus Amoebocyte Lysate Test" (S. W. Watson, J. Levin, and T. Novitsky, eds.), pp. 365–379. Alan R. Liss, Inc., New York.

122. Whittet, T. D. (1958). Ph.D. Thesis, University of London (quoted by Marcus and Nelson, 1977).

123. Wiblin, C. N. (1971). Ph.D. Thesis, University of London.

124. Wildfeuer, A., Heymer, B., Schleifer, K. H., and Haferkamp, O. (1974). Investigations on the specificity of the *Limulus* test for the detection of endotoxin. *Appl. Microbiol.* **28**, 867–871.

125. Williams, J. F., and Till, J. E. (1965). Formation of tumour colonies in the lungs of rats injected with rat embryo cells immediately after infection of the cells with polyoma virus. *Virology* **27**, 625–630.

126. Wolff, S. M. (1973). Biological effects of bacterial endotoxins in man. *J. Infect. Dis.* **128**, Suppl., S259–S264.

127. Wyke, J. A. (1970). Ph.D. Thesis, University of London.

128. Yin, E. T. (1975). Endotoxin, thrombin and the *Limulus* amoebocyte lysate test. *J. Lab. Clin. Med.* **86**(3), 430–434.

14

Field Tests

T. M. POLLOCK

Epidemiology Research Laboratory
Central Public Health Laboratory
Public Health Laboratory Service
London, United Kingdom

1. INTRODUCTION

The efficacy and safety of substances intended to treat or prevent human disease must ultimately be tested in man. The efficacy of treatment can be assessed with relative ease because patients can usually be readily identified, randomly allocated to one or more treatment groups and closely observed after treatment. Field tests of preventive methods, however, present two main difficulties. The assessment is made in healthy people whose response, unless they are selected and isolated in special units (1), is likely to be

Animal Cell Biotechnology, Vol. 2

influenced by many variables, environmental as well as individual, and the investigator may need to keep large groups of individuals under observation for long periods. Secondly, the ethics of trials in which healthy persons are given agents of unproven efficacy and potential hazard raise problems which set limits to the permitted boundaries of investigation. Experience of field tests has been largely gained by studies of immunising agents and the examples given in this section are taken from such studies, but the principles described apply also to field tests of other prophylactic agents,

The development of field trials has been slow. Typhoid vaccine, first produced before the turn of the century, was not adequately tested until more than 50 years had elapsed (2). Immunisation against tuberculosis, already available in 1922, remained unused in the United Kingdom at a time of great need for lack of an adequate field trial, and the efficacy of BCG vaccine was not confirmed intil 1953 (3). To the two problems of logistics and ethics was formerly added a third—ignorance of the statistical methods required. A major advance in the evolution of field trials was the publication in 1937 of Bradford Hill's "Principles of Medical Statistics" (4), which has been a formative influence on the large number of effective field trials made since the Second World War. The development of field trials has also been greatly assisted by advances in microbiology. Susceptible participants can now be identified at the outset and infections arising during the follow-up diagnosed with greater precision.

Three obvious errors—which are nevertheless still sometimes made— exemplify the need to base assessments of efficacy on valid comparisons of disease incidence in immunised and unimmunised groups. The first is to assume that the agent is necessarily ineffective because the disease is frequently reported in immunised patients. In fact, efficacy cannot be validly estimated unless it is known what proportion of the community is immunised, i.e. unless the assessment is based on rates, not on numbers. Consider as an example that 90% of 100,000 children are given a vaccine so effective that their attack rate is only one-fifth of that in the unimmunised group. If the attack rate in the unimmunised were 500 per 100,000, the number of cases expected in each group would be as shown in Table I. Clearly the figures alone give no indication of the efficacy of the vaccine.

The second error is to draw conclusions from a selected population without taking into account the mode of selection. This applies in a variety of situations. To take an obvious example:

In the case of children admitted to hospital with whooping cough the attack may be equally severe in the immunised and the unimmunised. From this it would be right to conclude that immunisation sometimes fails to reduce severity, but wrong to conclude that immunisation is always ineffec-

TABLE I   Expected Attack Rate for Immunised and Unimmunised Children

Group	No. of children at risk	Attack rate per 100,000	Expected no. of cases
Immunised	90,000	100	90
Unimmunised	10,000	500	50
Total	100,000		140

tive in doing so. Children admitted to hospital are selected on various grounds, usually because they are severely ill. The effect of immunisation in reducing severity cannot be shown by basing comparisons only on children who are severely ill, i.e. by taking only the immunisation failures. Mild cases must also be taken into account by comparing the proportions of mild and severe cases in each group.

The third error is to fail to take due account of factors which may affect the measurement on which the assessment of efficacy is founded. Estimates based on notifications are especially vulnerable. Notifications may be affected by a number of influences which are not easy to quantify. For example, the proportion of children immunised against pertussis fell by more than one-half after the outbreak of 1974 and the whooping cough notification rates in the next outbreak (in 1977) increased fourfold. This suggests that immunisation had been extremely effective in controlling whooping cough. However, a large part of the increase in notifications in 1977 was due not to an increase in the number of cases of whooping cough, but to an increase in the proportion of these cases which were notified. The effect produced by the fall in immunisation, therefore, was much less than a comparison of the notification rates indicated.

Field trials have two main functions—to assess the efficacy and to assess the safety of the agent under test. As regards efficacy, three basic methods are available:

1. The relative incidence of the disease concerned in an immunised and an unimmunised group can be compared.

2. The antibody titres in individual participants before and after immunisation can be examined.

3. The incidence of the relevant disease before and after the agent has been introduced can be assessed by community studies.

Safety assessments often necessitate comparisons between immunised and unimmunised groups, but other considerations may also be involved. These are dealt with later.

2. EFFICACY

2.1. Comparisons of Incidence During a Prospective Follow-up of an Immunised and Control Group

The basic feature of these investigations is the follow-up of immunised and unimmunised groups to compare the incidence of the disease in each. Because this incidence is often low and because the number of individual and environmental variables is large, many participants may need to be observed over a long period. The effort and the practical difficulties involved are such that studies which depend on incidence, although they play an important part in assessments of immunising agents, are not common. They are made only when other simpler methods are unsuitable. Examples from the United Kingdom include the trials of tuberculosis, and measles immunisation (3, 5).

The number of participants is decided at the outset on the basis of (1) probable environmental and individual variables, (2) the estimated incidence of the disease in the unimmunised controls and (3) the expected efficacy of the vaccine. Efficacy is calculated from the reduction in the attack rate in the immunised group as a percentage of the attack rate in the controls. For example, if the rate in the control group is 60 per 1000 and in the immunised group 10 per 1000, efficacy is 83%, i.e. $(60 - 10)/60 \times 100$

Such studies have four basic requirements:

1. The immunised and control groups under comparison must be similar in all relevant respects. The participants must therefore be allocated to an immunisation and control group by a random method.

2. The vaccine must be potent, i.e. capable of inducing an appropriate antibody response, when administered.

3. Both groups must be followed up impartially.

4. The diagnosis of the cases occurring in each group must be unbiased.

2.1.1. Similarity of Groups

To ensure that the immunised and control groups are similar, the participants in trials based on disease incidence are allocated by a random method. There are many methods of random allocation: random numbers, odd and even birth dates, odd and even record card numbers etc. Because of its importance, the similarity of the groups should be demonstrated from a sample. Table II, based on the Medical Research Council's report of a trial of measles vaccine (5), with about 50,000 participants, illustrates this.

Usually some participants will already have acquired immunity by natural infection. Although random allocation—provided there are sufficient participants—should ensure that the groups contain equal proportions of non-

TABLE II   Similarity of Immunisation Groups: Trial of Measles Immunisation

Immunisation group	No. in sample	Mean family size	Previous measles in siblings		Attending day nursery (%)	Previous immunisation with triple/ polio vaccine (%)
			Under 5 years (%)	Over 5 years (%)		
Killed/live vaccine	1127	2.07	17	76	2	94
Live vaccine	1015	2.09	19	78	2	94
Unvaccinated control	1732	2.17	21	74	2	90

susceptible immunes, it is an advantage to include as many susceptible participants as possible. This is easy to arrange when the study is concerned with the illnesses of childhood, as children in their second year of life have usually lost their passive maternal immunity but have not yet been actively immunised by natural infection. When dealing with few participants and older age groups, it may be important to discard the non-susceptibles identified by testing for circulating antibody or sometimes by cutaneous tests— before allocating the remainder to immunisation or control groups.

When the number of participants is limited and includes groups with different characteristics, e.g. age, race, place of residence, it may be necessary to make an initial stratification into sub-groups and make a random allocation within each sub-group. This will ensure that the total participants in each group include the same proportion possessing each of the characteristics concerned.

2.1.2. Potency of Immunising Agent

The potency of the agent at the time and place when the participants were immunised should be tested and the immunisation techniques standardised. Effectiveness cannot necessarily be assumed from the manufacturer's potency tests. Faulty transport and storage, for example will reduce the efficacy of a vaccine. The antibody response as an indication of potency can usually be simply tested in a sample of the participants.

2.1.3. Impartiality of Follow-up

Direct follow-up methods include home visits and postal and telephone enquiries. Participants may sometimes be followed up indirectly by searching hospital or clinical records for new cases of the disease. Whatever the method, the immunisation and control groups should be followed up with equal intensity. An example of checking impartiality at follow-up is shown in Table III.

TABLE III   Medical Research Council Trial of BCG Vaccine: Intensity of Follow-up[a]

Group	No. of participants	Returned postal enquiry (%)	Response Visited at home (%)	Response Chest radiograph (%)	Contacted by at least one method (%)
Control	13,200	83	77	51	96
Immunised	14,100	79	80	50	94

[a] From Medical Research Council (3).

2.1.4. Diagnosis

As diagnosis is often based on opinion rather than on ascertainable fact, it is not always easy to avoid bias. A physician may be prejudiced for or against the vaccine and may tend to reject or accept diagnoses accordingly. The most obvious way to preserve impartiality in diagnosis is to arrange a "blind" trial in which the controls are given a placebo and the immunised group is kept secret. This was done in the first MRC trials of whooping cough immunisation. Nowadays ethical considerations usually rule out inert injected placebos which are of no advantage to the control group, though this difficulty can sometimes be circumvented. For example, when the efficacy of immunisation against influenza was studied in school children, the pupils were given vaccine containing either influenza A or influenza B. This provided an effective control group during subsequent outbreaks of each influenza type. A different method of ensuring diagnostic impartiality was seen in the MRC trial of BCG vaccine (3). The diagnosis for each possible case of tuberculosis found in the follow-up was made by an independent assessor given all the radiographic, bacteriological and clinical evidence, but not the immunisation status, of each suspected case.

Inaccuracies in diagnosis may occur for reasons other than bias on the part of the physician. The immunisation in itself is a possible source of bias since it may make the illness more difficult to diagnose. For example, the symptoms of whooping cough in children immunised against pertussis may be replaced by a less characteristic cough common to many other viral and bacterial infections; an attack of measles in an immunised child does not necessarily include a readily identifiable rash; rubella, even in unimmunised persons, often presents clinical diagnostic difficulties, which may become insurmountable if the patient has been given immunoglobulin. Nor can inaccuracies invariably be avoided by eschewing diagnosis by clinical examination and insisting on laboratory confirmation. Thus it is more difficult to isolate *Bordetella pertussis* from an immunised child with whooping cough than from an unimmunised child. Immunisation, by inducing circulating antibody, also interferes with serological diagnosis. In practice, therefore, a

TABLE IV   Efficacy of Pertussis Immunisation in Whooping Cough of Varying Severity

No. of daily paroxysms	3 DTP Attack rate/1000	3 DT Attack rate/1000	Relative rate DTP : DT
None	0.4	1.1	1 : 2.8
1–9	3.8	17.3	1 : 4.6
≥10	4.6	30.8	1 : 6.7
Admitted to hospital	0.05	0.84	1 : 16.8

follow-up usually reveals cases of varying severity, from those whose characteristic symptoms cause few diagnostic difficulties, to those where a certain diagnosis is not possible. These varying grades of severity and their influence on the reliability of diagnosis have to be taken into account in assessing efficacy. This is illustrated in Table IV (6).

It is evident from Table IV that the efficacy of the pertussis vaccine under test varies widely according to the diagnostic criteria adopted. At one end of the scale are children with severe attacks who are admitted to hospital. In this group, in which an accurate diagnosis is reached after observation and investigation, efficacy is high. Assuming that all other factors affecting admission to hospital are the same in both groups, a comparison of hospital admission rates provides an accurate estimate of efficacy for a selected group of cases. At the other end of the scale are the cases without typical paroxysms. These may include illnesses due to other organisms as well as those due to *B. pertussis* in which severity might have been reduced by immunisation. There is no way of identifying these two types of case and in consequence an accurate assessment of the efficacy of the vaccine in preventing mild whooping cough is not possible.

Misdiagnosis, if frequent, can seriously affect estimates of the efficacy of any immunising agent. Influenza immunisation is especially vulnerable, since its symptoms are frequently indistinguishable from those caused by other respiratory viruses. Immunisation provides no protection against these other viruses, which in consequence occur with equal frequency in both immunised and control groups. The effect is to reduce the apparent efficacy. Consider the following hypothetical situation:

Group	Attack rates (per 10,000)		Combined attack rate
	Influenza	Misdiagnosed as influenza	
Control	150	75	225
Immunised	15	75	90

As stated previously, efficacy is calculated as the reduction in the attack rate expressed as a percentage of the rate in the control group. This gives a true efficacy in this situation of $(150 - 15)/150 \times 100$, or 90%. However, because of the misdiagnosis occurring in both groups, the apparent efficacy is the lower one of $(225 - 90)/225 \times 100$, or 60%.

Bias other than that due to misdiagnosis may affect the findings. Hospital admission rates, for example, are influenced by age and social class; attack rates by overcrowding, the number and age of siblings and by attendance at nursery groups. If differences in the proportions of these variables are discovered in the immunised and control groups, it may be possible to standardise for each variable and then compare the attack rates within each standardised group. This procedure requires an adequate number of participants and the availability of the relevant information.

Allocation by Groups.   The efficacy of immunising agents can be assessed by allocating groups at the outset rather than individuals as described above. To assess the commercial value of influenza immunisation by reducing sickness absence in telephone exchanges, immunisation was offered to the employees in some exchanges and withheld from others. Sickness absence was then monitored in both during outbreaks of influenza and also, as a second control, when influenza was absent (7).

2.2. Comparisons of Attack Rates in Vaccinated and Control Groups without Random Allocation of Controls

Ethical and practical considerations often prevent a random allocation of immunised and unimmunised individuals or groups. When whooping cough immunisation was re-assessed in 1978, strongly held views about immunisation—both for and against—were common enough to make randomisation impossible. Instead notification rates were compared in children who had been routinely immunised with either DTP or DT although, since these children were not randomly allocated, there could be no certainty at the outset that these two groups were similar. It was expected, for example, that social circumstances likely to affect liability to whooping cough might differ in the two vaccine groups. It was essential, therefore, to investigate such factors by visiting the homes of a random sample of DT and DTP immunised children from both groups. Social class, race, the number of siblings, attendance at a nursery group and the previous history of other infectious diseases (Table V) were recorded. In fact, the groups proved so similar with respect to the variables investigated that it was unnecessary to consider standardised sub-groups; the whole groups could properly be compared. Table V exemplifies this.

TABLE V Assessment of Pertussis Immunisation: History of Infectious Disease, According to Immunisation Group[a]

Group	Chickenpox (%)	Measles (%)	Mumps (%)
3 DTP	13	16	14
3 DT	15	14	13

[a] From Public Health Laboratory Service (6).

The efficacy of immunoglobulin in preventing viral hepatitis in travellers and rubella in pregnant contacts was also assessed in studies without randomly allocated controls. In the first of these, the estimate of efficacy was based on attack rates of jaundice in young people working overseas for periods of 1 or 2 years. Attack rates in young people given immunoglobulin and departing in the year when immunoglobulin was first introduced were compared with the rates among the unimmunised who had left the previous year. Despite the absence of randomly allocated controls, the findings showed that a single injection of immunoglobulin was highly effective—but only in the short term since its influence waned after 7 months (Table VI) (8).

In the second example, the infection rate in pregnant women given immunoglobulin as home contacts of rubella was compared with the rate in non-pregnant women contacts who received no immunoglobulin. Clearly there was at least one major difference between the women in these immunised and control groups. Despite this drawback, a comparison limited to susceptible women in contact with laboratory-confirmed cases showed that immunoglobulin did not prevent infection with rubella (9).

2.3. Advantages and Disadvantages of Field Trials Based on Attack Rates

Field trials based on attack rates provide a triple assessment—efficacy for the individual, the value of the vaccine to the community in the social and

TABLE VI Development of Infectious Hepatitis in Immunised and Unimmunised Groups[a]

						Months overseas							
Group[b]	1	2	3	4	5	6	7	8	9	10	11	12	13+	Total
Immunised	1							1	6	3	3	8	3	25
Unimmunised		2	1	1	2	1	1	5	3	4				20

[a] From Pollock et al. (8).
[b] 1079 immunised, 942 unimmunised.

TABLE VII   Annual Incidence per 1000
at 5-yearly Intervals during Follow-up

| | Interval (years) | | | |
Group	0–5	5–10	10–15	15–20
Control	2.50	1.06	0.26	0.08
Immunised	0.40	0.33	0.10	0.09

epidemiological circumstances of the time and its effect in reducing the severity of the disease. The value of immunisation in reducing the incidence and gravity of measles and whooping cough and in preventing tuberculous meningitis were clearly evident from the field trials concerned. However, such studies often are expensive and time-consuming, and their value in assessing long-term effects is limited. This is in part due to the practical difficulties of a long follow-up, but there are also obstacles to maintaining similarity of immunisation and control groups for long periods. Immunised and unimmunised groups, however, closely similar when first allocated, have an inherent tendency to differ as time goes on. The most vulnerable participants in the control group develop the disease and are removed, until the controls eventually tend to consist only of highly resistant individuals with a relatively low attack rate. Table VII, based on the MRC trial of tuberculosis vaccines, shows the changes in the relative attack rates of BCG immunised and unimmunised groups during a prolonged follow-up (10).

During the first 5 years the rates are about six times greater in the control than in the immunised group, whereas at 15–20 years the rates are almost the same. Because this difference may be due to a relative increase in the proportion of the more resistant individuals in the control group, the efficacy of the vaccine 20 years after immunisation cannot be assessed.

Another potential obstacle to prolonged field trials is the nature of the early findings. Should these indicate that immunisation is effective—at least in the short term—the controls too may need to be offered immunisation.

2.4. Assessment by Antibody

Viral immunising agents and toxoids can be readily assessed by comparing circulating antibody titres before and after immunization and such simple antibody studies are usually adequate after efficacy has been confirmed by a study based on incidence. American field trials of inactivated polio vaccine, for example, enabled subsequent studies in the United Kingdom to rely on antibody response alone. Antibody estimations are especially useful when examining the relative potency of different preparations of the same agent,

TABLE VIII   Measles Antibody 10 Years after Immunisation[a]

| Group | H.I. titre (%) | | | |
	<4	4–64	128	>256
Immunised	1	78	17	4
Unimmunised	32	41	19	8

[a] From Medical Research Council (14).

the efficacy of different doses and, sometimes, the duration of protection. Numerous examples are available (11–13).

Antibody comparisons have so far been less useful for bacterial immunising agents, where the relationship between serum antibody and protection is unclear. Tests for cutaneous sensitivity are used as evidence of successful immunisation against tuberculosis, but no relationship between the degree of sensitivity and degree of protection has been established.

Samples are usually tested for antibody before and 4–6 weeks after immunisation. To prevent interference by natural infection, such studies are made when the prevalence of the disease concerned is low.

The duration of protection in an unimmunised group can be assessed by antibody studies, but not the inherent effect of the agent, at least for diseases in which natural infection may play a part in boosting immunity (Table VIII) (14).

Table VIII shows the antibody titres in an immunised and control group 10 years after immunisation. The participants were 2 years old on entry, an age when children are likely to be without antibody to measles. It is clear that almost four-fifths of the control group, and therefore the immunised group also, have been exposed to natural measles infection. Clearly the efficacy of immunisation in conditions in which natural infection is absent cannot be estimated from this type of study.

2.5. Disease Incidence Before and After Immunisation as a Measure of Efficacy

If the disease is notifiable, a general assessment of the efficacy of the immunising agent is sometimes possible by comparing notification rates before and after its introduction. However, most diseases are affected by improvements in treatment and social changes and by cyclical variation. The validity of such comparisons is affected by several factors. One is the frequency of the disease and the uniformity of the cyclical pattern prior to immunisation. Measles, for example, was highly prevalent and, before the

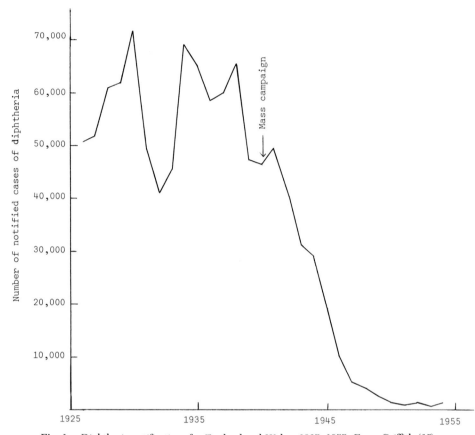

Fig. 1.   Diphtheria notifications for England and Wales, 1925–1955. From Griffith (*15*).

advent of immunisation, occurred in regular outbreaks every 2 years. The effect produced by immunisation of even 50% of the 1-year-old children annually was clearly evident from the notifications. Another factor is the extent of vaccine usage and the speed with which the prevalence of the disease declines; the more precipitous this decline, the easier to assess the part played by immunisation. The effect of mass diphtheria immunisation is shown in Fig. 1 (*15*). Yet another is the change in treatment. The advent of chemotherapy for tuberculosis, by preventing spread, made the part played by immunisation hard to discern, a difficulty greatly increased by the introduction of mass radiography, social changes and probable alterations in the proportion of cases notified.

Pertussis immunisation was introduced in the United Kingdom over a period of several years (the first half of the 1950s). While highly effective in preventing serious disease in the individual, mild cases in immunised children are common and the vaccine is not completely effective in controlling the dissemination of *B. pertussis* (*6*). Its true efficacy was not apparent until a marked decline in the proportion of children immunised was followed by a large increase in notifications (*16*). Even then the degree of efficacy was uncertain, since a relative increase in the proportion of cases notified exaggerated the difference in the notification rates before and after the fall in immunisation (*17*).

3. SAFETY

Studies to assess safety are affected by two factors. The first is the frequency of the hazard. it is easy to assess those reactions which are common and difficult to assess those which are rare. The second is the character of the reaction. Reactions which conform to a single pattern are readily identified. These include local pain and tenderness at the injection site and also those symptoms occurring after immunisation with live viral vaccines, which resemble those of the natural disease. Rash, sore throat and adenopathy after rubella immunisation and flaccid paralysis after immunisation with poliomyelitis virus are well-known examples. In contrast, the neurological disorders attributed to pertussis immunisation are difficult to assess; they are exceedingly rare, occur in unvaccinated as well as vaccinated children and exhibit no common clinical pattern.

Those common reactions which could be due only to the immunisation—pain and tenderness at the injection site, for example—can be assessed simply by following up immunised individuals, but a control group is essential when the symptoms of the reaction are less specific. In the example shown in Fig. 2 (*18*), the general illness attributed to measles vaccine in immunised children declines from 60% in immunised children to 26% when the illness in a control group is taken into account.

Among the less common reactions febrile convulsions have a special importance. Febrile convulsions have numerous causes and occur in immunised and unimmunised children. The part played by immunisation can only be assessed when the convulsion rate in both groups is compared. This is easily done when an unimmunised control group is available. In the MRC trial of immunisation against measles, for example, instances of convulsion were found in both the immunised children and unimmunised controls, but an excess—confined to a period 6–10 days after immunisation—could be distinguished in the immunised group (*5*). (Table IX).

However, control groups are not usually available, and in this case it is essential to establish the rate for febrile convulsions in the general population—the background rate. Figure 3 (*19*) shows the variation with age of the

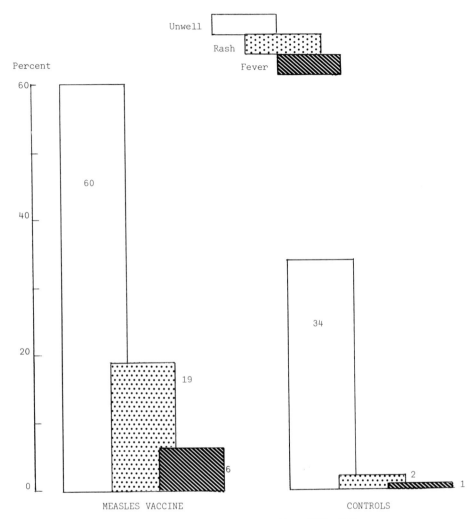

Fig. 2. Reactions to vaccination. From Miller (*18*).

TABLE IX Convulsions in a Measles Immunised and Control Group
According to Interval since Immunisation

Group	No. of children	Convulsion rate/1000	Interval (days)		
			0–5	6–9	10–21
Immunised	9,577	1.8	1	11	5[a]
Control	16,328	0.3	2	0	3

[a] Number excludes one child with epilepsy.

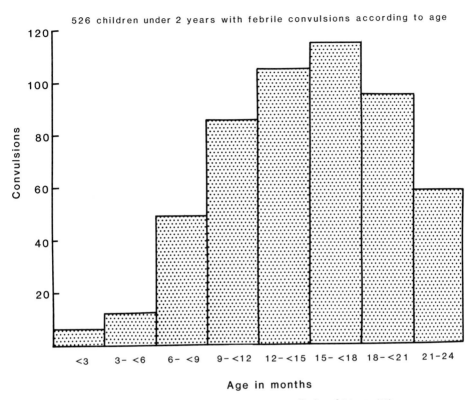

Fig. 3.   Hospital activity analysis 1979. From Pollock and Morris (*19*).

admission rate for febrile convulsions in children in a single health region in the United Kingdom. In the investigation for which this calculation was made, the 4-weekly rate for DTP (5.6 per 1000) was no different from the background rate for children of the corresponding age.

3.1. Case Control Studies

There are considerable difficulties in assessing the frequency of rare complications which for practical reasons cannot be assessed by following up hundreds of thousands of participants. Often all that can be done is to recognise that a rare hazard exists and that no reliable information about its true frequency is available. Recognition is based on reports of individual instances, as in the case of paralysis occurring within a week of immunisation with attenuated poliovirus. However, case control studies can be used to determine whether suspected symptoms are in fact vaccine induced. In these studies the frequency of immunisation in patients with the suspected symptoms is compared with the frequency of immunisation in matched controls. A good example is the National Childhood Encephalopathy Study of brain damage after pertussis immunisation (20). A major difficulty in case control studies is the matching of controls and cases. Valid matching should take into account all relevant variables in the cases and controls, an aim sometimes hard to achieve.

4. ETHICS

In the United Kingdom proposals for field trials with new immunising agents are considered by the Committee on Safety of Medicines, which takes into account the potential dangers involved. Permission should also be obtained from the relevant local Ethical Committee. Experiments in man are necessarily supervised by doctors, whose obligations to their patients may sometimes conflict with scientific objectives.

Field trials are made in varying circumstances and it is not easy to lay down precise directives, but the general medical approach is based on the Hippocratic oath originally restated by the World Medical Association as the Declaration of Geneva, and amended in 1968. The relevant statement is that the health of the patient must be the doctor's first consideration. Participants should not therefore be allocated to immunised and control groups if there are already good grounds for the belief that one group or the other is more likely to benefit. The sole exception to this rule occurs when a preparation is in short supply and the small amounts available can in good faith be reserved for a trial, as in the case of the early studies of immunisation with inactivated polio vaccine in the United Kingdom. As mentioned above, random allocation is not always essential and methods of assessment which fulfil accepted ethical standards can usually be devised. Doctors may also have to consider whether some participants—children with chronic illness, for example—should all be offered immunisation and excluded from the trial. If and when a benefit to the immunised group becomes apparent in the findings, immunisation must be offered to the controls despite the adverse effects on the scientific objectives of the investigation.

Other medical obligations include the maintenance of confidentiality, the prohibition of invasive tests, the taking of venous blood from infants, for example, unless these tests serve the patient's interest and not only the needs of the trial. A code of human experimentation incumbent on doctors to follow was drawn up by the World Medical Association in the Declaration of Helsinki which was revised in 1975: from a strictly legal point of view "subject to certain exceptions, consent must be freely and fully given by a person who understands the possible results of that which is proposed" (21). In the case of children, the consent of the parent or guardian must be obtained. In mental institutions, consent should be obtained from the patient if he or she is amenable to explanation; otherwise from the next of kin. Consent should be given in writing. When there are many participants it may not be possible to explain the issues verbally to every participant, but these issues should be fully described in a written statement which each must read and understand before signing the consent form. Finally, the form of consent should be part of, or be firmly attached to, the record card of each participant.

REFERENCES

1. Andrewes, C. H. (1949). The natural history of the common cold. *Lancet* 1, 71.
2. Yugoslav Typhoid Commission (1962). A controlled field trial of the effectiveness of phenol and alcohol typhoid vaccines. *Bull. W. H. O.* 26, 357.
3. Medical Research Council (1956). BCG and vole bacillus vaccines in the prevention of tuberculosis in adolescents. *Br. Med. J.* 1, 413.
4. Bradford Hill, A. (1937). "Principles of Medical Statistics." The Lancet, London.
5. Medical Research Council (1966). Vaccination against measles: A clinical trial of live measles vaccine given alone and live vaccine preceded by killed vaccine. *Br. Med. J.* 1, 441.
6. Public Health Laboratory Service (1982). Efficacy of pertussis vaccination in England. *Br. Med. J.* 285, 357.
7. Smith, J. W. G., and Pollard, R. (1979). Vaccination against influenza: A five-year study in the Post Office. *J. Hyg.* 83, 157.
8. Pollock, T. M., Reid, D., and Smith, G. V. (1969). Immunoglobulin for the prevention of infectious hepatitis in persons working overseas. *Lancet*, 2, 281.
9. Public Health Laboratory Service (1970) Studies of the effect of immunoglobulin on rubella in pregnancy. *Br. Med. J.* 2, 497.
10. D'Arcy Hart, P., and Sutherland I. (1977). BCG and vole bacillus vaccines in the prevention of tuberculosis in adolescence and early adult life. *Br. Med. J.* 2, 293.
11. Smith, J. W. G., Lee, J. A., and Fletcher, W. B. (1976). The response to oral polio vaccine in persons aged 16–18 years. *J. Hyg. (Cambridge)* 76, 235.
12. Collier, L. H., Polakoff, S., and Mortimer, J. (1979). Reactions and antibody responses to reinforcing doses of adsorbed and plain tetanus vaccines. *Lancet* 1, 1363.
13. Smith, J. W. G., Fletcher, W. B., Peters, M., and Perkins, F. T. (1975). Responses to influenza vaccine in Adjuvant 65-4. *J. Hyg.* 74, 251.
14. Medical Research Council (1977). Clinical trial of live measles vaccine given alone and live vaccine preceded by killed vaccine (4th report). *Lancet* 2, 571.
15. Griffith, A. H. (1979). International symposium on immunisation: Benefit versus risk factors. *Dev. Biol. Stand.* 43, 3–13.
16. Pollard, R. (1980). The relation between vaccination and notification rates for whooping cough in England and Wales. *Lancet* 1, 1180.
17. Miller, E., Jacombs, B., and Pollock, T. M. (1980). Whooping cough notifications (corresp.) *Lancet* 1, 718.
18. Miller, C. L. (1967). The evaluation of measles vaccine as a protective measure. *R. Soc. Health J.* 87, No. 5, 255.
19. Pollock, T. M., and Morris, J. (1982). Manifestations following vaccination in the North West Thames Region—a 7-year study. *Lancet* 1,753–757.
20. Department of Health and Social Security (1981). "Whooping Cough," p. 120. H. M. Stationery Office, London.
21. Thurston, G. (1966). Problems of consent. *Br. Med. J.* 1, 1405.

15

Regulatory Agencies: The Control of Medicinal Substances Intended for Use in Humans

J. P. JACOBS

National Institute for Biological Standards and Control
London, United Kingdom

1. INTRODUCTION

Regulatory authorities entrusted with the responsibility for controlling medicinal products have the duty to ensure that the manufacture, sale and

Animal Cell Biotechnology, Vol. 2

supply of such products are stringently controlled with specific reference to their safety, quality and efficacy. They are to be found in many countries functioning at various levels of complexity and competence ranging from the very advanced to the rudimentary. They have been established for some time in most developed countries; in some other countries they have been instituted only in recent times; and in many undeveloped countries there is little if any regulatory control.

Because of the virtually insuperable difficulty of obtaining detailed, reliable and up-to-date information, it is impossible to give an adequate account of the number and diversity of authorities established to control medicinal substances intended for use in man. Accordingly, the account that follows will be both general and selective. It will describe the regulatory system of the United Kingdom in reasonable detail and, in less detail, that of the United States of America. These particular systems have been chosen for description because they exemplify two effective regulatory bodies in use at present. They have been, and continue to be, prominent in influencing regulatory authorities in other developed countries and in the introduction and development of control systems in developing countries.

Other aspects that will be dealt with concerning the control of medicinal substances in general but with added regard to the control of biological substances are: the historical development of regulatory authorities, international co-operation and co-ordination, controversial opinions as to whether the measures enforced and the manner of their enforcement are satisfactory, excessive or inadequate, and trends and future prospects.

2. EARLY HISTORY

The really effective control of medicinal products began only in very recent times with the implementation of modern regulations enforced by legislative acts. Early control measures were very different from those of recent origin in that they were not concerned with present-day emphasis on safety and efficacy but simply with controlling quality and supply. In mediaeval times the manufacture of medicines in Muslim countries was subject to control. The original intention of such regulations was based on the desire to safeguard the faith and to inculcate and maintain a system of morals designed to prevent deceit and quackery. The regulatory codes included some which later extended into the area of medicine and, among others, applied to physicians, surgeons and syrup makers (whose responsibilities also included the preparation of medicines). The inspection of these medicines by officially appointed inspectors took place without warning, and a pharmacopoeia was drawn up which included in its regulations tests required to detect the adulteration of drugs that were common at the time. These included preparations such as henna, saffron, musk and acacia.

The existence of an Islamic regulatory system influenced thinking in Europe and contributed to the introduction of similar systems in several European countries. In the 12th century what is believed to have been the earliest law in Europe for controlling medical practice, requiring anyone wishing to practise medicine to pass an official examination, was proclaimed in Sicily. Corresponding restrictions were imposed in the 13th century in the nearby area of what is today part of southern Italy. In the pharmaceutical field apothecaries were subject to official control; they were compelled to ensure the quality of their drugs and to abide by a fixed scale of charges. In conjunction with these regulations, inspectors of apothecaries were also appointed. The notion of control along these lines and the gradual introduction of such systems spread to European countries but did not extend to the United Kingdom until later.

3. CONTROL IN THE UNITED KINGDOM

3.1. From the 14th to the 19th Century

In its very earliest form quality, as defined by its nature and the absence of contamination, was the one criterion demanded of a drug in Britain. The Ordinances of the Gild of Pepperers of 1316 are believed to be the first to attempt to effect a form of quality control in Britain. The Ordinances prohibited "the mixing of wares of different quality and price, the adulteration of bales of goods or falsifying their weight by wetting."

The picture for some time following the introduction of the first-proclaimed requirements is unclear. There were signs of a trend which led later to the apothecaries at first becoming preparers of various medicines and subsequently separating into two groups: the forerunners of medical practitioners, and a mixed group comprising apothecaries, druggists and chemists who founded the Pharmaceutical Society and eventually became known as pharmacists.

Throughout this long period considerable rivalry existed between the physicians and the apothecaries. Each of these groups attempted to advance its position at the expense of the other, the result of which was continued jealousy and hostility and the delayed emergence of effective control. The antagonism between the groups is reflected in the action of the physicians and surgeons of London who, in 1423, appointed two apothecaries to inspect the premises of pharmacists and empowered them to bring for trial anyone dispensing wares of unsatisfactory quality.

In 1518 Henry VIII founded the College of Physicians, and the earliest legislation for medicinal substances in the United Kingdom was enacted in 1540, which resulted in the physicians of London appointing four of their members to be inspectors of "Apothecary wares, Drugs, and Stuffs" (8). During the early part of the next century these inspectors were aided by the appointment of a number of representatives of the Society of Apothecaries whose duties were similar to those of the inspectors. In 1618, the first publication of a pharmacopoeia in the United Kingdom took place in London. This was the first of a series of 10 such publications to appear in London spanning a period of more than 200 years. The final pharmacopoeia of the series was published in 1851, by which time pharmacopoeias had also been published in Edinburgh and Dublin. In 1865, the first British Pharmacopoeia was issued. It was intended to be the official pharmacopoeia for the use of the United Kingdom, printed as a single publication to supersede those published separately in London, Edinburgh and Dublin.

During the 19th century additional important developments relating to the control of medicinal substances took place. The Pharmaceutical Society of Great Britain was founded in 1841. The Pharmacy Act of 1868 introduced legislation to control the retail supply of poisons; it also made the registration of pharmacists compulsory and imposed penalities for the adulteration of drugs. In 1874 the Society of Public Analysts was founded, and in the following year the Food and Drug Act of 1875 was passed, this being the final act of a number of similar acts introduced earlier. In parallel with these later developments, the pharmaceutical industry continued to grow and increasingly introduced self-imposed quality control of its products, and pharmacology was beginning to establish itself as a growing science.

3.2. The 20th Century

3.2.1. Medicines in General

Despite the prodigious amount of legislation introduced by the early part of the 20th century, the control of medicines was far from satisfactory, largely owing to lack of organisation. There were too many authorities with differing and sometimes far from clearly defined roles of authority. The government was well aware of this undesirable situation and of the confusing number of statutes in operation, and took measures to improve the anomalies. The Poisons and Pharmacy Act of 1908 was introduced to amend the Pharmacy Act of 1868, but poisons and dangerous drugs continued to be bought by individuals known to the pharmacist without the need to have them prescribed by a medical practitioner, and a wide range of proprietary medicines was freely available to the public. There was growing concern, especially by the British Medical Association, about the unrestricted sale of these nostrums, and in 1914 a report prepared by the Committee on Patent Medicines was published. The report made several recommendations, the most important of which were the following:

1. That there should be control of the sale and importation of patent medicines;
2. That the registration of manufacturers of such preparations should be mandatory;
3. That the ingredients of such products should be made known;
4. That an authority be appointed to permit or prohibit the sale of the products;
5. That an official inspectorate be formed.

Some of the recommendations were approved and put into practice; others either were not enforced until much later (following the introduction of the Medicines Act of 1968), or were disregarded. Further control was imposed by the introduction during World War I of the Defence of the Realm (Consolidation) Regulations of 1917, which restricted the supply of certain drugs such as morphine, barbiturates and cocaine to those issued on production of a physician's prescription. In 1920, the first Dangerous Drugs Act was introduced. It was the first of a series of such acts which were to appear over a period of years between 1920 and 1967 intended to introduce measures to control the supply and prescription of dangerous and narcotic drugs. Consolidation of these various acts was effected when the Misuse of Drugs Act was passed in 1971. A further measure which supported the control of medicines was achieved when pharmacists were recognised as members of a self-managed profession and the profession enforced its control measures through the Pharmaceutical Society of Great Britain.

In 1925, the government introduced the first of two acts designed for the control of therapeutic substances—the Therapeutic Substances Act of 1925. Additional regulations relevant to other substances considered coming within the scope of the act—mainly dealing with control over sale and supply—were introduced between 1925 and 1956. By 1956, it was clear that the act of 1925 needed to be brought up to date and, consequently, it was reviewed and revised and was superseded by the Therapeutic Substances Act of 1956. A year earlier an incident (the so-called Cutter incident) took place which emphasised the need for more rigid control. The incident resulted in the occurrence of more than 200 cases of poliomyelitis arising from the inadvertent use of what was meant to be an inactivated vaccine, but which contained traces of live virus.

The introduction of the Therapeutic Substances Act of 1956 was intended to consolidate the statutory position with respect to the control of medicinal products at the time. In large measure this attempted consolidation was achieved. However, the regulatory control imposed still left a conspicuous loophole in that only substances deemed to be of a biological nature or dangerous drugs were required to be tested by a manufacturer before they were released. This meant that any other medicinal product could be manufactured and sold without any obligation to test and control the product. It was clear that further legislation was needed, and in 1959 the government set up a Working Party to investigate the means of satisfying this need. Nevertheless, this unacceptable situation persisted until the Medicines Act of 1968 was passed following a thorough scrutiny of the control measures in existence at the time—and for some time following the thalidomide disaster in 1961. This appalling occurrence of the birth of thousands of deformed infants, because during pregnancy their mothers had taken this recently marketed sedative, highlighted dramatically the potential inadequacy of the existing control system and acted as a potent catalyst for the rapid overhaul of the measures for controlling medicinal substances.

In 1962, a sub-committee of the English and Scottish Standing Medical Advisory Committee evaluated the situation for the general control of drugs and recommended that an expert committee be formed to survey the facts regarding new drugs and to make recommendations about their toxicity. They further recommended that the committee's advice be given statutory support and that extensive revision of legislation should be carried out following a stage of review and consultation. Arising from these recommendations, the Committee on Safety of Drugs was formed in 1963. The committee was not legally sanctioned; it was intended to serve as an interim guiding force until backed by subsequent legislation. Its task was to assess new drugs before they were used and before being marketed for clinical trials, and to carry out the surveillance of drugs for the purpose of monitoring adverse reactions and to issue warnings if necessary.

The committee appointed three sub-committees to assist them in handling their prescribed responsibilities: a committee on toxicity and clinical trials, a committee on chemistry and pharmacy and a committee on adverse reactions to drugs. These sub-committees collectively further reviewed and consolidated the general position with regard to control measures.

In 1967 a white paper entitled "Forthcoming Legislation on the Safety, Quality and Description of Drugs and Medicines Act" was passed. This white paper formed the basis for the Medicines Act of 1968, by far the most important act passed up to that time and which, with subsequent amendments, is the act on which the present system of control of medicinal products in the United Kingdom is based.

The Medicines Act of 1968 confers the power for the introduction of a comprehensive set of regulations which supersede most of the earlier legislation for the control of various substances and articles including, in particular, substances for human and veterinary use. The act effectively embodies a compulsory system of licensing for the control of medicines (other than the different type of control required specifically for the possible misuse of narcotics and some other drugs), so that almost all control measures that had been carried out previously by voluntary means are consolidated on a statutory basis.

The main provisions of the act are:

1. The establishment of a compulsory procedure for control through the use of licences and certificates, applicable to the marketing, importation, manufacture and distribution of medicinal products

2. The creation of a Medicines Commission to act in a consultative capacity for the purpose of giving either the Health Minister or the Agriculture Minister advice on general matters relating to the act

3. The appointment of several expert committees with advisory responsibilities regarding matters of a specific nature

The Medicines Commission was formed in 1969 and is an independent body whose responsibilities are outlined above. Its members are mostly medically qualified doctors, veterinarians and pharmacists but it also includes lawyers and others. Administration of the act, insofar as human health matters are concerned, is carried out on behalf of the Health Ministers by the Medicines Division of the Department of Health and Social Security.

The division has the following wide range of responsibilities:

1. It acts on behalf of the Health Ministers as the authority for licensing, enforcement and inspection.

2. It is responsible for international relations on medicines.

3. It monitors adverse reactions to medicinal substances.

4. It functions as a centre for investigating reports of defective batches of medicines.

5. It keeps the Medicines Act under review, prepares new legislation and recommends amendments if considered necessary.

6. It is responsible for supplying the support services for the British Pharmacopoeia Commission, one of whose functions is to produce and keep up to date the British Pharmacopoeia.

The expert committees having advisory responsibilities comprise independent experts whose appointment was recommended by the Medicines Commission. There are five such committees.

These are:

1. The Committee on Safety of Medicines

2. The Committee on the Review of Medicines

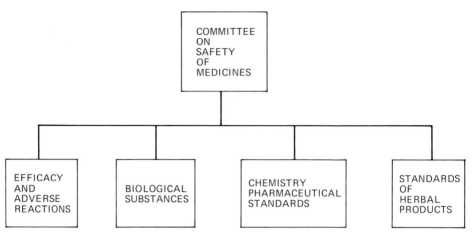

Fig. 1.   Sub-committees of the Committee on Safety of Medicines.

3. The British Pharmacopoeia Commission
4. The Comittee on Dental and Surgical Materials
5. The Veterinary Products Committee

With respect to governmental control of medicinal substances intended for human use, the most important of these committees is the Committee on Safety of Medicines. The relationship of this committee to the four sub-committees which assess licensing applications on behalf of the Committee on Safety of Medicines is shown in Fig. 1.

The advice of the appropriate committee on the safety, quality and efficacy of medicinal substances is considered by a licensing authority before a licence is issued or refused. Depending on the nature of the product, the ultimate licensing authority is either the Health Ministers or the Agriculture Ministers. In practice, however, a product for use in humans is licensed on behalf of the Health Ministers by the Medicines Division of the Department of Health and Social Security and a veterinary product is licensed on behalf of the Agriculture Ministers by the Animal Health Division of the Ministry of Agriculture, Fisheries and Food. The functions of the Medicines Division and its inter-relationship with a wide range of various bodies are shown in Fig. 2.

3.2.2. Licensing

Licensing applies to all materials in a pharmaceutical form intended for administration as a medicine and the Medicines Act provides powers to bring within the scope of licensing control substances used as ingredients in

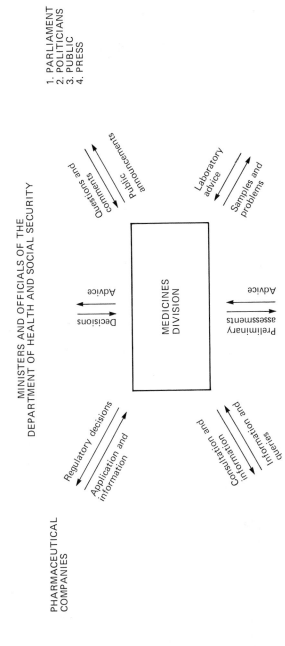

Fig. 2. The policy process of the Medicines Division.

medicinal products or substances which, if used without prior safeguards, may give rise to hazards to the health of the community.

Control measures based on compulsory licensing were implemented in 1971, since which time new medicinal products could neither be marketed nor used for clinical trial without the granting of the appropriate licence or certificate. With respect to products already on the market on the date that licensing control became mandatory, provided they satisfied particular pre-licensing requirements they were allowed to be licensed without assessment as to their safety, quality and efficacy. During the past decade products in this category have been further reviewed and have been renewed at the discretion of the licensing authority. In due course, all such products will be subjected to the same requirements as those applying to products for which ordinary licences are granted. Provided the product satisfies these requirements the Licence of Right applying to the product would be converted into an Ordinary Licence. There are four types of mandatory licences or certificates:

Manufacturer's Licence.   The manufacturer's licence is required for the manufacture or assembly of a medicinal product.

Product Licence.   The product licence is required for the importation, sale or supply of a medicinal product and is held by the person responsible for the composition of the product or by the importer. It authorises the holder freedom to act in any one of several ways with respect to the product. These are:

1. To sell, supply or export the product
2. To procure the sale, supply or exportation of the product
3. To procure the manufacture or assembly of the product for sale, supply or exportation
4. To import or procure the importation of the product

Clinical Trial Certificate.   The clinical trial certificate is required to supply a medicinal product for the purpose of a clinical trial in human beings. The product under investigation will be one which has shown evidence that it may be beneficial. It is administered to subjects to determine the extent to which it is beneficial and, concomitantly, to show evidence of freedom from any untoward effect it may have, such as adverse reactions.

Wholesale Dealer's Licence.   The wholesale dealer's licence is required for the sale of medicinal products to anyone who is not the ultimate user; that is, the buyer sells or supplies the product to someone else. The licensing of wholesalers is designed primarily for the purpose of identifying the distributor and the suitability of the conditions under which a product is stored.

In addition to these mandatory licences and certificates the Research Licence, which came into being as a result of the Therapeutic Substances Act of 1925, is still in use and there is an additional certificate connected with the licensing of medicinal products which is not compulsory but is issued on request. This is an Export Certificate. A licence holder in the United Kingdom who wishes to export a product can obtain this type of certificate, which signifies that the product to be exported satisfies the requirements of the United Kingdom legislation.

Procedures for considering applications for licence or clinical trial certificates are of a searching and demanding nature. Applications have to be accompanied by a wide range of information on all aspects of the nature of the product, its method of production and the results of tests conducted on the product including, when appropriate, animal tests and the findings on clinical studies and trials in man.

The application is normally referred to the Committee on Safety of Medicines, who, in turn, usually refer it to the appropriate sub-committees or perhaps several sub-committees for assessment. The reports of the sub-committee are considered by the Committee on Safety of Medicines, which then advises the licensing authority as to whether a licence should be granted.

When applying for Manufacturers' Licences and Wholesale Dealers' Licences, applicants must submit information which shows that manufacture of the product is consistent with respect to the suitability of the premises and equipment and the competence of the staff.

Enforcement of the act is a function of the Medicines Division. It applies this authority partly through the Medicines Inspectorate, which is an integral part of the division. The inspectorate visits and conducts inspections of manufacturers' premises both in the United Kingdom and overseas and the report they submit is considered very seriously when a decision is being made about the granting of licences.

In the event that an application for a licence is refused the applicant has the right to appeal against the decision. If the appropriate committee(s) advise the licensing authority not to grant a licence on the grounds of safety, quality or efficacy, the authority is obliged to inform the applicant of its opinion and allow him the opportunity of an oral hearing or of presenting a written case.

If the licensing authority is advised by the Committee on Safety of Medicines that the licence ought not to be granted despite the case made by the applicant, the authority is obliged to inform the applicant of the committee's advice and the reason for the advice. At this stage the applicant has the opportunity of making further representations, either by being heard or having his written representation referred to the Medicines Commission.

A number of other possibilities for redress are available to the applicant should any of the following eventualities develop:

1. The licensing authority is not in agreement with the recommendations of the Medicines Commission.

2. There has been no hearing or representation made to the commission which is different from the advice of the committee.

3. The licensing authority propose on grounds other than those of safety, quality and efficacy to refuse a licence be granted or to grant one different from that requested.

In any one of these situations the applicant is allowed an opportunity of presenting his case by a person appointed or the alternative of forwarding written representation concerning the application.

In certain circumstances the licensing authority may revoke, suspend or vary a licence or certificate without reference to the relevant advisory committee. However, before action of this kind is taken, when advising the licence holder of its intention the authority is obliged to allow him the opportunity of making either oral or written representation.

3.2.3. Biological Substances

Products categorised as biological substances in use at present include vaccines, toxins, antigens, sera, blood products and certain antibiotics, enzymes and hormones.

In the early part of this century a far more limited number of these substances was available but they were beginning to be produced in increasing quantities and in widening range. Concomitantly, it was becoming increasingly clear that the protection of the population against various diseases should be effected by immunisation and that, at the same time, adequate measures should be taken to control and standardise these immunising products.

These requirements presented a new problem. Hitherto, the medicines used were subjected to control tests based on chemical reactions, but the purity and potency of products of biological origin could not be measured by such means; instead, tests of a biological nature had to be devised and introduced.

In 1909, the British Pharmacopoeia Committee recommended that a governmental institution should be set up to standardise potent drugs and sera and to scrutinize manufacturing practices regarding the products. The recommendation was not acted upon until 1920, when an official Working Party was formed to advise on the measures required for the effective control of the quality and nature of therapeutic substances which could only be assessed by other than chemical means. The Report of the Working Party made several recommendations for the control of a wide range of biological substances. The main recommendations of the report were:

1. That manufacturers be licensed
2. That the premises and production procedures be subject to official inspection and approval
3. That a central laboratory under governmental control should be set up for testing and research, the testing to be carried out routinely on samples requested of and submitted by a manufacturer after removal from production batches
4. That imported products be subject to the same restrictions as those produced in the United Kingdom

The Therapeutic Substances Act of 1925 was of considerable importance for the control of biological substances. It followed closely the recommendations made by the Working Party, 1920, and was the first act introduced to make the licensing of medicines mandatory. It also made provision for the control of insulin (which had recently been discovered), for the control of as-yet-undiscovered biological substances, for describing standards of purity and the nature of tests to be used for measuring this property and for in-process control at the site of production. Among other rulings, the second Therapeutic Substances Act (that of 1956) authorized the issuing of three kinds of license : a Manufacturer's Licence, a Research Licence and an Import Licence. A Manufacturer's Licence entitled an individual or a company to manufacture for sale the therapeutic substances shown on the licence. A Research Licence allowed a named individual to import a scheduled or controlled substance to be used for research purposes. An Import Licence permitted an individual or company to import a substance or a substance listed from the source indicated. (With the exception of the Research Licence these licences were superseded in 1971 by those defined in the Medicines Act of 1968 referred to earlier in this chapter.) A Manufacturer's Licence and an Import Licence were granted only after inspection of the plant, premises, processes and personnel, whether within the United Kingdom or abroad, showed that the applicant was competent to produce the substance to the minimum specifications required. The inspections were carried out by professionally qualified staff of the Ministry of Health (now the Department of Health and Social Security), who were either medically qualified or pharmaceutically qualified, usually accompanied by a member of the scientific staff of the control laboratory. Following an official visit of this kind—during which points at issue would customarily be discussed with staff of the manufacturer—a licence would only be granted provided various requirements were met and after approval was given at several levels.

In addition to the control measures introduced by the Therapeutic Substances Acts, as a result of the recommendation contained in the Report of the Working Party of 1920 that a governmental laboratory be set up to control biological products, the Division of Biological Standards was established in 1923. It was set up as a laboratory within the National Institute for Medical Research for the purpose of testing manufacturers' products and establishing biological standards.

As time passed the responsibilities of the division grew as more biological substances, such as antibiotics and poliomyelitis vaccine, became available. These changes necessitated expansion of the work performed by the division and in 1960 resulted in the creation of a second division, the Division of Immunological Products Control.

During the 1960s the work of these laboratories continued to expand, and in 1967 a Working Party was formed to consider future changes and requirements. This study resulted in further reorganisation, which led to the establishment of the Division of Biological Standards and the Division of Immunological Products.

Control was accommodated in a single building in 1972 to form the National Institute for Biological Standards and Control (NIBSC). In 1975 the Biological Standards Act was passed and it came into force the following year. At the same time, control of the institute was transferred from the Medical Research Council to the National Biological Standards Board, which is a virtually autonomous body but is responsible through the Department of Health and Social Security to the Secretary of State.

The NIBSC has several functions, one of which is its essential role in the control of biological products. It prepares, characterises, stores and issues standard reference preparations of a number of therapeutic, prophylactic and diagnostic biological substances; it conducts control tests on samples of production batches of biological products submitted by manufacturers and scrutinizes documentary evidence of in-process control tests; it advises the licensing authoritory on measures for quality control; and it undertakes research related to the broad field of standardisation and control of biological products.

Members of the scientific staff of the NIBSC give advice to manufacturers regarding testing procedures on products. They also advise the licensing authority when application is made for a Product Licence or a Clinical Trial Certificate, and, in addition, they may give advice with respect to in-process control tests and if necessary accompany the inspector for inspection of the premises.

In a retrospective manner, the NIBSC also advises the licensing authority on problems that may arise relating to the quality control of a product. The detection or report of such a problem may result either from tests carried out at the NIBSC or from a report concerning a batch of a product already marketed.

In compliance with the requirements of the Medicines Act of 1968, which were enforced in 1971, all medicines in the United Kingdom are licensed today (2). If a new product is intended to be used for clincal trials or marketed, either of these steps can only be taken provided the appropriate licence or certificate has been issued.

With reference to the rationale employed for controlling medicines in the United Kingdom, the regulatory authority places much emphasis on the use of post-marketing surveillance (3). It does so in the belief that pre-marketing controls are in themselves not wholly adequate because clinical trials involve only a few hundred people and are therefore unlikely to provide evidence of serious adverse reactions since these are infrequent occurrences.

The general attitude of the regulatory authority to control is one of flexibility. It operates in close co-operation with the pharmaceutical industry, a respect for whose role in the regulatory system is reflected in the decision of the authority to provide guidelines rather than to issue stringent requirements.

4. CONTROL IN THE UNITED STATES

4.1. The Period Leading up to the Thalidomide Disaster

The early history of the use of drugs and related medicinal substances in the United States is an offshoot of the situation that existed in England with respect to the use of patent medicines; as already referred to, quackery and charlatanism relating to the sale and advertising of various remedies were rife for a long time in England.

When the earliest settlements of immigrants were being founded in New England in the early 17th century, patent medicines, at first imported from England but later prepared locally, were already being used by the settlers. However, the difference between the situation in England and in the newfound Colony was that by this time regulations relating to the sale and use of these potions were in force in England whereas in the United States no such control existed. Similarly, no restriction was imposed with regard to the practising of medicine; those who chose to, simply set themselves up with the title of physician, surgeon or apothecary.

The war of the American Revolution effectively very much reduced the importation of patent medicines and resulted in these substances being manufactured in the United States. With the growth of industrialisation and urbanization, problems of health associated with such changes also increased. Tuberculosis, typhoid and typhus were prominent as problems and provided impetus for the expansion of the output of nostrum remedies, most of which were purgatives, liniments or opiates.

Further momentum to the already widespread use of patent medicines was provided by the American Civil War. Among the combatants the incidence of diseases—particularly typhoid, dysentery and malaria—was much greater than casualties resulting from wounds and injuries. Panaceas flourished and their extensive use continued unrestrictedly with the return of peace, by which time an estimated 50,000 patent medicines were being marketed.

For some time the American Medical Association (AMA) had been campaigning to expose the uselessness, and in many cases the harmfulness, of these nostrums and actively to curb their use. In 1902, in collaboration with the American Pharmaceutical Association, the AMA established the National Bureau of Medicines and Foods, whose responsibility was that of certifying the identity, purity, quality and strength of medicines. Arising from the creation of this control body and subsequent developments with respect to its function, four years later the all-important Pure Food and Drugs Act of 1906 was passed. The act introduced the following important changes: if a preparation contained dangerous drugs their nature and quantity had to be shown on the label; falsification of claims for products became unlawful; the standards of pharmaceutical preparations set out in the United States Pharmacopoeia (USP) and the National Formulary (a supplement to the USP) were altered from publications of a public nature and became official standards. The act, however, made no provision for restricting the advertising of such preparations, which meant that incredible claims were frequently made. It was clear that such abuses had to be curtailed by introducing appropriate legislation, and in 1927 a new law was introduced which superseded the 1906 act.

The Bureau of Chemistry had been in existence for about 80 years, and the branch of the bureau which dealt with regulatory requirements was given the title "Food and Drug and Insecticide Administration." Four years later it was changed once again, this time to the Food and Drug Administration. At that time, however, its function was still limited almost entirely to matters concerning food supply.

In 1937, an almost unbelievable and horrific mistake occurred which rapidly led to the enactment of further legislation. The tragic deaths of 107 people resulted from the injection of a sulphanilamide preparation for which ethylene glycol had been used as the solvent (1). A number of the characteristics (e.g. flavour and fragrance) of the solvent had been determined but, astonishingly, it had not been tested for innocuity.

Arising from this appalling occurrence the Food, Drug and Cosmetics Act was enacted in 1938. This was a much more comprehensive statute than any of its predecessors and forms the framework of the system for the regulation of medicines that operates at present in the United States. The act demanded that drugs be shown to be safe prior to being marketed. In stipulating this requirement the United States government did so long before a similar requirement was introduced in any other country. The next country to do so was the United Kingdom, where this measure was not introduced until a generation later (7).

During the 1939–1945 war and in the years immediately following there was a rapid increase in the number of drugs available. This upsurge had a direct impact on the regulatory authority and resulted, in 1951, in the introduction of an amendment to the Food, Drug and Cosmetics Act. The amendment separated drugs into two distinct categories: those considered safe without medical supervision and available for purchase over the counter, and those deemed to be unsafe unless supervised and therefore only made available on prescription.

4.2. The Period since the Thalidomide Disaster

When the thalidomide disaster struck Europe the United States had the good fortune to escape the tragedy. However, the shock wave produced by this appalling event caused the United States government to review the entire matter of drug control and resulted in the introduction of the Drug Amendments Act of 1962. This was one of the most important acts passed in the United States concerning the regulation of medicinal products. It contained important amendments to the Food, Drug and Cosmetics Act, most of which were designed to strengthen the power of the regulatory agency. The most important innovation brought in by the act was the stipulation that any new medicinal substance should be not only safe but also effective, as demonstrated by clinical studies, before it was marketed. Other measures required by the act were that a manufacturer was compelled to inform the regulatory authority of any adverse reactions attributable to the use of a product and that he had to provide evidence of good manufacturing practice in compliance with official requirements.

Since 1962 regulations concerning the control of medicinal substances in the United States have not altered greatly, though several changes were introduced in the 1970s. One of the most important of these was the Biologicals Review, which introduced a requirement to re-examine the safety and efficacy of all biologicals licensed by the Public Health Service up to 1962 (authority for regulating these substances was by then vested in the Food and Drug Administration, which had been allocated the responsibility in 1972; the responsibility was entrusted to a body created within the Food and Drug Administration, designated the Bureau of Biologics). In this connection, it was decided that licence-holders would be allowed a period of 2 years in which to provide evidence of the efficacy of a product. The structure of the

Food and Drug Administration is shown in Fig. 3. The National Center for Drugs and Biologics was created in 1982 when the Bureau of Biologics and the Bureau of Drugs were merged because they have very closely related responsibilities.

In 1975 the Review Panel on New Drug Regulations was formed to study the rationale and procedures of the Food and Drug Administration (FDA) with respect to the approval of new drugs. The principal conclusions arising from the study were both supportive and critical.

It was concluded that the pre-marketing system of approving drugs based on evidence of safety and efficacy, as assessed by the FDA, was operating satisfactorily but that the procedure followed by the FDA, nevertheless, had several deficiencies. These were:

1. That the system did not allow sufficient participation by the public
2. That communications between the FDA and applicants for licences were too informal
3. That the scientific scope of the FDA was inadequate
4. That the standards it stipulated should be defined more clearly
5. That the procedure for licensing new drugs was unsatisfactory

With regard to the structure of the regulatory authority and the system through which it reaches decisions concerning the approval or disapproval of

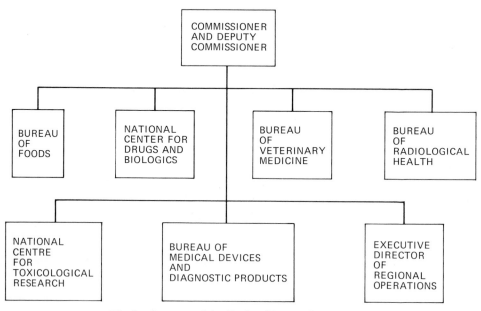

Fig. 3.   Structure of the Food and Drug Administration.

medicines, considerable controversy has existed in the United States for many years. Until very recently authority rested essentially with professional scientific civil servants, whose decisions were made strictly according to their own assessments and without the benefit of the additional opinions of outside experts. Compared with the systems operating in most countries, the method of control in the United States was very much an enclosed one and was discernibly rigid. The medical profession, the pharmaceutical industry and many groups representing the public were, in general, dissatisifed with, and critical of, the system for several reasons. Prominent among their accusations were the following:

1. That there was unnecessary delay between the time of discovery and the availability of new products
2. That regulations caused a reduction in the number of new products licensed
3. That because of delay, developmental costs for new medicines were increased

As a consequence of the 'tug-of-war' situation that existed between the regulatory agency and the other groups, for a long time relations were rarely friendly and at times verged on the hostile. However, by the mid-1970s the FDA had altered its stance in favour of a less rigid approach to control. It became increasingly convinced of the benefit, and began to accept the role, of external committees acting in an advisory capacity. Subsequently, it supported an official recommendation that the function of these committees be given legislative approval.

Since the mid-1970s there has been considerable activity involving exhaustive reviews of licenced pharmaceutical products. As a result of this situation two major reform bills, largely opposed in content, have been introduced for legislative approval.

The first of these bills is designed primarily to strengthen the position of the FDA by further extending its authority over the pharmaceutical industry and the medical profession concerning the production, licensing and use of medicinal substances. The second bill contains a number of counter-proposals because several of the measures proposed in the first bill have created strong opposition from both the pharmaceutical industry and the medical profession. The pharmaceutical industry objects to that part of the bill which it claims would affect adversely the proprietary position of safety and effectiveness. The medical profession opposes the bill because it maintains that if certain proposals were implemented its practice would be intruded upon and its authority undermined.

Very recently, there has been increased external pressure and a growing internal willingness for the FDA to relax some of its requirements regarding applications for new medicinal substances. At the same time, there have been reviews and recommendations at various levels designed to implement these changes. One important consequence of these developments is that the time required for the approval of a new drug has been noticeably shortened.

Whatever the disadvantages of the continuing debate and whatever the outcome, the unrestricted and protracted discussion that is taking place in proposals and counter-proposals contained in the bills should prove profitable. At the very least the extensive airing of opinions should serve to enhance the likelihood that issues may be more fully discussed and clarified than before and, therefore, should be beneficial for future discussion and legislation regarding the control of medicinal products.

5. SOME SIMILARITIES AND DIFFERENCES BETWEEN THE REGULATORY SYSTEMS OF THE UNITED KINGDOM AND THE UNITED STATES

The regulatory systems of the United Kingdom and the United States have some features that are similar and others that are different.

Two of the principal similarities are that both countries have an extensive and thorough system of control and that the control authorities share a stipulation, absent from the requirements of many other countries, that a product must be tested in animals and, further, that the protocol on the tests be submitted before authority is given for clinical studies to be conducted in humans.

The differences between the authorities are more evident than the similarities, as indicated by the following comparison:

1. The authority having the final power to license a product in the United Kingdom is the Minister of Health (who is responsible to Parliament), who acts on the advice of the externally appointed committees, assisted by a small staff of civil servants. The United States authority relies on a system in which decisions of professional civil servants continue to be influential, with the role of independent professional advisors being less important than in the United Kingdom. However, this situation has begun to change rapidly very recently; the authority is more flexible than hitherto and is more willing to accept the advice of outside committees.

2. The United Kingdom authority puts far greater stress on the safety than on the efficacy of a product and, therefore, demands less rigorous testing than the United States authority to establish efficacy as a criterion.

3. Although in keeping with its modified approach the United States regulatory agency has since the 1970s been more willing than earlier to

accept the results of tests performed in other countries, traditionally the United Kingdom control authority has accepted these data more readily. Acceptance is based largely on the premise that what is important is the quality of a study rather than its location or the status of the investigator.

4. The United Kingdom is more stringent with respect to post-marketing surveillance, although the United States control authority has in recent times put more emphasis on this aspect of control.

6. INTERNATIONAL AND REGIONAL CONTROL

The only body regulating medicinal products through international co-operation and whose function, as its name implies, is solely concerned with health matters is the World Health Organisation (WHO). A regional form of collaborative control is being attempted in a very limited way by one or two groups of countries situated in geographical proximity to each other such as the European Economic Community (EEC), whose main concern, however, is with economic problems.

6.1. The World Health Organisation

Among its various activities connected with the general problem of public health, the WHO provides advice and guidance on a wide range of problems to regulatory agencies in underdeveloped countries in particular but also to some developed countries. It also bring together representatives of control authorities from various countries with the object of endeavouring to advance harmonisation and it has direct links with about 150 countries who collaborate by exchanging information with the WHO on health problems in general.

As well as having the aim of promoting harmonisation, the WHO also performs a very valuable service in setting up programmes for monitoring, collating and reporting drug reactions on an international scale; in sponsoring programmes for combating some of the more important endemic diseases; in formulating agreed standards for evaluating medicinal products in order to reduce, or to avoid, duplication from country to country; and in assisting and encouraging countries who have difficulties in developing their own systems of effective policy and control for assessing medicinal substances either produced within the country or imported.

The WHO functions as a body whose intentions are seen as being consistent with an earnest desire to aid countries in their quest to improve the health of their people. Through its efforts, it has succeeded in fostering and establishing more congenial and frank discussion and in improving relations between the control authorities of many countries (5).

6.2. The European Economic Community

In contrast to the objective of the WHO, the co-ordination of matters relating specifically to health is of comparatively minor importance to the EEC. The main thrust of the consortium—and others such as the European Free Trade Association—is economic; the member countries have affiliated for the prime purpose of reducing, or even removing, trade barriers. Therefore, matters such as those concerning the regulatory control of medicines are considered relatively insignificant. Nevertheless, there has been some attempt to introduce measures relating to the control of medicines. Within the member countries minimum regulatory standards for the free movement of branded pharmaceutical products inside the Community have been prepared and issued. Approval of a product is shown when agreement is reached following discussion of a problem and final approval is indicated when it is published as a monograph in the European Pharmacopoeia. In this way, it is intended that if a substance is approved in any one of the member countries, it should be accepted without further assessment in the other constitutent countries. But this is intent, not realisation, at present.

Countries whose regulatory system is of a more stringent nature than those of other countries would be unlikely to follow recommendations meant to be accepted by the member countries. For example, the United Kingdom has very demanding criteria for safety and efficacy and would be unwilling to relax its requirements; instead, it would specify that a product approved by the country of its origin should also satisfy the requirements of the United Kingdom. A similar example can be given with regard to France. The French authority requires that even if a product they wish to import has been approved in another member country, tests based on pre-clinical, clinical and technical investigations must also be performed in France. In such cases as these, member countries submit details of their national requirements to the Community for approval.

7. ALTERNATIVE SYSTEMS: MULTIPLE SYSTEMS OR HARMONISATION

The diminished restrictive policy of the FDA in evaluating and approving pharmaceutical products means that the American system of control and those of a similar nature used in other countries are becoming more uniform. At the same time, there is a general trend in the direction of simplifying regulations towards the use of policies motivated more by medical and scientific considerations and a greater willingness to accept foreign data. The natural extension of this inclination to adopt a more common approach between countries controlling medicines is increased co-ordination and harmonisation. The alternative is a retention of the present pattern of the use of multiple control systems whereby each country chooses to use a system of regulatory control ostensibly best suited to its own particular circumstances. The opposing forces of these two systems base their preferences on a number of premises.

Those who support the use of regulatory agencies created on a national basis adopt their stance for a number of reasons. Problems concerned with public health, the treatment of diseases and the practice of medicine are very much influenced by the historical and cultural background. Therefore, the development of diverse regulatory authorities, each setting requirements for control that are relevant to the particular needs of the country, is more sensible than attempting to impose a uniform system which may prove inappropriate. For example, when changes of an epidemiological nature occur, as may be the case, for example, with infectious diseases for which vaccines are in use, a local regulatory authority with on-the-spot experience and not bound by international regulations would function advantageously. It would be able to assess and deal with problems using a more flexible approach than if it were compelled to follow regulations approved supranationally. Furthermore, there is the possibility that regulations prepared for international use may be based on the 'highest standard' approach and, accordingly, may be overstringent and unsuitable in certain cases. Another reason for not endorsing a uniform system of control is based on the contention that it would reduce the incentive to develop a local regulatory authority that might function more effectively than one imposed from outside.

The alternative approach supporting an alteration of the present system to one in which there would be increasing harmonisation, with the objective of achieving uniformity on a world-wide basis, or in very large measure, may be summarised by the following:

1. The realisation of the objective of harmonising procedures for regulating medicinal substances would be bound to have disadvantages, but those supporting the use of internationally agreed regulations believe the advantages would be far greater.

2. There would be standardisation of application, so that the pattern would not, as now, vary from country to country.

3. Duplication that is so common at present would be markedly diminished, and there would be widespread use of uniform regulations and automatic acceptance of a product once it had been approved.

In contrast to the system that exists at present, where foreign data are often not accepted, data from another country would be recognised except, perhaps, in rare instances where important differences of a genetic, racial or geographical nature might necessitate repeating certain studies on a different population (4). Despite the fact that legislation has been developed on a national basis, it seems natural to promote changes towards regionalisation and internationalisation because the major pharmaceutical companies have developed a structure that is international. Therefore, for the purpose of marketing worthwhile products on a world-wide scale, greater standardisation and simplicity is called for with respect to regulatory control procedures. If this change were introduced, it would benefit impoverished developing countries also, where lack of appropriately trained people to evaluate and study the substances they import is a problem.

8. THE REGULATORY POWER OF THE STATE: CONFLICTING OPINIONS

The role and power of the state in the control of pharmaceutical products varies enormously in different countries, ranging from total state involvement including state ownership of the pharmaceutical industry (as seen in Comecon countries) to the minimal or even perfunctory state control to be found in underdeveloped nations.

The extent to which medicinal substances should be subject to governmental control is a question that has been much discussed and debated in developed Western countries in which production is effected by privately owned companies while requirements for their control are decided and enforced by the state. This controversial matter has until recently been most hotly debated in the United States because although the American control authority has adopted a more flexible policy than hitherto, it still tends to be more cautious and bureaucratic than those of other countries who operate similar control systems. For this reason, in dealing with the contentious subject of the state control of medicinal products, it is appropriate to consider the opinions in the United States held in support of, or in opposition to, the extent or even the existence of state control.

The comparatively strict control measures that have been in force in the United States for about the past 20 years originated primarily from the thalidomide tragedy, but also from the Cutter incident. These occurrences had a galvanic effect on the government and were the principal reasons for the introduction of a much more stringent policy. Later, however, as in-process control procedures began to improve and surveillance on marketed products by the medical profession became more vigilant, the producers and users of products began to question the strictness of requirements. At the same time they also began to question the general role of the regulatory agency. The controversy grew, and by the 1970s it had reached its peak. It gave rise to considerable and heated discussion at various levels; at government level in Congress and in the courts, and at a number of conferences held by professional, industrial and academic groups.

The principal factors and opinions which either support or oppose the role of the government in controlling medicinal substances are summarised below.

As referred to earlier, the over-riding criticism levelled against the FDA is that since the introduction of much more stringent control measures in 1962 and later, and because of general over-cautiousness, there has been marked and unnecessary delay in the approval and marketing of new medicinal products (2). The criticism has been shown to be justified by a number of investigations. On the other hand, there is also evidence which indicates that while the number of drug applications submitted to the FDA is large, only a small proportion merit approval because many would have very limited use or are of doubtful worth. It has also been averred that the question as to how many products have been delayed from being marketed by FDA policy is far less relevant than the question of the number of important substances that have been developed and approved. Moreover, it is maintained that in this more meaningful context, on a year-by-year comparison, the number of products made available for use has remained unchanged during the past 25 years and the percentage of particularly beneficial substances has increased during the past decade. Other criticisms of FDA policy are that many poor countries cannot afford to buy many new products because the over-restrictive requirements create much higher development costs; that there is less incentive to invest in research and development compared to the attractiveness of investing in more profitable products; and that costs to the taxpayer are increased.

Those who oppose introducing major changes in the present regulatory system have given a number of reasons to support their case. They claim that because of the profit motive the pharmaceutical industry is too often unconcerned about medicines that would reduce diseases affecting poor or underdeveloped countries; that expenditure by the pharmaceutical industry in promoting their products is much higher than is spent on research; that on a risk-versus-benefit basis to change to a free market system would probably result in the use of medicines carrying greater risks than benefits because the consumers (prescribing doctors and patients) would be likely to be inadequately informed; and that pressures on a manufacturer, such as the urgency to market a product earlier than his competitor, the minimising of costs and the temptation to exaggerate the advantages and to understate the disadvantages of a product, could lead to inadequate pre-market testing and to disturbing social costs.

With regard to the pharmaceutical industry, whatever its real or putative shortcomings, in general it functions efficiently in terms of production, and the contribution it makes to controlling the quality of its products is considerable. One cannot, of course, affirm that an industry not operating on a profit motive would do less well; it might do better. A well-documented study completed about 10 years ago compared the output of pharmaceutical products in countries where the pharmaceutical industry is entirely, or in part, owned privately and in those countries where it is entirely under state ownership. The findings disclosed that the products introduced over a period of about a decade were developed and produced overwhelmingly in those countries where the industry was not owned by the state. Indeed, continued study has revealed that the USSR and other countries comprising the Eastern European bloc have produced scarcely a single new medicinal substance of genuine value (6).

The measure of the extent to which the pharmaceutical industry controls the quality of its products by a combination of voluntary and mandatory measures is shown by the fact that much of the industry applies controls which are beyond those required by law. If the thoroughness, and particularly the extent, with which the pharmaceutical industry generally controls its products were less, regulatory agencies would themselves be unable to do so because of shortages of staff and resources.

The controversies and the attendant struggle between groups in countries where the pharmaceutical industry is privately owned, whose aims are very similar but who differ in the control systems they employ, or would wish to employ, will doubtless continue. It is a discord comprising medical, economic, social and political interests which, collectively, decide the present shape of the control of medicinal products and influence their future.

9. CONTROL MEASURES FOR NEW PRODUCTS AND SUBSTANCES PRODUCED BY UNCONVENTIONAL METHODS

Major developments in technology are taking place which in some cases are supplementing and in others superseding the conventional methods by which medicinal substances are produced. These procedures are also making possible the manufacture of substances which cannot be produced by customary methodology.

New technological processes have already produced changes of a kind which until comparatively recently seemed very distant or unattainable. The preparation of antigenic components of whole viruses (those causing influenza and foot-and-mouth disease) have been synthesised by chemical means and are being used as experimental vaccines. In the field of biotechnology, developments of a revolutionary nature, which are increasingly finding many applications, have taken place; namely, recombinant DNA technology and the methodology for effecting the fusion of two different types of cell to produce hybrid cells, designated 'hybridoma cells' because one of the cells used originates from a tumour.

Modern recombinant DNA technology enables the isolation of genes (composed of nucleic acid molecules) coding for specific biologically active proteins, their insertion into prokaryotic or eukaryotic cells and subsequent expression. Substances with therapeutic potential, such as insulin, interferon and viral vaccines, are already being produced by means of this technology. Hybridoma cells have been used to produce specific monoclonal antibodies which are available commercially for use in diagnostic studies. They have at present several additional possible uses. These include: a reclaimed role for antibodies for producing passive immunisation against microbial infection; the purification of substances synthesised either by the use of conventional substrates or by prokaryotes or eukaryotes subjected to genetic manipulation; and, in the diagnosis of cancer, the selective recognition of tumour cell surface markers using antibodies labelled with cytotoxic tracers.

Substances produced by these new technological methods will have to be stringently controlled with regard to their quality, safety and efficacy. In some cases the criteria set by a regulatory authority for the acceptability of these products will be different from those in use at present; in others the requirements considered appropriate for substances produced by conventional methods will also be applied to substances produced by the use of recombinant DNA technology.

For new products, because of unfamiliarity and lack of experience concerning their nature and the procedures used in their production, the enforcement of rigid requirements would be both unwise and unwarranted. Instead, common sense counsels that it would be sensible to apply flexible requirements initially with the opportunity to introduce changes at any time later as experience regarding the production and the use of products accumulates. In support of this approach, experience has shown that if a binding requirement is introduced at the outset it is extremely difficult to remove it once it has become well established. Accordingly, control measures for new products in particular should not only be flexible but should also be subject to frequent review to ensure that at any stage requirements considered to be inappropriate or unnecessary should be altered or discontinued.

10. TRENDS AND PROSPECTS

A number of trends are evident with respect to the production and regulatory control of medicinal substances, some of which have already begun, others that seem likely to develop in the future.

Multi-national pharmaceutical companies may continue to expand and become larger in size and fewer in number. National regulatory authorities are likely to become more harmonised. Relationships between regulatory agencies, the pharmaceutical industry and consumer groups will probably continue to improve. Increased compatibility is already evident internationally and, should it continue, it ought to lead to added activity and productivity in research and development. The role of committees appointed externally to advise regulatory agencies is becoming more prominent and their influence and importance as part of regulatory systems are likely to continue to grow.

The production and control of medicinal substances are inseparably linked and, therefore, they have to be considered conjointly when speculating on the future control of these products. A number of trends seem possible with respect to such developments. One such likelihood is that the harmonisation process between the forces involved may be extended. Another possibility is that in the more distant future there may be increasing control and production by the state. But whatever form the forces responsible for production and control assume in the future, what is of paramount importance is that a responsible authority should be constantly motivated by the aim of introducing and maintaining a system that is intended to attain the maximum benefit and the minimum risk for the health of the people it serves.

REFERENCES

1. Abrams, W. B. (1976). Therapeutics and government: 1776 and 1976. *Clin. Pharmacol. Ther.* **20**, No. 1, 4.
2. Cuthbert, M. F., Griffin, J. P., and Inman, W. H. W. (1978). In "Controlling the Use of Therapeutic Drugs: An International Comparison" (W. M. Wardell, ed.), Vol. 14, p. 132. American Enterprise Institute for Public Policy Research, Washington, D.C.
3. Cuthbert, M. F., Griffin, J. P., and Inman, W. H. W. (1978). In "Controlling the Use of Therapeutic Drugs: An International Comparison" (W. M. Wardell, ed.), Vol. 14, p. 133. American Enterprise Institute for Public Policy Research, Washington, D.C.
4. Dukes, M. N. G. (1973). Law, medicines and the doctor: A critical look at drug regulation. *Curr. Med. Res. Opin.* **1**, No. 10, 623.
5. Dukes, M. N. G. (1973). Law, medicines and the doctor: A critical look at drug regulation. *Curr. Med. Res. Opin.* **1**, No. 10, 625.
6. Dunlop, D. (1978). The innovation, benefits, drawbacks and control of drugs. *J. R. Soc. Med.* **71**, No. 5, 329.
7. Dunlop, D. (1980). The growth of drug regulation in the United Kingdom. *J. R. Soc. Med.* **73**, No. 6, 407.
8. Penn, R. G. (1979). The state control of medicines: The first 3000 years. *Br. J. Clin. Pharmacol.* **8**, No. 4, 293–305.

FURTHER READING

Dorsey, R. (1979). The case for deregulating drug efficacy. *JAMA, J. Am. Med. Assoc.* **242**, No. 16, 1755–1760.

Douglas, R. D. (1979). National drug policies—more state intervention or less? *World Med.* **14**, No. 21, 29–36.

Helfand, W. H. (1975). The United States and international drug regulatory approaches. *J. Am. Pharm. Assoc.* **NS15,** No. 12, 702–704.

Krzyzyk, V. (1977). The history of drug regulations. *Bull. Parenter. Drug Assoc.* **31**, No. 3, 156–160.

Lasagna, L. (1976). Drug discovery and introduction: Regulation and over-regulation. *Clin. Pharmacol. Ther.* **20**, No. 5, 507–511.

Penn, R. G. (1979). The state control of medicines: The first 3000 years. *Br. J Clin. Pharmacol.* **8**, No. 4, 293–305.

Simmons, H. E. (1974). The drug regulatory system of the United States Food and Drug Administration: A defense of current requirements for safety and efficacy. *Int. J. Health Serv.* **4**, No. 1, 95–107.

Wardell, W. M., ed. (1978). "Controlling the Use of Therapeutic Drugs: An International Comparison," Vol. 14. American Enterprise Institute for Public Policy Research, Washington, D.C.

PART IV

PROSPECTS

16

Genetic Engineering of Animal Cells

F. COLBÈRE-GARAPIN
A. C. GARAPIN
Unité de Virologie Medicale
Institut Pasteur
Paris, France

Animal Cell Biotechnology, Vol. 2

1. INTRODUCTION

The cloning of eukaryotic genes in bacteria has allowed a determination of their primary structure. DNA sequencing has pointed out some regulatory regions which are usually found at the extremities of genes. However, the cloned cellular genes must be transferred back into animal cells to study their integrity, function and regulation, or to establish cell lines endowed with new properties.

The chromosomal gene, or the cDNA copy complementary to the messenger RNA (mRNA), of cellular or viral origin, can be cloned in bacteria or lower eukaryotes such as yeast. A variety of methods exist to transfer the cloned gene, or cDNA copy, back into the animal cells. These include transfection techniques, fusion of cells with liposomes or bacterial protoplasts, microinjection etc. A transient and/or permanent expression of the foreign gene can be induced in the recipient cells. The transient expression is induced in the first few days after gene transfer, and yields information concerning gene integrity and sequences involved in gene expression such as promoters, enhancers, polyadenylation signals etc. This kind of expression is strongly enhanced when the foreign gene is amplified by the use of viral vectors.

Stable expression of a gene after transfection is a rare event. Generally, it is necessary to select the cells which have stably acquired foreign DNA, which can be achieved by using a selective marker. Long-term expression of a gene following its cotransfer along with a selective marker is adequate to study gene induction, amplification etc.

Gene transfer into animals is possible at various stages of embryonic life or animal development, but the most promising experiments are performed at the one-cell stage.

2. CLONING OF VIRAL AND CELLULAR GENES

Chromosomal genes, or the cDNA copy complementary to the messenger RNA, can be cloned in plasmids, bacteriophages or cosmids.

Plasmids are double-stranded circular DNA molecules. The most widely used vector is pBR 322, which contains ampicillin- and tetracycline-resistance genes and several convenient restriction sites (6). This vector and derivatives are convenient for cloning DNA fragments smaller than 10 kilobase pairs (kb). Larger DNA fragments can be cloned in bacteriophage λ, which is a double-stranded DNA virus with a genome size of about 50 kb. The central third of the viral genome was shown to be nonessential for lytic growth. Bacteriophage vectors having a single restriction site at which foreign DNA is inserted are known as insertion vectors; those having a pair of sites spanning a segment that can be replaced by foreign DNA are known as replacement or substitution vectors (87).

The large DNA fragments (>20 kb) can be cloned in cosmids (Cos) (23). These cloning vectors contain a drug-resistance marker, a plasmid origin of replication and the ligated cohesive ends of bacteriophage λ. DNA fragments up to 45 kb in length can be cloned in such vectors. Some vectors carry selective markers for growth in bacterial and animal cells (54, 55). More details concerning vectors and cloning methods have been given in a review (87).

3. EUKARYOTIC VIRAL VECTORS

3.1. SV40 Vectors

3.1.1. *The Lytic System*

Eukaryotic viral vectors allow the amplification of foreign sequences in animal cells. For example, SV40 genomes are amplified up to 100,000-fold in permissive cells at the end of the lytic cycle due to replication [for review, see Gruss and Khoury (59)]. However, in lytic systems, the size of the foreign DNA which can be inserted into the viral genome is limited by the size of the virus capsid. No more than 2.5 kb of DNA can be inserted into SV40 vectors, and a helper virus is required for virus multiplication. On the other hand, small viral genomes such as the SV40 genome are easier to handle and contain fewer restriction sites than larger ones. Early or late viral sequences can be replaced by foreign DNA, and many results have been obtained using this vector.

For example, influenza virus haemagglutinin gene expression was studied extensively; it was shown that the polypeptide is correctly glycosylated, that it accumulates at the cell surface and that over 10^8 molecules are produced per cell (40, 65). The importance of the amino terminal signal of the molecule for glycosylation and cell surface localization was demonstrated by Gething and Sambrook (41). Haemagglutinin genes, lacking the hydrophobic carboxy-terminal sequences, were not accumulated at the cell surface, but excreted into the medium (144).

Gene splicing in a heterologous cell species, splicing signals and the necessity for sense orientation of the intron for splicing were studied by several groups working on rabbit or mouse β-globin gene expression (56, 61, 103, 139). Polypeptide synthesis, processing and excretion were studied in the case of human growth hormone genes (113) and human hepatitis virus B surface antigen (84, 102).

The regulation of gene expression can also be studied with SV40 vectors; for example Hamer and Walling (62) have shown that regulation of a mouse metallothionein gene by cadmium occurs primarily at the transcriptional level.

3.1.2. The 72-bp Enhancer Sequence

The genome of SV40 itself has been better studied and understood since deletion mutants and hybrid genomes could be constructed *in vitro* and transferred back into permissive cells. The role of tandem repeated sequences located on the late side of the SV40 DNA replication origin was examined by several groups (2, 58, 155). It has been shown that the largest of the repeated sequences, which is 72 base pairs (bp) long, forms an essential element in the early viral transcriptional promoter (58). It is a potentiator of initiation of transcription in either orientation. The 72-bp repeat potentiates RNA chain initiation preferentially from the proximal rather than the distal start site, which accounts for the so-called effect of distance, and it preferentially potentiates transcription in a direction away from rather than towards it (155).

Levinson *et al.* (81) have shown that the Moloney sarcoma virus (MSV) 72-bp repeat can substitute functionally for the SV40 repeats. However, the SV40 tandem repeats appear to activate gene expression to significantly higher levels in monkey kidney cells, while the MSV repeats are more active in two lines of mouse cells (79). Laimins *et al.* (79) suggest that the tandem repeat elements may constitute one of the elements determining the host range of these eukaryotic viruses.

BK virus, a human papovavirus, contains three copies of a 68-bp sequence to the late side of its replication origin (135). In the case of polyoma virus, de Villiers and Schaffner (28) have shown that a DNA fragment from the beginning of the late region strongly enhances the level of correct β-globin gene transcripts over a distance of at least 1400 bp.

3.1.3. The Cos Non-lytic Amplification System

A non-lytic cell system for SV40 vector amplification has been developed by Gluzman *et al.* (42). This was based on a previous observation: SV40 genomes having a functional origin of replication and large T antigen gene cannot stably transform permissive monkey cells because viral excision occurs, probably due to replication of integrated viral DNA (7). Gluzman *et al.* (42) constructed origin-defective mutants of SV40 which can stably transform monkey kidney cells. The transformed Cos cells express large T antigen and support the replication of pure populations of SV40 mutants with deletions in the early region. These cells are possible hosts for the propagation of populations of recombinant SV40 viruses (43). This system has several advantages: it allows the replication of molecules having a functional viral origin of replication, and the size of the DNA molecules is not strictly limited because they are not encapsidated. Several studies have already been made using these cells (1, 85, 94). Lusky and Botchan (86) found that the replication rates of various recombinant plasmids were 20- to 100-fold lower than those of wild-type SV40 DNA transfected into Cos cells, due to the presence of a "poison" sequence in pBR 322-derived vectors. However, Tsui et al. (148) have shown that removal of the pBR 322 inhibitory sequence is not necessary for stable maintenance of the recombinant molecules in Cos cells.

Maintenance of SV40 vectors in Cos cells is a rare event, since most of the DNA-transfected cells die. In fact, SV40 vector replication in Cos cells peaks about 2 days after transfection, and then declines. By using SV40 vectors carrying a selective marker, a few stably transformed clones, corresponding to approximately 0.01–0.1% of the initial transfected cell population, can be isolated (148).

Analogous to the Cos cells, COP lines of transformed mouse cells were established with a polyoma deletion mutant which was replication negative (149).

Polyoma virus vectors have been developed, which are similar to SV40 vectors (37).

3.2. Papilloma Vectors

Although the absence of a tissue culture system to propagate the papillomaviruses has precluded their analysis by standard virological procedures, these viruses are particularly interesting as vectors used to amplify foreign genes in animal cells. Mouse cells can be transformed by a plasmid carrying the transforming region of bovine papillomavirus (BPV) and the recombinant molecule of DNA propagates as a plasmid in mouse cells (as well as in bacteria) (5, 30, 132). There are 10 to 120 copies of circular DNA per cell. In this system, the rat pre-proinsulin gene is correctly expressed (132). Cells transformed with a BPV-linked gene that encodes a cell surface marker such as human or mouse histocompatibility antigens could be isolated by using a fluorescence-activated cell sorter. The BPV plasmid could then be recovered by transforming bacteria with plasmid DNA from the transformed mouse cells (45). This type of vector has several advantages: the size of the recombinant DNA propagated in mouse cells is not strictly limited, and this DNA has a selective marker in mammalian cells—the gene for oncogenic transformation.

3.3. Herpes Simplex Virus Vectors

Two types of herpes simplex virus (HSV) vectors have been developed, both of which are used under cell lytic conditions. In the first system the foreign gene, which can be up to 7.5 kb long, is inserted in the thymidine kinase (TK) gene of the HSV genome (98). Because the TK gene is not required for virus growth in dividing cells, TK-negative recombinant virus can replicate without helper, leading to the amplification of foreign sequences.

Speate and Frenkel (140) developed a cloning–amplifying vector which they called amplicon, which contains a herpesvirus origin of replication, a sequence required for DNA cleavage and encapsidation and bacterial plasmid DNA. Viral sequences are derived from defective genomes. The recombinant chimeric defective genomes are packaged into virus particles when cotransfected with helper virus DNA, and monomeric chimeric repeat units can be transferred back and forth between bacteria and eukaryotic cells (shuttle vector). The use of HSV amplicon allows the insertion of relatively large stretches of foreign DNA sequences. The chimeric genomes persist through undiluted virus propagation in the absence of selective pressure. Finally, virus populations containing the chimeric defective genomes can be used to infect a wide variety of host-cell species that are susceptible to HSV (98, 99, 140).

3.4. Retrovirus Vectors

Retroviruses are natural vectors for introducing foreign DNA into the vertebrate cell genome, because in their normal life cycle, retroviruses integrate their DNA into the cell genome. Wei et al. (157) and Shimotohno and Temin (137) have shown that a retrovirus containing the HSV TK gene can be propagated either as virus or provirus and can transform tk^- chicken and rat cells to a tk^+ phenotype with very high efficiency. These retrovirus vectors should have great utility for the insertion of genes into mammalian cells or into laboratory animals (146). They have also led to a better knowledge of retrovirus genomes, since the sequence required for encapsidation of genomic viral RNA was deduced from experiments using these vectors (156).

4. SELECTIVE MARKERS

Stable biochemical transformation of animal cells is a rare event. Thus, it is often necessary to select the cells which have acquired the foreign DNA. For this purpose, several selective markers have been cloned or constructed. The most commonly used is the thymidine kinase gene of HSV, which has been described in detail in several articles and reviews (18, 20, 21, 91, 116). This marker, which has a high transformation efficiency in tk^- mouse L cells, has, however, an important limitation: it can be used only in

tk^- cells. Other markers, which can be used in any cell type having no special enzymatic defect, have now been developed (20, 104, 105, 108, 125).

4.1. Xanthine–Guanine Phosphoribosyltransferase from *Escherichia coli* (Eco *gpt*)

The selection applied to animal cells after transfection with this marker is based on the possibility, unique to the bacterial enzyme, of using the xanthine efficiently as a precursor for the synthesis of guanylic acid. Only animal cells which have acquired the *E. coli* gene can survive in a medium containing mycophenolic acid and aminopterin, which inhibit the *de novo* synthesis of purines and xanthine (104, 105).

To make the Eco *gpt* gene function in animal cells, Mulligan and Berg (105) added to the coding region of the gene the transcription signals of a gene isolated from SV40. Monkey and mouse cells were transformed by the hybrid gene with a frequency of 10^{-5} to 10^{-4}. This marker has been used by Bourachot *et al.* (8) to study early and late control sequences of SV40 and polyoma virus.

4.2. Dihydrofolate Reductase (DHFR)

The DNA of hamster cells resistant to methotrexate is able to confer to mouse cells resistance to this drug (160). In transformed mouse cells, the hamster gene coding for DHFR and DNA sequences transferred along with the marker can be amplified upon passage in increasing concentrations of the drug. Such a strategy has been used by Christman *et al.* (16) to obtain the transfer, expression and amplification of the hepatitis B virus surface antigen (HBs) gene in mouse cells. However, the hamster gene coding for DHFR is not very convenient to use as a selective marker because of its large size (about 40 kb). For this reason, O'Hare *et al.* (108) have constructed a hybrid gene consisting of the coding region of the *E. coli* DHFR gene and eukaryotic transcription signals. This hybrid gene has been used as a selective marker in mouse L cells.

Another hybrid DHFR gene has been constructed by Ringold *et al.* (124), with the cDNA copy of the mouse DHFR mRNA and the transcription signals of the mouse mammary tumor virus. Although this gene has been used only in mutant cells, it is important to note that this hybrid gene is amplifiable like the eukaryotic chromosomal gene (39, 74, 124).

4.3. Aminoglycoside-3′-phosphotransferase (APH 3′)

Jimenez and Davies (72) have shown that yeast are sensitive to an antibiotic, G418, and that the bacterial transposon Tn601 is able to confer G418 resistance to yeast. These experiments suggested that a similar selection could be applied to higher eukaryotic cells.

All the animal cell lines that we have tested are sensitive to G418. We have constructed a hybrid gene in which the coding region is a bacterial gene coding for an aminoglycoside-3'-phosphotransferase (APH 3') which phosphorylates—and thus inactivates—G418. The APH 3' hybrid gene, under the control of herpes TK transcription signals, transforms mouse L cells to G418 resistance with an efficiency of 70 to 80 colonies per microgram of supercoiled plasmid DNA. Biochemical transformation of several human and monkey cell lines has also been successful, although with a lower efficiency. The cotransfer of a gene for which no selection exists, with the selective marker, has been demonstrated with various transferred genes and recipient cell systems (20, 22).

4.4. The Multifunctional CAD Enzyme

The multifunctional CAD enzyme catalyzes the first three steps of the *de novo* biosynthesis of uridine. One of these steps is specifically blocked by N-phosphonoacetyl-L-aspartate (PALA). When rodent cells are submitted to increasing concentrations of PALA, the CAD gene is amplified, and the cells synthesize large amounts of the CAD enzyme.

The CAD gene, which is at least 25 kb long, has been cloned in a cosmid (125) and transferred into hamster cells by protoplast fusion (see next section). Transformed cell colonies appear in the selective medium after the transfer. Some spontaneously PALA-resistant colonies are also detectable, but this marker, like the DHFR gene, has the advantage of being amplifiable. Other genes, such as the bacterial chloramphenicol acetyltransferase (CAT), could be used as dominant selective markers. Up to now, however, the CAT gene has only been used to compare the strengths of various promoters in mammalian cells (48, 49).

5. GENE TRANSFER METHODS AND RECIPIENT CELLS

5.1. Gene Transfer Methods

Several methods have been developed to transfer a gene into animal cells, ranging from simple DEAE–dextran transfection to the sophisticated microinjection technique. The DEAE–dextran method, first described by McCutchan and Pagano (90), has been improved and is still used routinely to transfect SV40 DNA into monkey cells (138). The technique is very simple since cells are only treated with a mixture of DEAE–dextran and DNA. As shown with the herpesvirus thymidine kinase gene, it seems, however, that this technique is more efficient for inducing a transient expression of the foreign gene than for obtaining stable transformants (97).

The calcium phosphate transfection technique developed by Graham and Van der Eb (53) and modified by Wigler *et al.* (159) has been the most commonly used method in the past few years. This is because it allows the biochemical transformation of tk^- mouse L cells to the tk^+ phenotype by the herpes TK gene with relatively good efficiency (1 colony per 20 pg of DNA per 10^6 cells). However, high efficiencies are only obtained when carrier DNA (salmon sperm, calf thymus or the recipient cell DNA) is included in the precipitate of calcium phosphate. Moreover, the transformation efficiency obtained with this technique is highly dependent on the cell line.

A transient gene expression can also be observed in as much as 70% of cells in the first few days after transfection when the calcium phosphate technique is followed by a polyethylene glycol (PEG)–sucrose shock (136).

Fraley *et al.* (35) have demonstrated the possibility of delivering viral DNA into recipient cells by the use of lipid vesicles or liposomes. However, the transfection efficiency, which is quite good in the case of RNA molecules (112), is lower than the efficiency obtained with the calcium phosphate technique in the case of DNA molecules (133).

The fusion of animal cells with bacterial protoplasts has been developed to transfer plasmids or cosmids into the cells (134). Biochemical transformation of tk^- cells with the TK gene has been obtained with a frequency of 1/300 to 1/500 (131). This technique, which has been improved (123), is thus very efficient, and the efficiency is not dependent on the cell type (125). A disadvantage of this technique, however, is the fact that all the bacterial DNA is transferred into the animal cell.

RNA or DNA microinjection into cells (51, 52) is the most efficient and precise technique. Molecules can be delivered to particular cell compartments—nucleus or cytoplasm—and the presence of carrier DNA is not required. Capecchi (15) has shown that transport of the DNA to the nucleus is required for its expression, and that the number of tk^- cells stably transformed by the TK gene is 100-fold lower than the number of cells which express the gene transiently. As with the calcium phosphate technique (18), stable transformants are obtained with a higher efficiency when linearized plasmid DNA is used for microinjection (34).

5.2. Recipient Cells

The TK selective marker and cotransferred genes have been, in most experiments, transfected into tk^- mouse LM cells (75). For example, the fact that cotransferred sequences are physically and genetically linked in the recipient cells has been established in these cells (117). The tk^- LM cells have several advantages: they never revert spontaneously to the tk^+ phenotype, and they have a high transfection efficiency with the calcium technique. Moreover, they are easily cloned and grown. Other cell lines, which have a much lower transfection efficiency, have been used for special purposes; for example, Robins et al. (127) have used a tk^- rat cell line to study the association of the transfected DNA with chromosomes, because this cell line is essentially euploid.

Teratocarcinoma cell lines have been used as recipient cells to study the expression of transfected genes at various stages of differentiation (83, 116). SV40 vectors which are amplified with a helper virus under lytic conditions have been transfected into permissive monkey kidney cells (103). The cos monkey kidney cells, which are transformed by an origin defective SV40 and express large T antigen permanently (42), have proved to be useful for obtaining amplification of SV40 vectors without helper virus (see Section 3). For potential applications, heteroploid monkey kidney cell lines (e.g. Vero) have an advantage over mouse cell lines, such as LM, since the former are not or are only weakly oncogenic for *nude* mice, and they do not harbour retroviruses (118).

Contact-inhibited mouse 3T3 cells have been used mainly to detect and isolate cellular oncogenic genes [for review, see Cooper (25)]. Other cells, such as bone marrow cells, have been transfected with viral genes, and the transformants have been selected *in vivo* after reimplantation in the animal (96). Among the most exciting experiments is the microinjection of fertilized mouse oocytes followed by reimplantation in a foster mother (see Section 8). *Xenopus* oocytes have been used mainly for short-term expression studies (92) and in a few cases for long-term expression studies (130).

6. TRANSIENT GENE EXPRESSION

Short-term expression studies, performed in the first few days after transfection, can provide much information about the functions encoded by DNA sequences. The infectivity of a DNA molecule is, of course, a typical short-term expression. When cloned genes began to be transferred back into animal cells, it was important to know whether a viral DNA, covalently linked to a vector, can initiate viral infection. As part of a risk-testing program, this question was investigated with polyoma virus DNA. It was shown that the only recombinant molecules which were infective as intact molecules were plasmids with dimeric polyoma DNA inserts (36).

Since then, it has been shown that cloned poliovirus complementary DNA is infectious in mammalian cells (122) and that cloned hepatitis B virus (HBV)

DNA causes hepatitis in chimpanzees (162). This kind of study demonstrates the integrity of genes and DNA sequences necessary for gene expression.

Short-term expression experiments have also been useful for identifying genes, for example genomic genes coding the human major histocompatibility antigens (3), or resolving the functions of overlapping genes (100).

Some genetic defects can now be analyzed at the molecular level and correlated with abnormal gene expression. For example, gene mapping and sequencing experiments have demonstrated a variety of causes for the thalassemias, which are hereditary defects in human globin chain synthesis. By studying the transient gene expression in the first few days after transfection, Busslinger et al. (13) and Felber et al. (32) showed that a β^+- and an α-thalassemia, respectively, were due to an abnormal RNA splicing.

Transcriptional control sequences, particularly those which are found upstream of the transcription start site, have been studied extensively by using short-term lytic or non-lytic expression systems. Point mutations or small deletions are first induced in the gene. In a second step, the mutated gene inserted in a vector is assayed for expression in animal somatic cells or Xenopus oocytes (4, 55, 93). Among somatic cells, the cos cells allow a high transient gene expression level under non-lytic conditions. Using this system, Humphries et al. (69) have shown that the human α-globin gene promoter functioned independently, but that the β- and δ-globin gene promoter was nearly totally dependent on the enhancing activity of the 72-bp direct repeats from the SV40 genome. Banerji et al. (2) had previously shown that the transcriptional enhancer element could act in either orientation at many positions.

To study the function of transcription signals, the HBV surface antigen (HBS) gene is a good model since the antigen is excreted into the cell culture medium (31), and the release of antigen can be detected as early as 48 hr after transfection (154). Using the HBS gene and the dominant selective marker APH 3' in monkey kidney cells, we found a late transient expression of the HBS gene in these cells (22), while a permanent expression of the gene was obtained in mouse cells under the same conditions. These results show that the type of gene expression obtained may depend on the recipient cells.

7. PERMANENT GENE EXPRESSION

7.1. Constitutive Expression

Permanent gene expression is generally obtained in a very few cells after the isolation of clones which have stably acquired a selective marker and the cotransferred sequences. Some results obtained in short-term expression systems can also be obtained in long-term expression systems: those concerning the identification of genes (95, 101), the localization of a gene (143), the promoter function (29, 120) or transcription enhancers (24, 73). The activity of a genomic gene is, however, more easily studied in clones of cells having stably acquired the foreign gene (9, 50, 78, 88, 121, 164). Long-term expression systems allow the study of interactions between the recipient cell and the transfected gene. For example, it was shown in biochemically transformed tk+ cells that a nuclease hypersensitivity of the TK gene promoter correlates with expression (145).

Long-term expression systems also allow the establishment of cell lines endowed with new properties. This may have potential applications in the case of cells permanently expressing a viral antigen (31), or for more fundamental studies, for example in the case of the establishment of cell lines containing nonsense mutations and functional suppressor tRNA genes (68).

The isolation and characterization of oncogenic genes present in chemically transformed or tumor cells by gene transfer is not the least interesting aspect of the possibilities offered by this technology. The oncogenic genes induce the morphological transformation of NIH 3T3 cells upon gene transfer [for review, see Weinberg (158) and Cooper (25)]. The cellular gene can then be analyzed by hybridization with homologous viral genes, cloned, sequenced and retransferred into animal cells.

7.2. State of the Transfected DNA

The state of the transfected gene in the recipient cell can be studied only in stable or semi-stable transformants. With the exceptions of SV40 vectors in cos cells and papilloma virus vectors, which can remain as stable episomes in transformed cells, the transfected DNA generally integrates into the recipient cell chromosome (127). There exists a single integration site, and the site differs in each independent transformant. When revertants are counterselected, most of them delete segments of transforming DNA (128).

Excision of the transfected genes can be induced in particular cases, for example, when mouse cells transformed by plasmids containing the SV40 origin of replication and early region are fused with permissive monkey kidney cells (63). With few exceptions, the recovery of a transforming plasmid from the animal cell DNA is not easy (76).

7.3. Regulation of Expression

The expression of several classes of cellular genes is inducible *in vivo*. Induction factors may be a viral infection, a heat shock, heavy metals, hormones, cell senescence or differentiation etc. Transcription of a given gene may also depend on the methylation of DNA, or its accessibility to enzymes (27). Finally, in response to specific inducers, the cell may amplify some genes or gene regions, which results in a very high level of gene expression. All these regulation mechanisms are now studied at the molecular level with gene transfer technology.

Human interferon (IFN) α, β and γ genes are induced in cell culture after infection with Newcastle disease virus or poly(rI)–poly(rC) treatment. It was found by Ohno and Taniguchi (109) and confirmed by others (14, 66, 165) that a cloned chromosomal DNA fragment which contains the human IFN β1 gene codes for bona fide IFN β when transferred into mouse cells, and that this expression is increased when cells are treated with IFN inducers. However, some authors (119) found that the cDNA coding sequence was sufficient to respond to the specific induction system, while others (147) found that specific IFN regulation implicated the region between nucleotides −144 and −186 from the transcription initiation site.

The 5′ side of the transcribed portion of genes has been implicated in regulation in many other systems (11, 17). The heat-shock proteins are synthesized in response to an increase in temperature in cells from a wide range of organisms. Deletion mutants of the *Drosophila* heat-shock protein 70 gene have been assayed in cos cells, using a SV40 vector. Pelham (114) has shown that residues −10 to −66 of this gene are sufficient for heat-inducible promotion.

Another example of such regulation is given by the metallothioneins (MT). Metallothioneins are ubiquitous proteins that bind heavy metals and are thought to be involved in resistance to heavy-metal toxicity. Analysis of a set of deletion mutants revealed that the minimum sequence required for cadmium regulation of the mouse metallothionein-I gene lies within 90 nucleotides of the transcription start site (11). Although the endogenous mouse MT-I gene is transcriptionally regulated by heavy metals and by glucocorticoid hormones, transcription of transfected cloned MT-I gene is not regulated by glucocorticoids (89).

On the contrary, hormonal regulation was observed after DNA transfer in clones of cells transfected with rat α2μ globulin genes (77), mouse mammary tumor virus long terminal repeat (67, 80) and human growth hormone genes (129).

Teratocarcinoma pluripotent stem cells of murine teratomas are a good model for the study of differentiation. When teratocarcinoma cells are transfected with the herpes TK gene, selected for TK expression and injected into syngeneic mice, most of the tumors derived from these cells maintain the HSV TK gene without loss or rearrangement (116). When the SV40 genome is cotransferred with the TK gene into the stem cells, the SV40 early gene products are detectable only if differentiation of the cells is induced (83).

Similarly, Moloney murine leukemia virus expression is blocked in embryonal carcinoma cells, in which proviral genomes are highly methylated (142).

DNA methylation has been correlated with reduced gene expression in a number of studies [for review, see Felsenfeld and McGhee (33)]. Undermethylation is a general phenomenon in all actively transcribed genes (106), and it has been shown by direct DNA methylation in vitro that methylation inhibits the expression of a gene transferred into an animal cell (141, 150, 151).

Gene amplification, which has been best studied in the case of the cellular dihydrofolate reductase gene (60, 107) can now be studied by gene transfer. Not only is the chromosomal DHFR gene amplified after transfection when the methotrexate concentration is progressively increased in the cell culture medium, but the cDNA copy can also be amplified (80). A model for the study of gene amplification has been developed by Roberts and Axel (126). These authors have selected tk^+ mouse cells that amplified a promoterless TK gene and have shown that these cells possess an efficient correction mechanism that maintains sequence homogeneity among repeated genetic elements.

8. TRANSGENIC ANIMALS

Introduction of specific gene sequences into animal embryos is a powerful tool for the study of differentiation and gene regulation. Transgenic animals should be useful in developing new strains for veterinary purposes, and they are also a good model for the correction of genetic defects. Very encouraging results have already been obtained with the mouse and Xenopus laevis (46, 130).

DNA is injected directly into the male pronucleus of fertilized oocytes and injected embryos are implanted into pseudo-pregnant females. In a high proportion of embryos, the foreign gene is stably acquired without any selective pressure. In some cases, the transferred gene is correctly transcribed, and the mRNA may even be translated (10, 12, 64, 152, 153). In the case of metallothionein DNA sequences, the gene copy number is constant in each tissue but the expression of the gene is tissue-specific and inducible by heavy metals (10). The stable germ line transmission of genes injected into mouse pronuclei has also been reported (26, 47). One of the most spectacular results in this domain has been the construction of "super mice" which are larger than normal mice because they have acquired and expressed the rat growth hormone gene (111). The mice contained variable numbers of the gene and, with one exception, the mice with the highest growth rate had the most copies of the gene. The use of animals to produce large amounts of commercially important proteins is a novel concept which may be applied to the production of proteins that are not correctly expressed in cultured eukaryotic cells (*111, 163*).

Gene regulation in early mouse embryos has also been extensively studied by microinjection of cloned DNA into morula stage pre-implantation mouse embryos (*71*), and it has been shown by Jaenisch *et al.* (*70*) that the chromosomal region at which a viral genome is integrated influences its expression during development and differentiation.

One of the main problems of such experiments which is still unresolved is how to integrate the microinjected gene at a specific locus. It is necessary not to inactivate an important cellular gene, not to activate silent genes and to integrate the foreign gene into a region of the cellular genome where it will be expressed.

9. DISCUSSION

During the past 4 years, great progress has been made in the field of gene transfer technology. Up to now, these results have been mainly of fundamental interest and concern the identification of genes or DNA sequences and the study of their function, regulation and interaction with the recipient cells. It is, however, possible to imagine that several applications will arise from these results. Although the advantage of bacterial cultures for synthesizing polypeptides of biological importance is obvious, some polypeptides may undergo post-translational modifications only in animal cells, or may be excreted into the cell culture medium only in the case of animal cells.

Before dominant selective markers were available, the recipient cells which could be used for gene transfer were mutant cells (except under conditions that were lytic or induced the morphological transformation of cells). This limited the recipient cells to a very small number of cell lines, among which are the well-studied tk^- LM cells (*75*). The LM cells are, however, oncogenic for the syngeneic species, harbour retroviruses, and thus cannot be used for medical purposes. Many more cell lines can now be studied and modified by gene transfer.

A leap forward was also made in the past few years when it became possible to microinject fertilized mouse oocytes with purified DNA and reimplant them into pseudo-pregnant females. At this one-cell stage, the gene transfer process is extremely efficient and the foreign gene is found in the embryos at a high frequency, without any selective pressure. It has been found with some hybrid genes (*10*) that the expression is tissue-specific, and that the hybrid gene can be transmitted vertically. Many problems remain to be solved: for example, how to induce the integration of the foreign gene at a specific locus of the chromosome, or to maintain the gene in an episomal form. However, the results are already extremely encouraging.

ACKNOWLEDGEMENTS

We are grateful to Mrs. N. Perrin and Miss V. Caput for typing the manuscript. Part of the work presented in this review was supported by grants from INSERM (124016 and 124006), CNRS (95.5097 and 033916) and MRI (82.L.1316).

REFERENCES

1. An, G., Hidaka, K., and Siminovitch, L. (1982). Expression of bacterial β-galactosidase in animal cells. *Mol. Cell. Biol.* **2**, 1628–1632.
2. Banerji, J., Rusconi, S., and Schaffner, W. (1981). Expression of a β-globin gene is enhanced by remote SV40 DNA sequences. *Cell* **27**, 299–308.
3. Barbosa, J. A., Kamarck, M. E., Biro, P. A., Weissman, S. M., and Ruddle, F. H. (1982). Identification of human genomic clones coding the major histocompatibility antigens HLA-A2 and HLA-B7 by DNA-mediated gene transfer. *Proc. Natl. Acad. Sci. U.S.A.* **79**, 6327–6331.
4. Benoist, C., and Chambon, P. (1980). Deletions covering the putative promoter region of early mRNAs of simian virus 40 do not abolish T-antigen expression. *Proc. Natl. Acad. Sci. U.S.A.* **77**, 3865–3869.
5. Binetruy, B., Meneguzzi, G., Breathnach, R., and Cuzin, F. (1982). Recombinant DNA molecules comprising bovine papilloma virus type 1 DNA linked to plasmid DNA are maintained in a plasmidial state both in rodent fibroblasts and in bacterial cells. *EMBO J.* **1**, 621–628.
6. Bolivar, F., Rodriguez, R. L., Greene, P. J., Betlach, M. C., Heyneker, H. L., Boyer, H. W., Crosa, J. H., and Falkow, S. (1977). Construction and characterization of new cloning vehicles. II. A multipurpose cloning system. *Gene* **2**, 95–113.
7. Botchan, M., Topp, W., and Sambrook, J. (1979). Studies on SV40 excision from cellular chromosomes. *Cold Spring Harbor Symp. Quant. Biol.* **43**, 709–719.
8. Bourachot, B., Jouanneau, J., Giri, I., Katinka, M., Cereghini, S., and Yaniv, M. (1982). Both early and late control sequences of SV40 and polyoma promote transcription of *Escherichia coli* gpt gene in transfected cells. *EMBO J.* **1**, 895–900.
9. Breathnach, R., Mantei, N., and Chambon, P. (1980). Correct splicing of a chicken ovalbumin gene transcript in mouse L cells. *Proc. Natl. Acad. Sci. U.S.A.* **77**, 740–744.
10. Brinster, R. L., Chen, H. Y., Trumbauer, M., Senear, A. W., Warren, R., and Palmiter, R. D. (1981). Somatic expression of herpes thymidine kinase in mice following injection of a fusion gene into eggs. *Cell* **27**, 223–231.
11. Brinster, R. L., Chen, H. Y., Warren, R., Sarthy, A., and Palmiter, R. D. (1982). Regulation of metallothionein–thymidine kinase fusion plasmids injected into mouse eggs. *Nature (London)* **296**, 39–42.
12. Bürki, K., and Ullrich, A. (1982). Transplantation of the human insulin gene into fertilized mouse eggs. *EMBO J.* **1**, 127–131.

13. Busslinger, M., Moschonas, N., and Flavell, R. A. (1981). β+ Thalassemia: Aberrant splicing results from a single point mutation in an intron. *Cell* **27**, 289–298.

14. Canaani, D., and Berg, P. (1982). Regulated expression of human interferon β1 gene after transduction into cultured mouse and rabbit cells. *Proc. Natl. Acad. Sci. U.S.A.* **79**, 5166–5170.

15. Capecchi, M. R. (1980). High efficiency transformation by direct microinjection of DNA into cultured mammalian cells. *Cell* **22**, 479–488.

16. Christman, J. K., Gerber, M., Price, P. M., Flordellis, C., Edelman, J., and Acs, G. (1982). Amplification of expression of hepatitis B surface antigen in 3T3 cells contransfected with a dominant-acting gene and cloned viral DNA. *Proc. Natl. Acad. Sci. U.S.A.* **79**, 1815–1819.

17. Cochet, M., Chang, A. C. Y., and Cohen, S. N. (1982). Characterization of the structural gene and putative 5'-regulatory sequences for human proopiomelanocortin. *Nature (London)* **297**, 335–339.

18. Colbère-Garapin, F., Chousterman, S., Horodniceanu, F., Kourilsky, P., and Garapin, A. C. (1979). Cloning of the active thymidine kinase gene of herpes simplex virus type 1 in *Escherichia coli* K12. *Proc. Natl. Acad. Sci. U.S.A.* **76**, 3755–3759.

19. Colbère-Garapin, F., Garapin, A. C., and Kourilsky, P. (1982). Selectable markers for the transfer of genes into mammalian cells. *Curr. Top. Microbiol. Immunol.* **96**, 145–157.

20. Colbère-Garapin, F., Horodniceanu, F., Kourilsky, P., and Garapin, A. C. (1981). A new dominant hybrid selective marker for higher eukaryotic cells. *J. Mol. Biol.* **150**, 1–14.

21. Colbère-Garapin, F., and Garapin, A. C. (1982). Transfert de gènes dans les cellules eucaryotes et marqueurs de sélection. *Bull. Inst. Pasteur (Paris)* **80**, 61–81.

22. Colbère-Garapin, F., Horodniceanu, F., Kourilsky, P., and Garapin, A. C. (1983). Late transient expression of human hepatitis B virus genes in monkey cells. *EMBO J.* **2**, 21–25.

23. Collins, J., and Hohn, B. (1978). Cosmids: A type of plasmid gene cloning vector that is packageable in vitro in bacteriophage λ heads. *Proc. Natl. Acad. Sci. U.S.A.* **75**, 4242–4246.

24. Conrad, S. E., and Botchan, M. R. (1982). Isolation and characterization of human DNA fragments with nucleotide sequence homologies with the simian virus 40 regulatory region. *Mol. Cell. Biol.* **2**, 949–965.

25. Cooper, G. M. (1982). Cellular transforming genes. *Science* **218**, 801–806.

26. Costantini, F., and Lacy, E. (1981). Introduction of a rabbit β-globin gene into the mouse germ line. *Nature (London)* **294**, 92–94.

27. Davies, R. L., Fuhrer-Krusi, S., and Kucherlapati, R. S. (1982). Modulation of transfected gene expression mediated by changes in chromatin structure. *Cell* **31**, 521–529.

28. de Villiers, J., and Schaffner, W. (1981). A small segment of polyoma virus DNA enhances the expression of a cloned β-globin gene over a distance of 1400 base pairs. *Nucleic Acids Res.* **9**, 6251–6264.

29. Dierks, P., Van Ooyen, A., Mantei, N., and Weissmann, C. (1981). DNA sequences preceding the rabbit β-globin gene are required for formation in mouse L cells of β-globin RNA with the correct 5' terminus. *Proc. Natl. Acad. Sci. U.S.A.* **78**, 1411–1415.

30. Di Maio, D., Treisman, R., and Maniatis, T. (1982). Bovine papillomavirus vector that propagates as a plasmid in both mouse and bacterial cells. *Proc. Natl. Acad. Sci. U.S.A.* **79**, 4030–4034.

31. Dubois, M.-F., Pourcel, C., Rousset, S., Chany, C., and Tiollais, P. (1980). Excretion of hepatitis B surface antigen particles from mouse cells transformed with cloned viral DNA. *Proc. Natl. Acad. Sci. U.S.A.* **77**, 4549–4553.

32. Felber, B. K., Orkim, S. H., and Hamer, D. H. (1982). Abnormal RNA splicing causes one form of α thalassemia. *Cell* **29**, 895–902.

33. Felsenfeld, G., and McGhee, J. (1982). Methylation and gene control. *Nature (London)* **296**, 602–603.

34. Folger, K. R., Wong, E. A., Wahl, G., and Capecchi, M. (1982). Patterns of integration of DNA microinjected into cultured mammalian cells: Evidence for homologous recombination between injected plasmid DNA molecules. *Mol. Cell. Biol.* **2**, 1372–1387.

35. Fraley, R., Subramani, S., Berg, P., and Papahadjopoulos, D. (1980). Introduction of liposome-encapsulated SV40 into cells. *J. Biol. Chem.* **255**, 10431–10435.

36. Fried, M., Klein, B., Murray, K., Greenaway, P., Tooze, J., Boll, W., and Weissmann, C. (1979). Infectivity in mouse fibroblasts of polyoma DNA integrated into plasmid pBR 322 or lambdoid phage DNA. *Nature (London)* **279**, 811–816.

37. Fried, M., and Ruley, E. (1982). Use of a polyoma virus vector. *In* "Eukaryotic Viral Vectors" (Y. Gluzman, ed.), pp. 67–70. Cold Spring Harbor Lab., Cold Spring Harbor, New York.

38. Garapin, A. C., Colbère-Garapin, F., Cohen-Solal, M., Horodniceanu, F., and Kourilsky, P. (1981). Expression of the herpes simplex virus type 1 thymidine kinase gene in *E. coli*. *Proc. Natl. Acad. Sci. U.S.A.* **78**, 815–819.

39. Gasser, C. S., Simonsen, C. C., Schilling, J. W., and Schimke, R. T. (1982). Expression of abbreviated mouse dihydrofolate reductase genes in cultured hamster cells. *Proc. Natl. Acad. Sci. U.S.A.* **79**, 6522–6526.

40. Gething, M. J., and Sambrook, J. (1981). Cell-surface expression of influenza haemagglutinin from a cloned DNA copy of the RNA gene. *Nature (London)* **293**, 620–625.

41. Gething, M. J., and Sambrook, J. (1982). Construction of influenza haemagglutinin in genes that code for intracellular and secreted forms of the protein. *Nature (London)* **300**, 598–603.

42. Gluzman, Y., Frisque, R. J., and Sambrook, J. (1979). Origin-defective mutants of SV40. *Cold Spring Harbor Symp. Quant. Biol.* **49**, 293–300.

43. Gluzman, Y. (1981). SV40-transformed simian cells support the replication of early SV40 mutants. *Cell* **23**, 175–182.

44. Goldfarb, M., Shimizu, K., Perucho, M., and Wigler, M. (1982). Isolation and preliminary characterization of a human transforming gene from T24 bladder carcinoma cells. *Nature (London)* **296**, 404–409.

45. Goodenow, R. S., McMillan, M., Orn, A., Nicolson, M., Davidson, N., Frelinger, J. A., and Hood, L. (1982). *Science* **215**, 677–679.

46. Gordon, J. W., Scangos, G. A., Plotkin, D. J., Barbosa, J. A., and Ruddle, F. H. (1980). Genetic transformation of mouse embryos by microinjection of purified DNA. *Proc. Natl. Acad. Sci. U.S.A.* **77**, 7380–7384.

47. Gordon, J. W., and Ruddle, F. (1981). Integration and stable germ line transmission of genes injected into mouse pronuclei. *Science* **214**, 1244–1246.

48. Gorman, C. M., Merlino, G. T., Willingham, M. C., Pastan, I., and Howard, B. H. (1982). The Rous sarcoma virus long terminal repeat is a strong promoter when introduced into a variety of eukaryotic cells by DNA-mediated transfection. *Proc. Natl. Acad. Sci. U.S.A.* **79**, 6777–6781.

49. Gorman, C. M., Moffat, L. F., and Howard, B. H. (1982). Recombinant genomes which express chloramphenicol acetyltransferase in mammalian cells. *Mol. Cell. Biol.* **2**, 1044–1051.

50. Gough, N. M., and Murray, K. (1982). Expression of the hepatitis B virus surface, core and e antigen genes by stable rat and mouse cell lines. *J. Mol. Biol.* **162**, 43–67.

51. Graessmann, A., Wolf, H., and Bornkmann, G. W. (1980). Expression of Epstein–Barr virus genes in different cell types after microinjection of viral DNA. *Proc. Natl. Acad. Sci. U.S.A.* **77**, 433–436.

52. Graessmann, M., and Graessmann, A. (1976). Early simian virus 40 specific RNA contains information of tumor antigen formation and chromatin replication. *Proc. Natl. Acad. Sci. U.S.A.* **73**, 366–370.

53. Graham, F. L., and Van der Eb, A. J. (1973). A new technique for the assay of infectivity of human adenovirus 5 DNA. *Virology* **52**, 456–467.

54. Grosveld, F. G., Lund, T., Murray, E. J., Mellor, A. L., Dahl, H. H. D., and Flavell, R. A. (1982). The construction of cosmid libraries which can be used to transform eukaryotic cells. *Nucleic Acids Res.* **10**, 6715–6732.

55. Grosveld, G. C., de Boer, E., Shewmaker, C. K., and Flavell, R. A. (1982). DNA sequences necessary for transcription of the rabbit β-globin gene in vivo. *Nature (London)* **295**, 120–126.

56. Gruss, P., and Khoury, G. (1980). Rescue of a splicing defective mutant by insertion of an heterologous intron. *Nature (London)* **286**, 634–637.

57. Gruss, P., Efstratiadis, A., Karathanasis, S., König, M., and Khoury, G. (1981). Synthesis of stable unspliced mRNA from an intronless simian virus 40–rat preproinsulin gene recombinant. *Proc. Natl. Acad. Sci. U.S.A.* **78**, 6091–6095.

58. Gruss, P., Dhar, R., and Khoury, G. (1981). Simian virus 40 tandem repeated sequences as an element of the early promoter. *Proc. Natl. Acad. Sci. U.S.A.* **78**, 943–947.

59. Gruss, P., and Khoury, G. (1981). Gene transfer into mammalian cells: Use of viral vectors to investigate regulatory signals for the expression of eukaryotic genes. *Curr. Top. Microbiol. Immunol.* **96**, 159–170.

60. Haber, D. A., and Schimke, R. T. (1981). Unstable amplification of an altered dihydrofolate reductase gene associated with double-minute chromosomes. *Cell* **26**, 355–362.

61. Hamer, D. H., and Leder, P. (1979). Expression of the chromosomal mouse β-maj.-globin gene cloned in SV40. *Nature (London)* **281**, 35–40.

62. Hamer, D. H., and Walling, M. J. (1982). Regulation in vivo of a cloned mammalian gene: Cadmium induces the transcription of a mouse metallothionein gene in SV 40 vectors. *J. Mol. Appl. Genet.* **1**, 273–288.

63. Hanahan, D., Lane, D., Lipsich, L., Wigler, M., and Botchan, M. (1980). Characteristics of an SV40–plasmid recombinant and its movement into and out of the genome of a murine cell. *Cell* **21**, 127–139.

64. Harbers, K., Jähner, D., and Jaenisch, R. (1981). Microinjection of cloned retroviral genome into mouse zygotes: Integration and expression in the animal. *Nature (London)* **293**, 540–542.

65. Hartman, J. R., Nayak, D. P., and Fareed, G. C. (1982). Human influenza virus hemagglutinin is expressed in monkey cells using simian virus 40 vectors. *Proc. Natl. Acad. Sci. U.S.A.* **79**, 233–237.

66. Hauser, H., Gross, G., Bruns, W., Hochkeppel, H. K., Mayr, U., and Collins, J. (1982). Inducibility of human β-interferon gene in mouse L-cell clones. *Nature (London)* **297**, 650–654.

67. Huang, A. L., Ostrowski, M. C., Berard, D., and Hager, G. L. (1981). Glucocorticoid regulation of the Ha-MuSV p21 gene conferred by sequences from mouse mammary tumor virus. *Cell* **27**, 245–255.

68. Hudziak, R. M., Laski, F. A., Raj Bhandary, U. L., Sharp, P. A., and Capacchi, M. R. (1982). Establishment of mammalian cell lines containing multiple non sense mutations and functional suppressor tRNA genes. *Cell* **31**, 137–146.

69. Humphries, R. K., Ley, T., Turner, P., Davis Moulton, A., and Nienhuis, A. W. (1982). Differences in human α-, β- and δ-globin gene expression in monkey kidney cells. *Cell* **30**, 173–183.

70. Jaenisch, R., Jähner, D., Nobis, P., Simon, I., Löhler, J., Harbers, K., and Grotkopp, D.

(1981). Chromosomal position and activation of retroviral genomes inserted into the germ line of mice. *Cell* **24**, 519–529.

71. Jähner, D., Stuhlmann, H., Stewart, C. L., Harbers, K., Löhler, J., Simon, I., and Jaenisch, R. (1982). De novo methylation and expression of retroviral genomes during mouse embryogenesis. *Nature (London)* **298**, 623–628.

72. Jimenez, A., and Davies, J. (1980). Expression of a transposable antibiotic resistance element in *Saccharomyces*. *Nature (London)* **287**, 869–871.

73. Joyner, A., Yamamoto, Y., and Bernstein, A. (1982). Retrovirus long terminal repeats activate expression of coding sequences for the herpes simplex thymidine kinase gene. *Proc. Natl. Acad. Sci. U.S.A.* **79**, 1573–1577.

74. Kaufman, R. J., and Sharp, P. A. (1982). Construction of a modular dihydrofolate reductase cDNA gene: Analysis of signals utilized for efficient expression. *Mol. Cell. Biol.* **2**, 1304–1319.

75. Kit, S., and Dubbs, D. R. (1963). Acquisition of thymidine kinase activity by herpes simplex infected mouse fibroblast cells. *Biochem. Biophys. Res. Commun.* **11**, 55–59.

76. Kretschmer, P. J., Bowman, A. H., Huberman, M. H., Sanders-Haigh, L., Killos, L., and Anderson, W. F. (1981). Recovery of recombinant bacterial plasmids from *E. coli* transformed with DNA from microinjected mouse cells. *Nucleic Acids Res.* **9**, 6199–6217.

77. Kurtz, D. T. (1981). Hormonal inducibility of rat α2μ gobulin genes in transfected mouse cells. *Nature (London)* **291**, 629–631.

78. Lai, E. C., Woo, S. L. C., Bordelon-Riser, M. R., Fraser, T. H., and O'Malley, B. W. (1980). Ovalbumin is synthesized in mouse cells transformed with the natural chicken ovalbumin gene. *Proc. Natl. Acad. Sci. U.S.A.* **77**, 244–248.

79. Laimins, L. A., Khoury, G., Gorman, C., Howard, B., and Gruss, P. (1982). Host specific activation of transcription by tandem repeats from simian virus 40 and Moloney murine sarcoma virus. *Proc. Natl. Acad. Sci. U.S.A.* **79**, 6453–6457.

80. Lee, F., Mulligan, R., Berg, P., and Ringold, G. (1981). Glucocorticoids regulate expression of dihydrofolate reductase cDNA in mouse mammary tumor virus chimaeric plasmids. *Nature (London)* **294**, 228–232.

81. Levinson, B., Khoury, G., Van de Woude, G., and Gruss, P. (1982). Activation of SV40 genome by 72-base pair tandem repeats of Moloney sarcoma virus. *Nature (London)* **295**, 568–572.

82. Lindenmaier, W., Hauser, H., Greiser de Wilke, I., and Schütz, G. (1982). Gene shuttling: Moving of cloned DNA into and out of eukaryotic cells. *Nucleic Acids Res.* **10**, 1243–1256.

83. Linnenbach, A., Huebner, K., and Croce, C. (1980). DNA transformed murine teratocarcinoma cells: Regulation of expression of simian virus 40 tumor antigen in stem versus differentiated cells. *Proc. Natl. Acad. Sci. U.S.A.* **77**, 4875–4879.

84. Liu, C. C., Yansura, D., and Levinson, A. D. (1982). Direct expression of hepatitis B surface antigen in monkey cells from a SV40 vector. *DNA* **1**, 213–221.

85. Lomedico, P. T., and McAndrew, S. J. (1982). Eukaryotic ribosomes can recognize preproinsulin initiation codons irrespective of their position relative to the 5' end of mRNA. *Nature (London)* **299**, 221–226.

86. Lusky, M., and Botchan, M. (1981). Inhibition of SV40 replication in simian cells by specific pBR322 DNA sequences. *Nature (London)* **293**, 79–81.

87. Maniatis, T., Fritsch, E. F., and Sambrook, J. (1982). "Molecular Cloning, a Laboratory Manual." Cold Spring Harbor Lab., Cold Spring Harbor, New York.

88. Mantei, N., Boll, W., and Weissman, C. (1979). Rabbit β-globin mRNA production in mouse L cells transformed with cloned rabbit β-globin chromosomal DNA. *Nature (London)* **281**, 40–46.

89. Mayo, K. E., Warren, R., and Palmiter, R. D. (1982). The mouse metallothionein-1 gene is transcriptionally regulated by cadmium following transfection into human or mouse cells. *Cell* **29**, 99–108.

90. McCutchan, J. H., and Pagano, J. S. (1968). Enhancement of the infectivity of simian virus 40 deoxyribonucleic acid with diethylaminoethyl dextran. *J. Natl. Cancer Inst. (U.S.)* **41**, 351–357.

91. McKnight, S. L. (1980). The nucleotide sequence and transcript map of the herpes simplex thymidine kinase gene. *Nucleic Acids Res.* **8**, 5949–5964.

92. McKnight, S. L., Gavis, E. R., and Kingsbury, R. (1981). Analysis of transcriptional regulatory signals of the HSV thymidine kinase gene: Identification of an upstream control region. *Cell* **25**, 385–398.

93. McKnight, S. L. (1982). Functional relationships between transcriptional control signals of the thymidine kinase gene of herpes simplex virus. *Cell* **31**, 355–365.

94. Mellon, P., Parker, V., Gluzman, Y., and Maniatis, T. (1981). Identification of DNA sequences required for transcription of the human α1-globin gene in a new SV40 host-vector system. *Cell* **27**, 279–288.

95. Mellor, A. L., Golden, L., Weiss, E., Bullman, J., Hurst, J., Simpson, E., James, R. F. L., Townsend, A. R. M., Taylor, P. M., Schmidt, W., Ferluga, J., Leben, L., Santamaria, M., Atfield, G., Festenstein, H., and Flavell, R. A. (1982). Expression of murine H-2Kb histocompatibility antigen in cells transformed with clones H-2 genes. *Nature (London)* **298**, 529–534.

96. Mercola, K. E., Stang, H. D., Browne, J., Salser, W., and Cline, M. J. (1980). Insertion of a new gene of viral origin into bone marrow cells of mice. *Science* **208**, 1033–1035.

97. Milman, G., and Herzberg, M. (1981). Efficient DNA transfection and rapid assay for thymidine kinase activity and viral antigenic determinants. *Somatic Cell Genet.* **7**, 161–170.

98. Mocarski, E. S., and Roizman, B. (1982). Herpes virus dependent amplification and inversion of cell associated viral thymidine kinase gene flanked by a sequences and linked to an origin of DNA replication. *Proc. Natl. Acad. Sci. U.S.A.* **79**, 5626–5630.

99. Mocarski, E. S., and Roizman, B. (1982). Structure and role of the herpes simplex virus DNA termini in inversion, circularization and generation of virion DNA. *Cell* **31**, 89–97.

100. Montell, C., Fisher, E. F., Caruthers, M. H., and Berk, A. J. (1982). Resolving the functions of overlapping viral genes by site-specific mutagensis at a mRNA splice site. *Nature (London)* **295**, 380–384.

101. Moore, K. W., Sher, B. T., Sun, Y. H., Eakle, K. A., and Hood, L. (1982). Identification of a Balb/c H-2Ld gene by DNA-mediated gene transfer. *Science* **215**, 677–682.

102. Moriarty, A. M., Hoyer, B. H., Shih, J. W. K., Gerin, J. L., and Hamer, D. H. (1981). Expression of the hepatitis B virus surface antigen gene in cell culture by using a simian virus 40 vector. *Proc. Natl. Acad. Sci. U.S.A.* **78**, 2606–2610.

103. Mulligan, R. C., Howard, B. H., and Berg, P. (1979). Synthesis of rabbit β-globin in cultured monkey kidney cells following infection with a SV40 β-globin recombinant genome. *Nature (London)* **277**, 108–114.

104. Mulligan, R. C., and Berg, P. (1980). Expression of a bacterial gene in mammalian cells. *Science* **209**, 1422–1427.

105. Mulligan, R. C., and Berg, P. (1981). Selection for animal cells that express the *Escherichia coli* gene coding for xanthine–guanine phosphoribosyl transferase. *Proc. Natl. Acad. Sci. U.S.A.* **78**, 2072–2076.

106. Navey-Many, T., and Cedar, H. (1981). Active gene sequences are undermethylated. *Proc. Natl. Acad. Sci. U.S.A.* **78**, 4246–4250.

107. Nunberg, J. H., Kaufman, R. J., Chang, A. C., Cohen, S. N., and Schimke, R. T. (1980).

Structure and genomic organization of the mouse dihydrofolate reductase gene. *Cell* **19**, 355–364.

108. O'Hare, K., Benoist, C., and Breathnach, R. (1981). Transformation of mouse fibroblasts to methrotrexate resistance by a recombinant plasmid expressing a prokaryotic dihydrofolate reductase. *Proc. Natl. Acad. Sci. U.S.A.* **78**, 1527–1537.

109. Ohno, S., and Taniguchi, T. (1982). Inducer-responsive expression of the cloned human interferon β1 gene introduced into cultured mouse cells. *Nucleic Acids Res.* **10**, 967–977.

110. Oi, V. T., Morrison, S. L., Herzenberg, L. A., and Berg, P. (1983). Immunoglobulin gene expression in transformed lymphoid cells. *Proc. Natl. Acad. Sci. U.S.A.* **80**, 825–829.

111. Palmiter, R. D., Brinster, R. L., Hammer, R. E., Trumbauer, M. R., Rosenfeld, M. G., Birnberg, N. C., and Evans, R. M. (1982). Dramatic growth of mice that develop from eggs microinjected with metallothionein growth hormone fusion genes. *Nature (London)* **300**, 611–615.

112. Papahadjopoulos, D., Fraley, R., and Heath, T. (1980). Optimization of liposomes as a carrier system for the intracellular delivery of drugs and macromolecules. *In* "Liposomes and Immunobiology," pp. 151–164. Elsevier/North Holland, Amsterdam.

113. Pavlakis, G. N., and Hamer, D. H. (1983). Regulation of a metallothionein–growth hormone hybrid gene in bovine papilloma virus. *Proc. Natl. Acad. Sci. U.S.A.* **80**, 397–401.

114. Pelham, H. R. B. (1982). A regulatory upstream promoter element in the *Drosophila* Hsp70 heat shock gene. *Cell* **30**, 517–528.

115. Pellicer, A., Robins, D., Wold, B., Sweet, R., Jackson, J., Lowy, I., Roberts, J. M., Sim, G. K., Silverstein, S., and Axel, R. (1980). Altering genotype and phenotype by DNA-mediated gene transfer. *Science* **209**, 1414–1422.

116. Pellicer, A., Wagner, E. F., El Kareh, A., Dewey, M. J., Reuser, A. J., Silverstein, S., Axel, R., and Mintz, B. (1980). Introduction of a viral thymidine kinase gene and the human β-globin gene into developmentally multipotential mouse teratocarcinoma cells. *Proc. Natl. Acad. Sci. U.S.A.* **77**, 2098–2102.

117. Perucho, M., Hanahan, D., and Wigler, M. (1980). Genetic and physical linkage of exogenous sequences in transformed cells. *Cell* **22**, 309–317.

118. Petricciani, J. C., Kirschstein, R. L., Hines, J. E., Wallace, R. E., and Martin, D. P. (1973). Tumorigenicity assays in non-human primates treated with antithymocyte globulin. *J. Natl. Cancer Inst. (U.S.)* **51**, 191–196.

119. Pitha, P. M., Ciufo, D. M., Kellum, M., Raj, N. B. K., Reyes, G. R., and Hayward, G. S. (1982). Induction of human β-interferon synthesis with poly(rI–rC) in mouse cells transfected with cloned cDNA plasmids. *Proc. Natl. Acad. Sci. U.S.A.* **79**, 4337–4341.

120. Post, L. E., Mackem, S., and Roizman, B. (1981). Regulation of α genes of herpes simplex virus: Expression of chimeric genes produced by fusion of thymidine kinase with α gene promoters. *Cell* **24**, 555–565.

121. Pourcel, C., Louise, A., Gervais, M., Chenciner, N., Dubois, M.-F., and Tiollais, P. (1982). Transcription of the hepatitis B surface antigen gene in mouse cells transformed with cloned viral DNA. *J. Virol.* **42**, 100–105.

122. Racaniello, V. R., and Baltimore, D. (1981). Cloned poliovirus complementary DNA is infectious in mammalian cells. *Science* **214**, 916–919.

123. Rassoulzadegan, M., Binetruy, B., and Cuzin, F. (1982). High frequency of gene transfer between bacteria and eukaryotic cells. *Nature (London)* **295**, 257–260.

124. Ringold, G., Dieckmann, B., and Lee, F. (1981). Co-expression and amplification of dihydrofolate reductase cDNA and the *Escherichia coli* XGPRT gene in Chinese hamster ovary cells. *J. Mol. Appl. Genet.* **1**, 165–175.

125. Robert de Saint-Vincent, B., Delbrück, S., Eckhart, W., Meinkoth, J., Vitto, L., and Wahl, G. (1981). Cloning and reintroduction into animal cells of a functional CAD gene, a dominant amplifiable genetic marker. *Cell* **27**, 267–277.

126. Roberts, J. M., and Axel, R. (1982). Gene amplification and gene correction in somatic cells. *Cell* **29**, 109–119.

127. Robins, D. M., Ripley, S., Henderson, A. S., and Axel, R. (1981). Transforming DNA integrates into the host chromosome. *Cell* **23**, 29–39.

128. Robins, D. M., Axel, R., and Henderson, A. S. (1981). Chromosome structure and DNA sequence alterations associated with mutations of transformed genes. *J. Mol. Appl. Genet.* **1**, 191–203.

129. Robins, D. M., Pack, I., Seeburg, P. H., and Axel, R. (1982). Regulated expression of human growth hormone genes in mouse cells. *Cell* **29**, 623–631.

130. Rusconi, S., and Schaffner, V. (1981). Transformation of frog embryos with a rabbit β-globin gene. *Proc. Natl. Acad. Sci. U.S.A.* **78**, 5051–5055.

131. Sandri-Goldin, R., Goldin, A. L., Levine, M., and Glorioso, J. C. (1981). High frequency transfer of cloned herpes simplex virus type 1 sequences to mammalian cells by protoplast fusion. *Mol. Cell. Biol.* **1**, 743–752.

132. Sarver, N., Gruss, P., Law, M. F., Khoury, G., and Howley, P. M. (1981). Bovine papillomavirus deoxyribonucleic acid: A novel eukaryotic cloning vector. *Mol. Cell. Biol.* **1**, 486–496.

133. Schaefer-Ridder, M., Wang, Y., and Hofschneider, P. H. (1982). Liposomes as gene carriers: Efficient transformation of mouse L cells by thymidine kinase gene. *Science* **215**, 166–168.

134. Schaffner, W. (1980). Direct transfer of cloned genes from bacteria to mammalian cells. *Proc. Natl. Acad. Sci. U.S.A.* **77**, 2163–2167.

135. Seif, I., Khoury, G., and Dhar, R. (1979). The genome of human papovavirus BKV. *Cell* **18**, 963–977.

136. Shen, Y. M., Hirschhorn, R. R., Mercer, W. E., Surmacz, E., Tsutsui, Y., Soprano, K. J., and Baserga, R. (1982). Gene transfer: DNA microinjection compared with DNA transfection with a very high efficiency. *Mol. Cell. Biol.* **2**, 1145–1154.

137. Shimotohno, K., and Temin, H. (1981). Formation of infectious progeny virus after insertion of herpes simplex thymidine kinase gene into DNA of an avian retrovirus. *Cell* **26**, 67–77.

138. Sompayrac, L. M., and Danna, K. J. (1981). Efficient infection of monkey cells with DNA of simian virus 40. *Proc. Natl. Acad. Sci. U.S.A.* **78**, 7575–7578.

139. Southern, P. J., Howard, B. H., and Berg, P. (1981). Construction and characterization of SV40 recombinants with β-globin cDNA substitutions in their early regions. *J. Mol. Appl. Genet.* **1**, 177–190.

140. Spaete, R. R., and Frenkel, N. (1982). The herpes simplex virus amplicon: A new eukaryotic defective virus cloning–amplifying vector. *Cell* **30**, 295–304.

141. Stein, R., Razin, A., and Cedar, H. (1982). *In vitro* methylation of the hamster adenine phosphoribosyltransferase gene inhibits its expression in mouse L cells. *Proc. Natl. Acad. Sci. U.S.A.* **79**, 3418–3422.

142. Stewart, C. L., Stuhlmann, H., Jähner, D., and Jaenisch, R. (1982). *De novo* methylation, expression and infectivity of retroviral genomes introduced into embroyonal carcinoma cells. *Proc. Natl. Acad. Sci. U.S.A.* **79**, 4098–4102.

143. Summers, W., Grogan, E., Shedd, D., Robert, M., Liu, C. R., and Miller, G. (1982). Stable expression in mouse cells of nuclear neoantigen after transfer of a 3.4 megadalton cloned fragment of Epstein–Barr virus DNA. *Proc. Natl. Acad. Sci. U.S.A.* **79**, 5688–5692.

144. Sveda, M. M., Markoff, L. J., and Laĭ, C. J. (1982). Cell surface expression of the influenza virus hemagglutinin requires the hydrophobic carboxy-terminal sequences. *Cell* **30**, 649–656.

145. Sweet, R., Chao, M. V., and Axel, R. (1982). The structure of the thymidine kinase gene promoter: Nuclease hypersensitivity correlates with expression. *Cell* **31**, 347–353.

146. Tabin, C. J., Bradley, S. M., Bargmann, C. I., Weinberg, R., Papageorge, A. G., Scolnick, E. M., Dhar, R., Lowy, D. R., and Chang, E. H. (1982). Mechanism of activation of a human oncogene. *Nature (London)* **300**, 143–149.

147. Tavernier, J., Gheysen, D., Duerinck, F., Van der Heyden, J., and Fiers, W. (1983). Deletion mapping of the inducible promoter of human IFN-β gene. *Nature (London)* **301**, 634–636.

148. Tsui, L. C., Breitman, M. L., Siminovitch, L., and Buchwald, M. (1982). Persistence of freely replicating SV40 recombinant molecules carrying a selectable marker in permissive simian cells. *Cell* **30**, 499–508.

149. Tyndall, C., La Mantia, G., Thacher, C. M., Favaloro, J., and Kamen, R. (1981). A region of the polyoma virus genome between the replication origin and late protein coding sequences is required in cis for both early gene expression and viral DNA replication. *Nucleic Acids Res.* **9**, 6231–6250.

150. Vardimon, L., Kressmann, A., Cedar, H., Maechler, M., and Doerfler, W. (1982). Expression of a cloned adenovirus gene is inhibited by in vitro methylation. *Proc. Natl. Acad. Sci. U.S.A.* **79**, 1073–1077.

151. Waechter, D. E., and Baserga, R. (1982). Effect of methylation on expression of microinjected genes. *Proc. Natl. Acad. Sci. U.S.A.* **79**, 1106–1110.

152. Wagner, E. F., Stewart, T. A., and Mintz, B. (1981). The human β-globin gene and a functional viral thymidine kinase gene in developing mice. *Proc. Natl. Acad. Sci. U.S.A.* **78**, 5016–5020.

153. Wagner, T. E., Hoppe, P. C., Jollick, J. D., Scholl, D. R., Hodinka, R. L., and Gault, J. B. (1981). Microinjection of a rabbit β-globin gene into zygotes and its subsequent expression in adult mice and their offspring. *Proc. Natl. Acad. Sci. U.S.A.* **78**, 6376–6380.

154. Wang, Y., Schäfer-Ridder, M., Stratowa, C., Wong, T. K., and Hofschneider, P. H. (1982). Expression of hepatitis B surface antigen in unselected cell culture transfected with recircularized HBV DNA. *EMBO J.* **1**, 1213–1216.

155. Wasylyk, B., Wasylyk, C., Augereau, P., and Chambon, P. (1983). The SV40 72 bp repeat preferentially potentiates transcription starting from proximal natural or substitute promoter elements. *Cell* **32**, 503–514.

156. Watanabe, S., and Temin, H. (1982). Encapsidation sequences for spleen necrosis virus, an avian retrovirus, are between the 5' long terminal repeat and the start of the *gag* gene. *Proc. Natl. Acad. Sci. U.S.A.* **79**, 5986–5990.

157. Wei, C. M., Gibson, M., Spear, P., and Scolnick, E. M. (1981). Construction and isolation of a transmissible retrovirus containing the *src* gene of Harvey murine sarcoma virus and the thymidine kinase gene of herpes simplex virus type 1. *J. Virol.* **39**, 935–944.

158. Weinberg, R. (1981). Use of transfection to analyse genetic information and malignant transformation. *Biochim. Biophys. Acta* **651**, 25–35.

159. Wigler, M., Pellicer, A., Silverstein, S., and Axel, R. (1978). Biochemical transfer of single-copy eukaryotic genes using total cellular DNA as donor. *Cell* **14**, 725–731.

160. Wigler, M., Perucho, M., Kurtz, D., Dana, S., Pellicer, A., Axel, R., and Silverstein, S. (1980). Transformation of mammalian cells with an amplifiable dominant acting gene. *Proc. Natl. Acad. Sci. U.S.A.* **77**, 3567–3570.

161. Wigler, M., Levy, D., and Perucho, M. (1981). The somatic replication of DNA methylation. *Cell* **24**, 33–40.

162. Will, H., Cattaneg, R., Koch, H. G., Darai, G., Schaller, J., Schellekens, H., Van Eerd, P. M., and Deinhardt, F. (1982). Cloned HBV DNA causes hepatitis in chimpanzees. *Nature (London)* **299**, 740–742.

163. Williams, J. G. (1982). Mouse and supermouse. *Nature (London)* **300**, 575.

164. Wold, B., Wigler, M., Lacy, E., Maniatis, T., Silverstein, S., and Axel, R. (1979). Introduction and expression of a rabbit β-globin gene in mouse fibroblasts. *Proc. Natl. Acad. Sci. U.S.A.* **76**, 5684–5688.

165. Zinn, K., Mellon, P., Ptashne, M., and Maniatis, T. (1982). Regulated expression of an extrachromosomal human β-interferon gene in mouse cells. *Proc. Natl. Acad. Sci. U.S.A.* **79**, 4897–4901.

17

The Biotechnological Future for Animal Cells in Culture

R. E. Spier
Department of Microbiology
University of Surrey
Guildford, Surrey
United Kingdom

F. Horaud
Unité de Virologie Medicale
Institut Pasteur
Paris, France

Animal Cell Biotechnology, Vol. 2

1. INTRODUCTION

During the late 1970s a reading of the popular scientific press would have led one to believe that proteins, which until that time could only be produced in animal cells, would in the future be produced in prokaryotic bacterial cells or in the primitive eukaryotic yeast cell (77). The features that would have been highlighted would have been those which demonstrated that bacteria and yeasts

1. Are more highly productive of biomass
    a. Per unit volume of system
    b. Per unit capital expenditure
    c. Per unit operating cost
2. Are more reliable in production operations
    a. Less fastidious medium requirements
    b. Simple control of exogenous contaminants
3. Can make materials which could not otherwise be made
4. Produce immunogens free from infectious particles

Indeed Katinger Chapter 7, (Volume 1) and Spier (73) have concluded that bacterial biomass productivities are over 40 times that of a rapidly doubling animal cell (see Tovey, Chapter 8, Volume 1). However, the wary would have had some reservations based on an equally impressive list of difficulties faced by generating products from genetically engineered bacteria and (to a lesser extent) yeasts:

1. The bacterial proteins lack the appropriate post-translational modifications
    a. Protein cleavage sites
    b. Glycosylations
    c. Disulphide bridges
    d. Environment which promotes the emergence of the exact three-dimensional structure of the native protein
    e. Amidations
    f. Carboxsylations
    g. Phosphorylations
2. Bacteria and yeasts do not generally excrete proteinaceous materials
    a. Need to break cells
    b. Separation of products problems
    c. Contamination by endotoxins
3. As the expressed yeast proteins are "foreign" to the cell they are subject to intracellular proteases and can be endotoxic.
4. The bacteria and yeast which have been "turned on" to express a foreign protein are at a competitive disadvantage vis-à-vis their untrans-

formed variants. This leads to lower rates of growth, more complex media and susceptibility to contamination.

In recent times a more sober appraisal of the potential of genetically engineered prokaryotic systems is emerging. It has been accepted that it is unlikely that the antigens of either foot-and-mouth disease virus or poliovirus will be prepared in a sufficiently immunogenic form from engineered bacteria.

While the effort in producing such antigens from bacteria is on the wane, new interest has been stimulated by the demonstration of the protective efficacy of a chemisynthetic 20-mer polypeptide whose sequence mimicked that of amino acids 141 to 160 of the VP1 virus protein of foot-and-mouth disease virus (FMDV) (4). This material protects guinea pigs after a single vaccination, yet it fails to protect the larger animal, bovines, following a single dose of vaccine. A similar situation appears to hold for polypeptides which mimic a portion of the hepatitis B surface antigen (HB_sAg) (74).

As a result of such developments and following a considerable amount of work on the genetic engineering of animal cells (21, 22, 65) a situation has developed whereby it is now considered practicable to produce materials from engineered animal cells or from animal cells in culture which have been selected, stimulated or modified to yield commercially viable quantities of saleable products. The context of the later sections of this chapter will therefore focus on these products, which will be selected so that it is unlikely that they would be made by any of the alternative routes described above (genetically engineered bacteria, simple eukaryotes such as yeasts or chemisynthetically). During this forward look, it should not be forgotten that the traditional processes in which cultured animal cells have been and are being successfully used will continue to operate. The large-scale production of foot-and-mouth disease vaccines, human viral vaccines (measles, rubella, mumps, polio) and other veterinary vaccines (rabies, Marek's, Newcastle's disease, rinderpest) will continue to depend on conventional methods. These methods have been built up and refined over three decades. Also, the acceptance of such products for mass vaccination by the regulatory authorities, a process which requires an expenditure of £6–40 million (54), impedes the manufacture of new products which would have to carry this high overhead, which could result in their being "priced out of the market." However, developments in these areas are to be expected, some of them involving the novel use of genetically modified animal cell cultures.

Virus vaccines have been the mainstay of the animal cell biotechnology effort until the most recent times. Now, following the discovery of Kohler and Milstein in 1975 (32) that it is possible to continuously maintain an antibody-secreting cell in culture, a new industry has been founded to produce such cells and to use them to generate products. (See also Osterhous, Chapter 3, this volume.)

This new industry of monoclonal antibody production, and the other exciting areas of development defined by the exploitation of the discovery of a wide variety of immunobiological regulator materials whose potential use can only be guessed at, will provide a considerable impetus to further technological developments in cell culture. Such major departures should not occlude a description of the significant activities in the areas of hormone production (the large, glycosylated hormones; see below), insecticide development, animal cell enzyme applications and the use of whole cells as a bioproduct. These major product areas will be discussed below, placing particular emphasis on the future and as yet unrealised prospects of this area of biotechnology.

2. IMMUNOBIOLOGICAL MATERIALS

The explosive expansion of immunologically associated activities (leading to the existence of over 30 journals with "immuno" in the title) has resulted in new ways of looking at vaccines and in the methods for the potentiation of their effects. The consequence of such applications is the production of antibodies which, following the demonstration of the utility of "hydridoma techology" (32), can now be made on the large scale in bioreactors.

2.1. Vaccines Produced from Animal Cells in Culture

The vaccines which are conventionally produced for human and veterinary applications have been summarised in Table I and discussed in Chapter 1 of this volume. While such vaccines have achieved much in the remission of disease there are significant improvements yet to be made. Starting from the set of criteria for an ideal vaccine:

1. Administered once
2. Life-long immunity
3. Safe (high benefit/risk ratio)
4. Cheap to produce (preferably locally with simple equipment)
5. Easy to apply
6. Contains a minimum of inessentials
7. Simple to assay and characterise
8. One vaccine to cover as wide a variety of virus strains as possible

it is possible to appreciate how much has yet to be done. The present vaccines suffer from one or more deficiencies in a number of the listed

TABLE I Current Human and Veterinary Vaccines
Produced from Cultured Animal Cells

Human vaccines	Veterinary vaccines
Mumps	Foot and mouth (FMD)
	Marek
Measles	Newcastle
	Intestinal bovine rhinotracheitis
Rubella	Parainfluenza 3
	Rinderpest
Rabies	Bovine viral diarrhoea
	Rabies
Yellow fever[a]	Calf scours
	Canine distemper
Polio	Canine and feline panleucopenia
	Ibaraki
Influenza[a]	Ephemeral fever
	Fowl pox
Adenovirus	African horse sickness
	Hog cholera
Smallpox[a]	Canine contagious hepatitis
	Transmissible gastroenteritis
Swine influenza	Contagious pustular dematitis
	Blue tongue
	Feline infectious enteritis
	Contagious ecthyema
	Infectious bursal disease
	Louping ill
	Laryngotracheitis
	Rift Valley fever
	Pseudorabies (Aujeszky's)

[a] Made in whole egg or whole animal systems also.

criteria. With present-day techniques of molecular biology it is possible to insert the genes coding for the immunogenic proteins of a difficult-to-produce virus (hepatitis B) into a vector which is derived from a strongly immunogenic virus (smallpox) (72). The resulting product has been shown to have considerable promise as a vaccine. (The risks of contracting encephalitic side effects from the smallpox vaccine have been measured at 3 per 10^6 applications; the risk of more debilitating effects of hepatitis B in rapidly developing countries is many times more.) The development of such an immunogenic vector for the vaccines of rabies, herpes simplex II and malaria is presently in hand (51). The contribution of an immunogenesis vector from herpes virus and measles virus is also under active investigation. Were such vaccines to emerge as major elements in vaccination programs, the demand for animal cell cultures would increase significantly.

An alternative approach is to genetically engineer animal cells to produce immunogenic materials which can be used for FMD vaccines or polio vaccine. For both of these situations, the whole virus provides the most effective immunogen. Were it possible to produce the complete capsid structure (containing all four viral proteins) assembled in the native configuration yet lacking the virus nucleic acid, then it is conceivable that such a material could be made at high yield (low cost) and safely. Clearly, the animal cell would have the genes for the polyprotein and viral protease and would provide the environment for the appropriate post-translational modification to the virus capsid (50). While viral genes (i.e. positive strand RNA) will be present in cells making viral proteins, the absence of the viral gene for the RNA-dependent RNA polymerase will seriously impede the replication of the RNA carried by the virus. Thus it would be unlikely that infectious, pathogenic, virus would be formed in such genetically engineered cells.

Speculating that an animal cell, burdened by having to service a gene which did not contribute to its survival, would be difficult to grow, it is possible to consider the insertion of the virus genes in a construction that would contain elements which would enable the production of the proteins foreign to the animal cell to be confined to the end of the cell growth phase. Such regulator genes have been found to be transposable between animal cells and have been shown to be subject to such controlling stimuli as glucocorticoid hormone, heat, heavy metals and polyribosyl I–polyribosyl C (22). The combination of the correct genetic elements and their phased expression in cultured animal cells offers much.

2.1.1. Anticancer Vaccines

Following the demonstration that there are a number of genes which are present in cancerogenic viruses and which are also present in native mammalian cells (15), it is possible to consider the development of vaccines which would stimulate the production of antibodies to proteins that are made if and when the cancer gene changes its intrachromosomal location. Thus it is conceivable that cell lines could be described which express the products of such viral proteins. Such expressed products could then be used as vaccines to generate antibodies so that when a normal body cell which suddenly begins to make such a protein is induced, the cell is recognised as foreign by the previously formed antibodies and is thereby eliminated. One such protein could be a component of the histocompatibility complex, as this material is made in response to a transition in the position of a virus oncogene normally resident and quiescent in the untransformed mammalian cell (6).

These ideas would be difficult to check as one would have to challenge a putatively protected subject with a known cancer-inducing material, an act which would probably not be allowed. However, it may be that when offered the possibility of unproven material the public may choose to evaluate a material which could not have been tested and would therefore only have a "theoretical" prophylactic quality. Statistical surveys of the consequences of such actions would then (over many years) indicate the justification, or lack of it, for the putative vaccine.

2.1.2. Other New Vaccines

There are a number of types of new vaccines which are presently under investigation. Such new human vaccines as cytomegalovirus vaccine (82), hepatitis B_sAg vaccine prepared from geneticaly engineered animal cells (62, 74), varicella–zoster and herpes simplex (8) are under test in human subjects, while others (see Table II) are still under investigation at the research stage. One such vaccine is that of yellow fever, which since 1939 has been made in embryonated chick eggs. At present there is a movement supported by the World Health Organisation (WHO) towards the development of a new cell culture vaccine.

On the veterinary side, the vaccines are held up because of technical difficulties resulting from a lack of reliable immunogens or methods of application. Cost–benefit analysis is also a significant factor which determines the availability of these vaccines.

There is yet much progress to be made in this area, the result of which will be a renewed demand for the exploitation of animal cells in culture and their genetically engineered counterparts.

TABLE II Some Virus Vaccines in Various States of Preparation

Human	Veterinary
Hepatitis B_sAg	African swine fever
Hepatitis A	Equine influenza
Herpes simplex I and II	Fish rhabdovirus
Cytomegalovirus	Equine encephalitis
Varicella–zoster	Rotavirus diarrhoea
Respiratory syncytial virus	Corona virus
Rift Valley fever	
Rhinovirus	
Rheovirus	
Rotavirus	
Epstein–Barr virus	
Leukemia viruses	
Dengue	

2.2. Immunoregulatory Materials

Since the discovery of the existence of different subsets of T cells (7) which are held to (1) be specifically differentiated to promote or regress the formation of antibodies by B cells and (2) have functions in the areas of generating cell cytotoxicity or cell killing, there has been an immense volume of work which has sought to define and produce the regulatory factors controlling these differentiations. The number of factors discovered or proposed is large and it is possible that the same factor masquerades under several different names (Table III).

2.2.1. The Interferons

Animal cell systems are widely used to produce the interferons. There are 18 or so varieties of the α-leucocyte interferon, one version of which can be

TABLE III   Immunoregulators

Material	Effect
Thymosin	Reverses effects of thymectomy
	Causes T-cell differentiation
	Produces MIF (macrophage migration inhibitory factor)
	Increases production of antibody-forming cells
Thymopoietin	Generation of cytotoxic T cells
	Prevention of autoimmunity
	Rejection of carcinoma
THF (thymic humoral factor)	Reverses effects of thymectomy
Serum thymic factor	Induces T-cell markers
Lymphocyte activation factor—LAF—interleukin I	B-Cell mitogen
T-Cell growth factor—TGGF	Cytotoxic T-cell mitogen
Interleukin II	Aids T-cell differentiation
Interleukin III	Maintains natural cytotoxic cells in culture
Suppressor factors (immune interferon type II)	Non-specific suppression of antibody response via macrophage soluble factor
	Suppresses cytotoxic T-cell response
Interferon type α	Potentiation of cytotoxic T-cell and natural killer cell response
Interferon type β	
Interferon type γ	
Immunotranquilliser factor	Suppresses antibody production to new antigens but not existing antigens
B-Cell growth factor (BGF)	
Colony-stimulating factor (CSF)	
Migration-inhibition factor (MIF)	Inhibits the migration of macrophages
Macrophage cytotoxicity factor (MCF)	
T-Cell replacing factor (TRF)	

made from Epstein–Barr virus-transformed lymphoid cells or lympho-blastoid cells (*10*). The β-(fibroblastic) version of interferon is generally asso-ciated with the fibroblastic, anchorage-dependent cells and unit processes for its large-scale production from animal cells have been described (*41*). However, both such types of interferon may be made as unglycosy-lated, yet active, proteins in genetically engineered bacteria (*75*) and indeed it is easier to produce a wider variety of types of α-interferons in this way. It is also practicable to produce "hybrid" interferon molecules having proper-ties intermediary to any two particular α-interferons (*75*). A third type of interferon, referred to as γ - or immune interferon, is a more glycosylated version of interferon and therefore a preferred production method is from genetically engineered animal cells. Indeed, a methodology based on the use of Chinese hamster ovary cell lines whose inserted γ-interferon genes have been amplified by using methotrexate is under test as a practicable system for producing γ-interferon in its native and most active form (*1, 31*).

Although how useful the interferons will be is a moot point at this time (*46, 67*) there is little doubt that such materials have been shown to be of significant value in particular situations (*26*) and, whereas the universal, antiviral wonder cure is probably not at hand, it is probable that very useful therapeutic adjuncts can be derived from these materials, used either singly or more likely in combinations (*17*). Should this be the case then it is not unlikely that one or other of the production methodologies will be founded on cultured animal cells.

2.2.2. *Lymphocyte Growth and Differentiation Promotors*

The three interleukins 1, 2 and 3 (*13, 20*) and an assortment of other factors (Table III) are involved in the control of proliferation of the cells of the immune system. The latter has been likened to a network (*28*) in which a change in one element causes a distortion in the system, which seeks to regain its equilibrium position. The present goal is one which would enable the administration of a prophylactic or therapeutic agent to exploit the po-tential of the immune system to the maximum extent. By current theories this would involve, for example, increasing the number or proportion of T helper cells, or correspondingly decreasing the T-suppressor cell effect. Such action could apply at the cell number level (via growth factors) or at the point of control of the concentration of T helper cell "factors" or T-sup-pressor cell excretions (*53, 71*). However, at the time of writing, it is uncer-tain which particular products will emerge. Whether they will be factors which promote the growth (negatively or positively) of particular cell types or the control materials which such controlling cells produce, the method of their production will depend on the nature of the molecules concerned. At present the animal cell biotechnologist has the task of producing sufficient cells and factors to enable the molecular biologists to isolate the genes, amplify them and use them to produce materials in sufficient quantity to determine their applicability in beneficially controlling the immune system. The final production methodology could also rest with the animal cell system, for such systems produce materials identical to those in use naturally and thus they do not generate immune reactions to themselves, nor would they be expected to stimulate an allergic response on repeated application.

2.3. Antibodies

Antibodies are complex molecules (Fig. 1) which are unlikely to be chemically synthesised or secreted in their native configuration from either engineered prokaryotes or yeasts (cf. ref. 30). They may, however, be excreted in prolific amounts [a mature plasma cell secretes 10^5 antibody molecules/hr (29)] from appropriately tuned animal cells and there are prospects for engineering animal cells to enhance production further. Thus as the mechanisms which lead to the assembly of the mRNA, from which the various

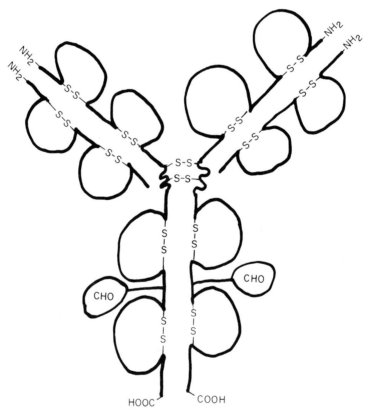

Fig. 1. Schematic of IgG type of antibody molecule.

strains of the antibody molecule are made, become known in more detail it becomes possible to augment the promoter and enhancer sequences artificially and thereby increase the rate of antibody synthesis (22).

Although antibodies with high titres (reactive at dilutions of $> \frac{1}{10}^5$) are available, such materials are made from sera and contain a mixture of antibody molecules in proportions which differ from one induced animal to another or from one induction to the next. As the reactivity of such mixtures is the sum of the reactions of the individual molecules, it is difficult to obtain consistency, reliability and the same degree of specificity between uniquely induced preparations. For this reason the discovery of Kohler and Milstein (32) that it is possible to produce a cell line which is capable of secreting a single species of antibody molecule was a breakthrough in this area. The cell line formed from the fusion of a myeloma and a single antibody-excreting lymphocyte produces material referred to as a monoclonal antibody (MCA). Such antibodies may be produced in two ways. One way which is commonly used is to inject the mouse-derived myeloma–lymphocyte hybrid (the hybridoma) into the peritoneum of a mouse, where it replicates and produces an ascites tumour, a loose collection of cells which fills the peritoneal cavity and which can be harvested as an antibody-rich cell suspension. The cells and non-antibody proteins are removed and the remaining material, which is mainly (but not completely) antibody, is used as a product. The second method for producing MCA is by growing the hybridoma *in vitro* in an animal cell culture vessel. When it is possible to use a serum-free medium for the antibody production phase in such systems, then the material produced is relatively pure and is thereby easier to purify further.

While the former, mouse-based technology is highly productive of a concentrated antibody preparation and therefore constitutes the most widely used procedure, the method is not without its problems. The scale-up of the mouse-based technology involves thousands of mice, which becomes an expensive, labour-intensive operation. The materials produced require considerable purification and the ethical considerations involved in subjecting a large number of animals to the discomfort of an ascites tumour become important. For such reasons emphasis is returning to the improvement of the productivity of animal cells in culture for such materials.

There is a pronounced impetus to further develop the animal cell culture method because of the burgeoning uses for which the product MCAs are becoming indispensable components. Such uses are in diagnostic medicine, purification procedures, epidemiology, new vaccines, targeted drugs and basic research. These areas will be considered separately below.

2.3.1. Monoclonal Antibodies in Diagnostic Medicine

Perhaps the most widespread use of monoclonal antibodies is in the "diagnostics area." Conventional antibody-based systems whose reactions have been amplified by developments in immunological techniques such as immunofluorescence, enzyme-linked immunosorbent assay (ELISA) (cheap to use and commonly found in research labs) and radioimmunoassay (RIA) (more expensive, sensitive and reliant on expensive γ-counters).

With the development of packaged kits for diagnosis it has become increasingly important to use stable, consistent elements for the reactions. This provides more guaranteed reactivity and specificity. Also, by a judicious choice of reagent it is possible to improve the specificity of the reactions considerably. These developments have led to the emergence of a new MCA-based industry (48).

In addition to the *in vitro* diagnostic applications it becomes possible to attach radioactive, X-ray opaque or magnetically active materials to the monoclonal antibodies which have been raised to attach to diseased cells, organs or tissues. Such materials can be used to locate, by non-invasive techniques, sites of pathology.

Further developments in the area of specific disease diagnosis may be expected. In particular, as sensors and computers become more developed, smaller and less power-consuming it becomes possible to conjecture that the heat-evolving interaction of a few unique antibody molecules with the appropriate antigens would result in an electrical signal in a defined physical sensor. Such a signal could then be incorporated in a form suitable for diagnostic analysis. Many such antibody sets may be located at the end of an array of sensors; the whole assembly may be implanted under the skin and provide a telemetered signal to a mobile receiver, which would in turn supply information back to the sensed individual or to a public health system computer. While such ideas may be "far-fetched" today, in 20–50 years they could be commonplace.

2.3.2. Monoclonal Antibodies for Purification Procedures

Modern technology used for the purification of immunoactive materials requires the covalent attachment of antibodies to a solid support matrix [e.g. Sepharose (Pharmacia)], which complex is then used to remove only those molecules which react with the area of the antibody molecule that confers its specificity. Clearly the use of an antibody preparation containing a single molecular species presents considerable advantages for such a procedure. Once the required molecules have been attached to the matrix and the contaminants have been washed away, it is possible to recover the product by disassociating the antigen–antibody complex by using either acids (pH 2.5–4.5), chaotropic agents (KSCN, NH_4SCN) or hydrogen bond breakers (urea, guanidine).

The use of such chromatographic procedures can involve the recycling of the column many times (20–80×), and industrial processes for the production of hepatitis B_sAg (74) and interferon (68) have been reported.

2.3.3. Epidemiology

The specificity of the monoclonal antibody can be so selected that it reacts with a single epitope on a virus. Thus a panel of such antibody preparations can define uniquely a particular viral material and can distinguish it from closely related viruses. By such means it is possible to determine the spread of a disease [number of changes in a virus capsid protein being directly proportional to the distance (in time and space) of that virus from other viruses of the same basic strain] and thereby locate the origin of an epidemic. This proves useful with such diseases as FMD (40) and influenza where virus differences occur readily. The early location of the origins of the disease could lead to stifling the spread of the disease. Also, the availability of distributed, defined and consistent reagents would lead to a better world-wide standardised disease surveillance program.

2.3.4. New Vaccines

Antibodies which combine with the reactive site of other antibodies can stimulate the production of more of the latter antibody molecules. Such anti-antibody antibodies are called anti-idiotypic antibodies and are prepared as in the schedule described in Fig 2. It can be inferred that such anti-idiotypic antibodies act on the immune system as a pseudo-antigen and therefore may be used to stimulate the immune system to produce antibodies to disease-causing agents. Hence their use would be like that of the antigen in a vaccine preparation (27, 63) (see Chapter 2, this Volume). Such effects have been observed in a murine system with anti-idiotypic antibodies to rabies virus glycoprotein (57).

The value of passive, antibody-based vaccines has been exploited many times. Mixtures of antibody and tetanus toxin or toxoid are commonly ad-

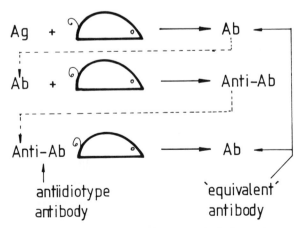

Fig. 2.   Production of antiidiotype antibody from mice.

ministered as it has been held that such antibody–antigen complexes are capable of stimulating the immune system. Also, it may be thought necessary to store quantities of antibodies as emergency reagents to combat noxious agents involved in such diseases as Lassa fever, Marburg and Ebola as well as insect or snake bites. Panels of monoclonal antibodies might also be assayed against the allergens present in pollens, dusts and hairs.

2.3.5. *Therapeutic Use of Monoclonal Antibodies*

For monoclonal antibodies to be used effectively in humans the antibody molecules should be of the type normally found in the species. However, work to date has shown that this is a difficult aim to achieve (33), yet it may be expected with some confidence that clones producing human monoclonal antibodies reliably and profusely will soon become available. In spite of such difficulties, the short-term application of foreign antibodies can elicit therapeutic responses. Indeed, work with rat monoclonal antibodies to human tumour cell antigens has rendered such tumour cells more susceptible to complement-mediated lysis (11, 23).

2.3.6. *Monoclonal Antibodies and Targeted Drug Therapy*

It is becoming increasingly apparant that a wide variety of cancers result from the activation (by chemical agents or inter- or intrachromosomal translocations) of endogenous genes (59). Such genes have been shown to code for proteins (6). This gives credence to hypotheses which posit that cancer cells may be differentially delineated by unique proteins which could be expressed at the cell surface (6). Several groups have been engaged in such investigations, which seem to center on the use of the plant lectin, ricin, as the toxic agent attached to the monoclonal antibody which has been selected for attaching itself to pathological tissue (37, 48, 80). The use of the A-chain subunits of ricin or different moieties of the diphtheria toxin coupled with an MCA has also been assayed.

The further development of this technology can augment or substitute for the application of immunoregulators as potentiators of the immune response. As it would appear that the immune system is a balanced network of interacting molecular and cellular species, the elimination of an agonist enables the more extensive expression of an antagonist. Thus antibodies to suppressor cells, suppressor cell factors or cells which generate growth factors for suppressor cells could also result in immunopotentiation effects.

2.3.7. *Basic Research and Monoclonal Antibodies*

The identification of the immunoactive site on a virus by using panels of MCA has been used widely (78). The sequence followed is to (1) determine the base sequence of the gene coding for the viral capsid proteins, (2) identify likely areas for immunogenic sites on the basis of hydrophilicity, proximity to loop or hinge structure, hypervariability or sensitivity to enzymic attack, (3) synthesise the amino acid chains which span such areas of interest, (4) raise monoclonal antibodies to the polypeptides so formed and then (5) determine which of the antibodies so raised will neutralise virus preparations *in vitro* and hence *in vivo* (one can expect differences between extra- and intracorporeal activities and differences between target and test animal systems). Having identified the putative immunogenic sites, one can proceed to chemisynthetic polypeptides as potential vaccines (*43*).

As a result of their specificity, MCAs have been extensively used as probes for the determination of the precise nature of cell surface molecules (*24, 76*).

A further use of these materials is in the development of an understanding of the complex interaction between an antibody and an antigen. Such events may now be studied with monomolecular species of reactants, thus simplifying experimental procedures.

3. INSECTICIDES

Insecticides derived from viruses lethal for the insects they uniquely and specifically infect fulfill two functions. They can remove the defined target insect pest and leave helpful insects unharmed, and they do not constitute a hazardous pollutant (*25, 42*). The baculoviruses have been selected for this function as they infect moths and butterflies (*16*). In recent years several examples of the deliberate use of such viruses for the elimination of plant-debilitating insects have been observed. It has been estimated that some 80% of the pesticides used in agriculture could be replaced by Baculoviridae at about 40% of the cost of the chemical materials.

There are two ways of producing such materials. The one *in vivo* requires the labour-intensive raising of insects and insect larvae. The product could be contaminated with bacteria and/or other unknown viruses. On the other hand, production of the virus in insect cells grown *in vitro* [which presently can be effected at moderate scales (tens of litres) to high cell densities] leads to a decrease in the production of the virus in the most desirable form (that of the polyhedra), so that although cells may be grown to $1–6 \times 10^6$/ml, they are most productive at 2×10^5/ml (see Chapter 4, this volume) (*79*). This results in an uneconomic process. A second problem is the maintenance of the polyhedral form and yield of polyhedra on continuous passaging of the virus. Although this problem can be circumvented by management of seeding techniques etc., it is an interesting question in biology whose solution could lead to significant increases in the commercial prospects for the system (*35*).

This area where animal cells in culture can be exploited is of a scale and a value (already established in relation to the chemical pesticide industry) that warrant significant additional scientific research and technological development to yield cost-effective and socially acceptable products.

4. ENZYMES FROM ANIMAL CELLS

Animal cells, like bacteria, produce several thousand different enzymes, of which fewer than ten may become commercial products. Although these enzymes may be made in bacteria, the problems of glycosylation (with consequences for the stability of the enzyme once injected), post-translational modification (renin) and excretion (plasminogen activator) may well result in such products being most profitably made from cultured animal cells. Table IV gives a list of enzymes which could be of such value. These, unlike the enzymes which can effect the equivalent reactions but which can be expressed in non-human cells, do not act as foreign material to the human immune system; they can therefore be used at higher concentrations and dosed more often with little fear of allergic side reactions.

At present clinical assay systems are calibrated by using standard enzymes. Such enzymes are normally produced from bacteria or fungi; however, a better standard for human enzyme activity would be an enzyme preparation derived from a human cell source, as such an enzyme would be expected to have characteristics identical to those of the enzyme under investigation.

TABLE IV   Enzymes and Proteins from Cultured Animal Cells

Enzyme	Reference
Plasminogen activator	39, 44
Urokinase	52
Asparaginase	
Tyrosine hydroxylase	
Hyaluronidase	
Cytochrome P-450	
Renin (chymosin)	39, 40
Pepsin	
Trypsin	
Collagenase	39
Factor VII	
Factor VIII	
Factor X	
Enzyme standards for clinical laboratories	
Angiogenic factors	45

Other enzymes, represented by factors VII and VIII (83) of the human blood-clotting system, might be preferentially made in genetically engineered animal cell systems, particularly in view of the problem expressed under the name acquired immune deficiency syndrome (AIDS), which has been associated with the use of materials derived from fractionated donor blood (66). Such considerations have resulted in a sharp increase in the intensity of efforts to produce a hepatitis B_sAg material for a vaccine from a non-blood source (39).

5. HORMONES

While many animal hormones are relatively small polypeptides (up to 20–30 amino acids long) and are most economically produced by chemical synthesis (19), other hormones have 50–200 amino acids and some of them are glycosylated, which implies that they may be best produced in cultured animal cells (12). Erythropoietin is such a hormone (MW 46,000). It can be used to promote the generation of red blood cells. Other glycoprotein hormones are shown in Table V. Animal cell cultures of pituitary or kidney cells could produce such hormones, which might then be used for the amelioration of infertility states or in the controlled production of ova or spermatozoa for use in *in vitro* fertilisation operations. Hormones may also be generated by cells from tissues not normally associated with hormone excretions. Hormones which are so produced are called ectopic hormones. It is of interest to note that in the late 1970s much interest was focussed on the production of both human and animal growth hormones from genetically engineered bac-

TABLE V Glycoprotein Hormones

Hormone	Molecular weight	Action	Reference
Intestinal cell		Development of	49, 50
Stimulating hormone or luteinizing hormone (ICSH or LH)	40,000 (sheep) 100,000 (pig)	tissues of testes and follicle	49, 50
Follicle-stimulating hormone (FSH)	70,000 (sheep)	Development of ovarian follicles and maintenance of spermatogenesis	49
Chorionic hormone (HCG)	100,000 (man) 46,000 (man)	Placental development	49, 50
Erythropoietin		Increases red blood cell numbers	

teria (*69*). Presently there are indications that work is in hand to produce such materials from genetically engineered Vero cells (*7*).

6. OTHER PROTEINS DERIVABLE FROM ANIMAL CELLS IN CULTURE

Proteins such as albumin or collagen do not easily fall into the above categories, yet they can be produced from animal cells in culture (*56, 58*). Other proteins, glycoproteins and sulphated mucopolysaccharides have also been shown to be produced by animal cells in culture (*14, 34, 52*). Similarly, there are a variety of growth factors which could find widespread application in cell culture and which may be formed from cell cultures. For example, attachment factors such as cold-insoluble globulin (CIG or fibronectin) (*52*), fibroblast growth factor, nerve cell growth factor (*49*), epidermal growth factor and transferrin may also be produced from cells in culture. The above do not constitute a definitive list but do serve to indicate the range of possibilities (see Chapter 14, this volume). The possibility of genetically engineering a cell line to produce its own growth factor should not be overlooked.

7. CELLS AS PRODUCTS

Cells produced in suspension cultures grow to about 2×10^6/ml (although it is possible to get up to 10×10^6/ml (*30*). At these concentrations productivity is about 1–5 g/litre of cellular material (wet weight). This can be compared to the productivity of bacterial cultures of 10–50 g/litre (w/w). While this is one measure of "productivity" another measure is the amount of biomass produced per litre per hour in continuous culture. When compared at their minimal doubling times (12 hr for animal cells and 20 min for bacteria), the bacteria produce about 50 times more biomass per unit time.

However, there are other uses of animal cells which are becoming more important as a result of (1) cost-cutting exercises and (2) increased social and ethical pressures to decrease the number of live animals used for test and experimental purposes. Such animals serve three main functions:

1. Toxicology testing
2. Assay of bioproducts
3. Screening systems for potentially useful materials (antiviral agents, anticancer agents)

In each of the above three activities it has become more practicable to use animal cells (*2*). Clearly, for large-scale use it would be preferable to produce cells in bulk (biotechnology) and to effect the necessary quality control procedures *en masse* before using the cells for evaluative work. It is not difficult to envisage a technology of production—quality control testing—storage and use, as part of an operation which can grow and control the masses of animal cells needed for such work.

A further application of animal cells as products is their use in specifically designed environmental chambers implanted in the body and their expression in this form as artificial organs or hormone-secreting glands. Under such conditions the secretions of the implanted cells would come under the normal control mechanisms and would therefore be particularly suitable for human applications.

7.1. Screening Systems for Chemotherapeutics Based on Animal Cells in Culture

It is possible in this review of opportunities for cultured animal cells only to indicate the nature of the type of screening that has been suggested and is being implemented presently. The ends to which such screening systems can be put are many. The action of drugs which are presently on the market on the shape, metabolism and reproduction of a wide variety of cells in culture can be observed. Indications of the way in which the drug acts can then be used to screen other chemicals for a more extensive reactivity or an activity which may cause the appropriate biochemical change but which does not affect growth or shape of ability to grow in soft agar (a test for a transformed or potentially cancerogenic cell). Antidotes for drugs may also be found in such systems and it could be that the antidote itself could be used as a drug in cases of disease with the "opposite" symptoms.

One should also be aware that during such investigations it is possible to find materials which have unsuspected effects leading to spin-offs into other areas of therapy or cell technology.

7.2. Toxicity Testing

It is not uncommon for a batch of human *vaccine* to be subjected to a large number of separate tests (four to six) in a variety of tissue culture systems as part of a regulatory agency process control requirement. This kind of testing may also be applied to all other pharmaceuticals and even food stuffs and food additives. Further applications in the area of environmental contaminants can be envisaged. In the event that such tests prove valuable, the development of *fully automated test systems* could lead to a useful commercial spin-off in the area of packaged equipment sales. Models for such systems are extant in the bacteriological field, where screening for antibiotics has been the objective of intense study and has led to highly mechanised methods.

7.3. Assay Developments

The whole live animal-based bioassay system has numerous drawbacks. It is unethical when used for human cosmetic evaluation and when the data derived from a test are often *not* highly indicative of product potency. A classic example is the cattle test for the evaluation of foot-and-mouth disease vaccine; 26 steers are used in a test which takes 3–5 weeks to complete; it costs about £15,000 and results in an answer which has a two- standard-deviation error of 300%! That is with a good test; some such tests (about 5%) lead to results which are without any significance. A good bioassay for immunogenesis based on cell cultures would be of value for FMD vaccines but also for other vaccines: the development of such a test could also lead to the more rapid development of new vaccines and immunogens and tell us more about how the body responds to foreign substances.

7.4. Encapsulated, Implanted Hormone-excreting Animal Cells (9, 38, 70)

Hormone therapy can be effected in three ways: by regular or mechanically timed and automated injections of the depleted hormone, by presenting the human/animal with a slow-release type of preparation or by setting up an artificial organ or gland within the body. Much attention has been focussed on the use of insulin-secreting cells for people who are insulin-deficient. The problems of cellular rejection may be overcome by the manipulations of the lymph system (antilymphocyte antibodies) or by siting the cells in an immunologically privileged environment (the testes or the brain). An alternative strategy is to grow the cells in a slowly pulsating semi-permeable chamber which prevents contact of the host lymph cells with the histocompatibility antigens of the implanted cells whilst keeping the foreign cells provided with nutrients and a method of voiding their waste materials. It would be a function of the animal cell biotechnologist to produce the appropriate cells in their properly differentiated form, and to encapsulate such cells in a suitable chamber. It may also be necessary to replace the cells with fresh cells periodically without, necessarily, altering the position of the implantation chamber. A further, and major, advantage of such a system over the alternatives referred to above is that the rate of release of hormone would be controlled by the signals normally operating within the body. One would expect therefore that the rate of release of insulin would be related to blood glucose level in the normal way.

8. FURTHER DEVELOPMENT OF THE BIOTECHNOLOGY WHICH PRODUCES AND EXPLOITS ANIMAL CELLS IN CULTURE

While the previous sections of this chapter have concentrated on the products which can be derived from the development of animal cells in culture, the *technology as such* is an area which, if improved further, would yield considerable benefits to those who could exploit such advances.

Technology or the art, craft or capability of achieving ends (colloquially "doing things") in the area of animal cell biotechnology can be divided into three fields: (1) equipment, (2) process fluids and (3) process operations and their timing.

8.1. Equipment

Two basic approaches to equipment development are beginning to consolidate themselves into an accepted modus operandi. The first approach is to design dedicated equipment specifically for the particular job it has to do. The second is to design multipurpose equipment to do any job asked of it. An example of the former is the production of an immobilised bed of animal cells which are capable of generating products and from which one can expect such advantages as: (1) less involved separation (cells and products) operations; (2) "milking" stationary (non-growing) cells, i.e. most of the nutrients supplied can be diverted to product generation as opposed to cell biomass; (3) application of biochemical regulatory materials can be done rapidly and removed as quickly, which gives the operator close control of the cell's chemical environment; and (4) operations may be effected at higher cell concentrations.

The demonstrated growth of cells (and virus in those cells) on packed beds of 3-mm glass spheres [developed to the 100-litre scale (*81*)] renders this kind of system a prime candidate for development of those applications where milking product from stationary non-growing cells is the approved way of material manufacture.

An alternative methodology to the immobilisation of cells on the surface of the elements of a packed bed is to contain the cells within the structure of a porous matrix (*47, 61, 64*). Such systems have been assayed for the production of antibodies from hybridoma cells but their performance relative to alternative systems (air lift, stirred tank reactor) has yet to be critically evaluated.

Equipment which can produce homogeneous or quasi-homogeneous cultures of animal cells (i.e. microcarriers) on 100- to 1000-litre scales has also been developed (*44*). Such equipment is also capable of growing animal cells in suspension (which includes hybridomas) as well as cells growing on microcarriers (the worst case). Minor modifications to the equipment (air inlet and gearing of stirrer drive) should lead to adequate productivities of bacteria, yeasts and other fungi. There are no a priori reasons which would prevent the production of *plant* cell cultures in such equipment also!

However, it is important to establish such equipment on production performance criteria and to properly compare it with alternative systems [air lift (36), tower fermenter, waldorf tube etc.], an exercise in technology evaluation. In addition, it is still unclear for most (if not all) animal cell lines what factors are involved in "shear" sensitivity as well as the mechanism whereby "bubbles" interfere with the viability and productivity of cultured animal cells. Such questions are best answered off the shop floor, using different vessels and exploiting the controllable environment which can be obtained in fully continuous culture systems (chemostat or turbidostat).

As in other areas of biotechnology there is a basic inability to scale up the size of process equipment without altering basic geometric and transfer relationships (surface area/volume, power/volume and $K_L a$). We need to know how best to scale up animal cell culture systems in which bubbles (and small bubbles, which can generate high surface areas) are not welcome features. We also need more information on the factors which limit the growth of cells in batch and continuous culture.

For process optimisation on-line sensors, local and supervisory computers (5) and host computers are essential and have for the most part been developed. Continuous (fully automated) assay systems for immunoactive materials are still needed. *Steam-sterilisable* sensors for biological monomers [amino acids, sugars, ammonia, lactic acid immunoactive materials (55)] do not as yet exist in robust and reliable forms; this of necessity constitutes an area in which valuable developments can be anticipated. Also, schemes for process optimisation of continuous culture systems have yet to be implemented (they exist on paper but not on the shop floor).

In addition, suitable and *useful* models have to be developed from data which can now be obtained and stored in great quantities but which at present few (if any) investigators have used productively (other than as an acceptable record of a controlled process from the point of view of regulatory authorities).

8.2. Process Media

Media for animal cell systems can be as complex as one with more than 50 components to one with as few as 5. This means that there is considerable scope for development as not only is each component a variable, but the sum total of the interaction variables is enormous. Two main objectives, however, stand out. One involves the development and successful use of media in which foetal calf serum (£50–£100 per litre) can be replaced with cheaper sera (<£5 per litre) or development media which are serum-free or, better yet, completely defined chemically, and which do not require that the cells become transformed to grow in them (3, 18, 60). A second objective is to design media in which cell growth does not occur but cell viability is maintained and into which product is excreted. An objective of Japanese investigators is to produce a "non-saline" medium for the production of animal cells (60).

The question of the medium is potentially endless as each individual cell system may require a medium with a unique composition. However, with investigation it could prove that we presently know very little about what makes a cell grow and produce outside the body and that new media can be developed which are cheaper, easier to use and formulate (fewer components with higher definition) and which can be applied with greater universality. Again, the incentive to achieve success in this area has come from the requirement of the MCA production system, which is effected preferentially in serum-free medium.

8.3. The Process

A process consists of a defined series of operations with specified pieces of equipment and particular process materials at designated times. Each of these features is often a long and painstaking operation which is the culmination of years of work. Clearly, the procedures which can be used for vessel optimisation can be used for the process as a whole, but it is the author's experience that once a process has arrived on the shop floor there is little opportunity for change. This is a reason for the exploration of as many of the process variables as possible (and, if practicable, their interactions) at the developmental stage. In such a development, areas of "sensitivity" should be sought, as these will ultimately control productivity and lead to further yield gains if suitably massaged. A further method of developing exploitable culture systems is to derive a model off-line and then attempt to implement those conditions which the model has shown to be most productive (see Chapter 7, Volume 1)

9. CONCLUSION

Animal cell biotechnology began as an area of activity when antibiotics became available. Media were developed in the 1950s and vaccines in the 1960s. In the 1970s new forms of unit process technology were designed and demonstrated and the developments described in the introduction to this chapter were made. It can be anticipated with confidence that in the remaining years of this century and well into the next millenium, these developments will find expression in new products for direct use, new equipment for product generation and new processes for making those products.

REFERENCES

1. Axel, R., Wigler, M. H., and Silvestein, S. J. (1983). Processes for inserting DNA into eucaryotic cells and for producing proteinaceous materials. U.S. Patent 4,399,126.
2. Balls, M., and Horner, S. (1982). Cell and organ cultures as alternatives in toxicity testing. *Alternatives Lab. Anim.* **10,** 16–26.
3. Barnes, D., and Sato, G. (1980). Serum-free cell culture: A unifying approach. *Cell* **22,** 649–655.
4. Bittle, J. L., Lerner, R. A., Rowlands, D. J., and Brown, F. (1982). Protection against foot-and-mouth disease by immunisation with a chemically synthesized peptide predicted from the viral nucleotide sequence. *Nature (London)* **298,** 30–33.
5. Breame, A. J., and Spier, R. E. (1981). Down market computers for research. *Compu. Appl. Ferment. Technol. Soc. Chem. Ind. Publ.,* pp. 13–22.
6. Brickwell, P. M., Latchman, D. S., Murphy, D., Wilson, K., and Rigby, P. W. J. (1983). Activation of Qa/Tla class I major histocompatibility antigen gene is a general feature of oncogenesis in the mouse. *Nature (London)* **306,** 756–760.
7. Chemistry and Industry (1984). *Chem. Ind. (London)* (7), 233.
8. Chen, M-H., Dong, C. Y., Lui, Z. H., Skinner, G. R. B., and Hartley, C. E. (1983). Prevention of type 2 herpes simplex virus induced cervical carcinoma in mice by prior immunization with a vaccine prepared from type 1 herpes simplex virus. *Vaccine* **1,** 13–16.
9. Chick, N. L., Perna, J. J., Low, D., Lewis, V., Galletti, P. M., Panol, G., Whittemore, A. D., Like, A. A., Colton, C. K., and Lysaght, M. J. (1977). Artificial pancreas using living beta cells: Effects of glucose homeostasis in diabetic rats. *Science* **197,** 780–81.
10. Christofinis, G. J., Steel, C. M., and Finter, N. B. (1981). Interferon production by human lymphoblastic cell lines of different origins. *J. Gen. Virol.* **52,** 169–171.
11. Cobbold, S. P., and Woldmann, H. (1984). Therapeutic potential of monovalent monoclonal antibodies. *Nature (London)* **308,** 460–462.
12. Cox, R. P., and Day, D. G. (1981). Regulation of glycopeptide hormone synthesis in cell culture. *Adv. Cell Cult.* **1,** 15–65.
13. Djen, J. Y., Lanza, E., Pastore, S., and Hapel, A. J. (1983). Selective growth of natural cytotoxic but not natural killer effector cells in interleukin-3. *Nature (London)* **306,** 788–791.
14. Dorlyman, A., and Ho, P. L. (1970). Synthesis of acid mucopolysaccharides by glial tumor cells in tissue culture. *Proc. Natl. Acad. Sci. U.S.A.* **66,** 495–499.
15. Duesberg, P. H. (1983). Retroviral transforming genes in normal cells. *Nature (London)* **304,** 219–226.
16. Falcon, L. A. (1980). "Economical and Biological Importance of Baculoviruses as Alternatives to Chemical Pesticides." Bundesministerium fur Forschung und Technologies, Bonn.

17. Finter, N. B. (1981). Comment made at the 2nd European Congress of Biotechnology Meeting at Eastborne, April 1981.
18. First European Conference on Serum-Free Cell Culture, Heidelberg (1982). *Conf. Ges. Biol. Chem.* **42** (abstr.).
19. Genetic Technology (1982). "A New Frontier," p. 80. West View Press Groom/Helm. Office of Technology Assessment. Washington, D.C.
20. Gillis, S., Watson, J., and Mochizaki, D. (1982). Biochemical characterization of interleukin-2 (T-cell growth factor). *In* "Isolation, characterization and utilization of T lymphocyte clones" (C. G. Fathman and F. W. Fitch, eds.), pp. 24–41. Academic Press, New York.
21. Gluzman, Y., ed. (1982). "Eukaryotic Viral Vectors." Cold Spring Harbor Lab., Cold Spring Harbor, New York.
22. Gluzman, Y., and Shenk, T., eds. (1983). "Enhancers and Eukaryotic Gene Expression." Cold Spring Harbor Lab., Cold Spring Harbor, New York.
23. Haspel, M. V., Hoover, H. C., McCabe, R. P., Pomato, N., Knowlton, J. V., Jonesch, N. J., and Nanna, M. G. (1984). Human colon cancer: Generation of tumor-specific human monoclonal antibodies. *Nature (London)* (submitted for publication).
24. Hughes, E. N., and August, J. T. (1981). Characterization of plasma membrane proteins identified by monoclonal antibodies. *J. Biol. Chem.* **256**, 664–671.
25. Ignoffo, C. M., and Anderson, R. F. (1979). Bioinsecticides. *In* "Microbial Technology" (H. J. Peppler and De Perlman, eds.), Vol. 1, Chapter 1, pp. 1–28. Academic Press, New York.
26. Ikic, D., Manicic, Z., Oresic, V., Rode, B., Nola, P., Smudj, K., Krezevic, M., Jusic, D., and Soos, E. (1981). Application of human leucocyte interferon in patients with urinary bladder papillomatosis, breast cancer and melanoma. *Lancet*, 1022–1030.
27. Janeway, C. A. (1981). Manipulation of the immune response by anti-idiotype. *In* "Immunology 80: Progress in Immunology IV" (M. Fougereau and J. Dausset, eds.), Vol. 3, pp. 1149–1159. Academic Press, London.
28. Jerne, N. K. (1973). The immune system. *Sci. Am.* **229**, 52–60.
29. Karp, G. (1979). "Cell Biology," p. 799. McGraw-Hill, New York.
30. Katinger, H. W.-D., and Scheirer, W. (1979). Mass cultivation of mammalian cells in an airlift fermenter. *Dev. Biol. Stand.* **42**, 11 (abstr.).
31. Kaufman, R. J., and Sharp, P. A. (1982). Amplification and expression of sequences contransfected with a modular dihydrofolate reductase complementary DNA gene. *J. Mol. Biol.* **159**, 601–621.
32. Kohler, G., and Milstein, C. (1975). Continuous cultures of fused cells secreting antibody of predefined specificity. *Nature (London)* **256**, 495–497.
33. Kozbor, D., and Roder, J. C. (1983). The production of monoclonal antibodies from human lymphocytes. *Immunol. Today* **4**, 72–79.
34. Kraemer, P. M. (1968). Production of heparin related glycosaminoglycans by an established mammalian cell line. *J. Cell. Physiol.* **71**, 109–120.
35. Krieg, A. (1980). "Production and Application of Baculoviruses in Germany." Bundesministerium fur Forschung und Technologies, Bonn.
36. Lab Equipment Digest (1983). The best laid schemes of mice and men. *Lab. Equip. Dig.*, p. 101.
37. Laurent, J. C., Jansen, F. K., and Gros, P. (1983). Potential therapy of monoclonal antibodies coupled with toxins. *Dev. Biol. Stand.* **55**, 181. (Abstr.)
38. Lim, F., and Sun, A. M. (1980). Microencapsulated islets as bioartificial endocrine pancreas. *Science* **120**, 908–910.
39. McAleer, W. J., Buynack, E. B., Maigetter, R. Z., Wampler, D. E., Miller, W. J., and

Hilleman, M. R. (1984). Human hepatitis B vaccine from recombinant yeast. *Nature (London)* **307**, 178–180.

40. McCullough, K. C., and Butcher, R. (1982). Monoclonal antibodies against foot-and-mouth disease virus 146S and 12S particles. *Arch. Virol.* **74**, 1–9.

41. Merck, W. A. M. (1981). Large scale production of human fibroblast interferon in cell fermenters. *Dev. Biol. Stand.* **50**, 137–140.

42. Miltenburger, H. G., ed. (1980). "Safety Aspects of Baculovirus as Biological Insecticides." Bundesministerium fur Forschung und Technologies, Bonn.

43. Minor, P. D., Schild, G., Bootman, J., Evans, D. M. A., Ferguson, M., Reeve, P., Spitz, M., Stanway, G., Cann, A. J., Hauptmann, R., Clarke, L. D., Moutford, R. C., and Almond, J. W. (1983). Location and primary structure of a major antigenic site for poliovirus neutralization. *Nature (London)* **301**, 674–679.

44. Montagnon, B., and Fanget, B. (1983). Thousand litres scale microcarrier culture of Vero cells for killed poliovirus vaccine: Promising results. *Dev. Biol. Stand.* **55**, 37–42.

45. Neville, D. M., and Youle, R. J. (1982). Anti-thy 1,2 monoclonal antibody–ricin hybrid utilized as a tumor suppressant. U.S. Patent 4,359,457.

46. Newmark, P. (1981). Interferon: Decline and stall. *Nature (London)* **291**, 105–106.

47. Nilsson, K., and Mosbach, K. (1980). Preparation of immobilised animal cells. *FEBS Lett.* **118**, 145–150.

48. Nowinski, R. C., Tam, M. R., Goldstein, L. C., Strong, L., Kuo, C. C., Corey, L., Stamm, W. E., Handsfield, H. H., Knapp, J. S., and Holmes, K. K. (1983). Monoclonal antibodies for diagnosis of infectious diseases in humans. *Science* **219**, 637–644.

49. Orger, J., Arnason, B. G. W., Pantazis, H., Lehrich, J., and Young, M. (1974). Synthesis of nerve cell growth factor by L and 3T3 cells in culture. *Proc. Natl. Acad. Sci. U.S.A.* **71**, 1551–1558.

50. Palmenberg, A. C. (1982). In vitro synthesis and assembly of picornaviral capsid intermediate structures. *J. Virol.* **44**, 900–906.

51. Panicali, D., Davis, S. W., Randall, L., Weinberg, L., and Paoletti, B. (1983). Construction of live vaccines by using genetically engineered poxviruses: Biological activity of recombinant vaccinia virus expressing influenze virus hemagglutinin. *Proc. Natl. Acad. Sci. U.S.A.* **80**, 5364–5368.

52. Parry, G., Soo, W.-J., and Bissell, M. J. (1979). The uncoupled regulation of fibronectin and collagen synthesis in Rous sarcoma virus transformed avian tendon cells. *J. Biol. Chem.* **254**, 11763–11766.

53. Pick, E., ed. (1981). "Lymphokines—A Forum for Immunoregulatory Cell Products," Vol. 3. Academic Press, New York.

54. Powledge, T. M. (1984). Interferon on trial. *Biotechnology* **2** (3), 214–228.

55. Preston, K., and Spier, R. E. (1985). A method for the automated semi-continuous on-line measurement of immunocative materials produced in animal cell culture system. (In preparation).

56. Quarori, A., and Trelstad, R. L. (1980). Biochemical characterisation of collagens synthesised by intestinal epithelial cell cultures. *J. Biol. Chem.* **255**, 8351–8361.

57. Reagan, K. J., Wunner, W. H., Wiktor, T. J., and Kaprowski, H. (1983). Anti-idiotype antibodies induce neutralizing antibodies to rabies virus glycoprotein. *J. Virol.* **48**, 660–666.

58. Richardson, U. I., Tashjian, A. H., Jr., and Levine, L. (1969). Establishment of a clonal strain of hepatoma cells which secrete albumin. *J. Cell Biol.* **40**, 236–246.

59. Robertson, M. (1983). Paradox and paradigm: The message and meaning of *myc. Nature (London)* **306**, 733–736.

60. Rogers, M. D. (1982). The role of the Japanese Government in biotechnology research and

development. 6th Report from the Education Science & Arts Committee of the House of Commons (HC 289, pp. 47–56). H. M. Stationery Office, London.

61. Rosevear, A., and Lambe, C. A. (1982). Immobilised plant and animal cells. *Top. Enzyme Ferment. Biotechnol.* **7**, 13–37.

62. Rutter, W. J., Laub, O., and Rall, L. (1982). Virus protein synthesis. European Patent Appl. 82 400,564.9.

63. Sacks, D. L., Esser, K. M., and Sher, A. (1982). Immunisation of mice against African trypanosomiasis using anti-idiotypic antibodies. *J. Exp. Med.* **155**, 1108–1119.

64. Scheirer, W., Nilsson, K., Merten, O. W., Katinger, H. W. D., and Mosbach, K. (1983). Entrapment of animal cells for the production of biomolecules such as monoclonal antibodies. *Dev. Biol. Stand.* **55**, 155–162.

65. Schimke, R. T., ed. (1982). "Gene Amplification." Cold Spring Harbor Lab., Cold Spring Harbor, New York.

66. *Science* **220**, 806–871.

67. Scott, G. M., and Tyrrell, D. A. J. (1980). Interferon: Therapeutic fact or fiction for the 80's *Br. Med. J.* 1558–1562.

68. Secher, D. S., and Burke, D. C. (1980). A monoclonal antibody for large-scale purification of human leucocyte interferon. *Nature (London)* **285**, 446–450.

69. Seeburg, P. H., Shine, J., Martial, J. A., Ivarie, R. D., Morris, J. A., Ulliman, J. A., Baxter, J. D., and Goodman, H. M. (1978). Synthesis of growth hormone by bacteria. *Nature (London)* **276**, 795–798.

70. Sefton, M. V. (1982). Encapsulation of live animal cells. U.S. Patent 435,388.

71. Sercarz, E. S., and Cunningham, A. J., eds. (1980). "Strategies of Immune Regulation." Academic Press, London.

72. Smith, G. L., Mackett, M., and Moss, B. (1983). Infectious vaccinia virus recombinants that express hepatitis B virus surface antigen. *Nature (London)* **302**, 490–495.

73. Spier, R. E. (1982). Animal cell or genetically engineered bacteria for the manufacture of particular bioproducts. *Dev. Biol. Stand.* **30**, 311–321.

74. Spier, R. E. (1983). Modern approaches to vaccines—A report of a Cold Spring Harbor conference in modern approaches to vaccines Aug 1983. *Vaccine* **1**, 53–54.

75. Streuli, M., Hall, A., Boll, W., Stewart, E. W., II, Nagata, S., and Weissmann, C. (1981). Target cell specificity of two species of human interferon-α produced in *Escherichia coli* and of hybrid molecules derived from them. *Proc. Natl. Acad. Sci. U.S.A.* **78**, 2848–2852.

76. Taylor, D. L., and Wang, Y.-L. (1980). Fluorescently labelled molecules as problems of the structure and function of living cells. *Nature (London)* **284**, 405–410.

77. Time (1981). Shaping life in the lab. *Time Mag.* March 9, pp. 36–43.

78. Van der Werf, S., Wychowski, C., Bruneau, P., Blondel, B., Crainic, R., Horodniceanu, F., and Girard, M. (1983). Localisation of a poliovirus type 1 neutralisation epitope in viral capsid polypeptide VPI. *Proc. Natl. Acad. Sci. U.S.A.* **80**, 5080–5084.

79. Vaughn, J. L. (1981). Insect cells for insect virus production. *Adv. Cell Cult.* **1**, 281–295.

80. Vitetta, E. S., Krolick, K. A., Miyama-Inaba, M., Cushley, W., and Uhr, J. W. (1983). Immunotoxins: A new approach to cancer therapy. *Science* **219**, 644–650.

81. Whiteside, J. P., and Spier, R. E. (1981). The scale-up from 0.1 to 100 litres of a unit process system based on 3mm diameter glass spheres for the production of four strains of FMDV from BHK monolayer cells *Biotechnol. Bioeng.* **23**, 2551–2565.

82. Yamane, Y., Furukawa, T., Plotkin, S. A. (1983). Supernatant virus release as a differentiating marker between low passage vaccine strains of human cytomegalovirus. *Vaccine* **1**, 23–25.

83. Zachavski, L. R., Bowie, E. J. W., Titus, J. L., and Owen, C. A. (1969). Cell-culture synthesis of a factor VIII-like activity. *Mayo Clin. Proc.* **44**, 784–792.

FURTHER READING

Almond, J. W. (1983). Location and primary structure of a major antigenic site for poliovirus neutralization. *Nature (London)* **301**, 674–679.

Barnes, D., and Sato, G. (1980). Methods for growth of cultured cells on serum-free medium. *Anal. Biochem.* **102**, 255–270.

Birch, J. R., Cartwright, T. B., and Ford, J. A. (1981). Verfahren zum Zuchten von tierischen und mit einem Plattenstapel versehene Sellzuchtungsapparatur. German Patent De 30 31 674 A1.

Bodeker, B. G. D., and Muhlradt, P. F. (1982). Inhibitors regulate interleukin-2 synthesis of stimulated lymphocytes: Consequence for production procedures. *Dev. Biol. Stan.* **55**, 247–254.

Falkner, F. G., and Zachau, H. G. (1982). Expression of mouse immunoglobulin genes in monkey cells. *Nature (London)* **298**, 286–288.

Feder, J., and Tolbert, W. R. (1983). The large scale cultivation of animal cells. *Sci. Am.* **248**, 24–31.

Gershon, A. A. (1978). Varicella–Zoster virus: Prospects for active immunisation. *Am. J. Clin. Pathol.* **70**, 170.

Henney, C. S., Kuribayashi, K., Kerm, D. E., and Gillis, S. (1981). Interleukin-2 augments natural killer cell activity. *Nature (London)* **291**, 335–338.

Hull, R. N., and Huseby, R. M. (1976). Enhanced production of plasminogen activation. U.K. Patent 1,443,189.

Kadouri, A., and Bohak, Z. (1983). Production of plasminogen activator in cultures of normal human fibroblasts. *Biotechnology* **1**, 354–358.

Katinger, H. W. D., and Bliem, R. Production of enzymes and hormones by mammalian cell culture (in press).

Paul, W. E., and Benacerraf, B. (1977). Functional specificity of thymus-dependent lymphocytes. *Science* **195**, 1293–1300.

Pye, E. K., Macaig, T., and Iyengar, R. (1977). Production enzymes and proteins in tissue culture. *In* "Biotechnological Applications of Proteins and Enzymes" (Z. Bohak and N. Sharon, eds.), Chapter 5, pp. 63–79. Academic Press, New York.

Robertson, A. L., Smeby, R. R., Bumpus, F. M., and Page, I. H. (1965). Renin production by organ cultures of renal cortex. *Science* **149**, 650–651.

Tashjian, A. H., Jr. (1969). Animal cell cultures as a source of hormones. *Biotechnol. Bioeng.* **11**, 109–126.

Thurman, G. B., Marshall, G. D., Low, T. L. K., and Goldstein, A. L. (1980). Thymosin— Structural studies and immunoregulatory role in host immunity. *In* "Thymus, Thymic Hormones and T Lymphocytes" (F. Aiuti and H. Wigzell, eds.), pp. 175–185. Academic Press, London.

Index

459